Advanced Carbon-Based Materials for Next-Generation Batteries and Supercapacitors

Advanced Carbon-Based Materials for Next-Generation Batteries and Supercapacitors

Guest Editors

Binghui Xu
Haichao Chen

MDPI

Basel • Beijing • Wuhan • Barcelona • Belgrade • Novi Sad • Cluj • Manchester

Guest Editors
Binghui Xu
Institute of Materials for
Energy and Environment
Qingdao University
Qingdao
China

Haichao Chen
Institute of Materials for
Energy and Environment
Qingdao University
Qingdao
China

Editorial Office
MDPI AG
Grosspeteranlage 5
4052 Basel, Switzerland

This is a reprint of the Special Issue, published open access by the journal *Batteries* (ISSN 2313-0105), freely accessible at: https://www.mdpi.com/journal/batteries/special_issues/JA7116YR6X.

For citation purposes, cite each article independently as indicated on the article page online and as indicated below:

Lastname, A.A.; Lastname, B.B. Article Title. *Journal Name* **Year**, *Volume Number*, Page Range.

ISBN 978-3-7258-6296-2 (Hbk)
ISBN 978-3-7258-6297-9 (PDF)
https://doi.org/10.3390/books978-3-7258-6297-9

Cover image courtesy of Binghui Xu

Contents

batteries

MDPI

Article

Sustainable Synthesis of a Carbon-Supported Magnetite Nanocomposite Anode Material for Lithium-Ion Batteries

Hui Zeng, Jiahui Li, Haoyu Yin, Ruixin Jia, Longbiao Yu, Hongliang Li and Binghui Xu *

Institute of Materials for Energy and Environment, College of Materials Science and Engineering, Qingdao University, Qingdao 266071, China
* Correspondence: xubinghuiqdu@qdu.edu.cn or xubinghuicsu@163.com

Abstract: Transition metal oxide magnetite (Fe_3O_4) is recognized as a potential anode material for lithium-ion batteries owing to its high theoretical specific capacity, modest voltage output, and eco-friendly character. It is a challenging task to engineer high-performance composite materials by effectively dispersing Fe_3O_4 crystals with limited sizes in a well-designed supporting framework following sustainable approaches. In this work, the naturally abundant plant products sodium lignosulfonate (Lig) and sodium cellulose (CMC) were selected to coprecipitate with Fe^{3+} ions under mild hydrothermal conditions. The Fe-Lig/CMC intermediate sediment with an optimized microstructure can be directly converted to the Lig/CMC-derived carbon matrix-supported Fe_3O_4 nanocomposite sample (Fe_3O_4@LigC/CC). Compared with the controlled Fe_3O_4@LigC material, the Fe_3O_4@LigC/CC nanocomposite provides superior electrochemical performance in the anode, which has inspiring specific capacities of 820.6 mAh g^{-1} after 100 cycles under a current rate of 100 mA·g^{-1} and 750.5 mAh g^{-1} after 250 cycles, as well as more exciting rate capabilities. The biomimetic sample design and synthesis protocol closely follow the criteria of green chemistry and can be further developed in wider scenarios.

Keywords: magnetite; sodium lignosulfonate; sodium cellulose; green chemistry; lithium-ion batteries

1. Introduction

Rechargeable lithium-ion batteries are predominantly utilized as an energy storage apparatus in modern society, and they play an extremely important role in relieving the pressure caused by the depletion of fossil fuels [1–4]. Upgrading the electrode materials with significantly improved lithium-ion storage capabilities is becoming pivotal in developing next-generation lithium-ion battery products. Compared with the current insertion-type graphitic anode materials, the conversion-type transition metal oxide magnetite (Fe_3O_4) has a much higher specific capacity as well as a modest voltage plateau and thus receives increasing research attention [5–7]. On the other hand, it is commonly recognized that the heavy volume expansion after lithiation and the poor inherent electric conductivity for the Fe_3O_4 anode are among the most critical problems before its commercialization.

In recent research, it has been confirmed that synthesizing Fe_3O_4 crystals with reduced particle sizes and further constructing a supportive carbon skeleton can effectively enhance their lithium-ion storage performance [8]. Moreover, these materials with attractive physical and chemical properties are being widely investigated in emerging fields, such as energy harvesting, photocatalysis, biosensing, biomedicine, and so on [9–12]. From the green chemistry viewpoint, it is particularly meaningful and challenging to engineer the corresponding composite materials with unique microstructures via simplified and sustainable approaches. In the natural plant kingdom, plant stems are generally recognized for their good mechanical strength and high flexibility, which enable them to stand up in a variety of environments. It has been revealed that the ordered crossing arrangement of the long-chain cellulose and smaller polymer lignin jointly contribute to the favorable microstructure of these plant stems [13–15]. Moreover, cellulose and lignin are among the three most abundant available natural polymers, which means these materials have high potential for large-scale conversion and utilization. Therefore, they offer an alternative material design and engineering strategy following the bionic principles [16,17]. Sodium lignosulfonate (Lig) and sodium cellulose (CMC) are available products derived from natural lignin and cellulose, respectively. Both Lig and CMC feature good hydrophilicity due to their rich functional groups. Therefore, the structural advantages enable Lig and CMC to have good processibility in aqueous conditions, which means they can easily complex with transition metal ions to synthesize precursor samples [18]. In our previous work, CMC and Fe^{3+} ions were first employed as starting materials to synthesize a final porous carbon-supported Fe_3O_4 nanocomposite (Fe_3O_4@PC) using a two-step method [19]. However, due to the ineffective stabilization from the PC domains derived from the 1D CMC, the Fe_3O_4@PC exhibited unsatisfactory electrochemical behaviors. Therefore, optimizing sample design and synthesis are challenging tasks in developing a sustainable strategy.

On the foundation above, as shown in Scheme 1, both Lig and CMC with different sizes and microstructures were jointly selected as carbon precursors to interact with Fe^{3+} ions in mild hydrothermal conditions. Taking advantage of the coordination capability with Fe^{3+} ions of Lig and CMC, an Fe-Lig/CMC intermediate sample with improved microstructure can be engineered. After a simple inert thermal treatment, the Fe-Lig/CMC can finally be converted to the Lig/CMC-derived carbon matrix-supported Fe_3O_4 nanocomposite sample (Fe_3O_4@LigC/CC). It is not surprising that the Fe_3O_4@LigC/CC sample delivers significantly elevated lithium-ion storage performance compared with the controlled Fe_3O_4@LigC sample. This work demonstrates a sustainable and scalable sample synthesis protocol using abundant, natural, plant-derived products as raw materials. Moreover, the Fe_3O_4@LigC/CC sample can be considered for more feasible applications beyond the energy storage field.

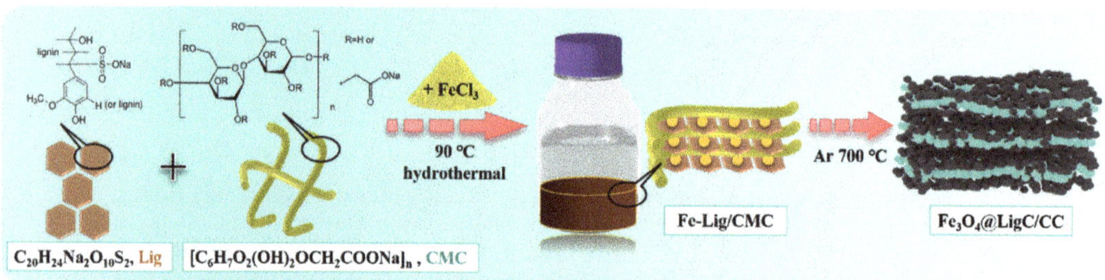

Scheme 1. Schematic diagram of the engineering of the Fe_3O_4@LigC/CC nanocomposite.

2. Experimental

2.1. Sample Synthesis

The chemicals were of analytical purity; they were purchased from the Sinopharm Company and used as received.

CMC (0.5 g) and Lig (0.5 g) were added to deionized water with uninterrupted stirring until a homogeneous suspension was formed in the first 4 h. After gradually adding the $FeCl_3$ $6H_2O$ (2.0 g) aqueous solution into the reaction system, brown flocculent Fe-Lig/CMC sediment was gradually generated in the subsequent 6 h. The Fe-Lig/CMC intermediate sample was collected after water washing and following freeze-drying treatment, and then it was thermally treated in Ar at 700 °C for 2 h. Finally, the black Fe_3O_4@LigC/CC powder sample could be synthesized.

The controlled Fe_3O_4@LigC sample was fabricated following the same procedures used for the Fe_3O_4@LigC/CC sample without using CMC in the raw materials.

2.2. Sample Characterization and Electrochemical Measurement

The detailed sample characterization and electrochemical measurement can be found in the supporting information file.

3. Results and Discussion

3.1. Engineer the Fe_3O_4@LigC/CC Nanocomposite from Natural Products

The coprecipitation between Lig, CMC molecules, and Fe^{3+} ions is first triggered in mild hydrothermal conditions utilizing the coordination interaction, and the Fe-Lig/CMC intermediate sediment sample can be engineered. The well-dispersed Fe^{3+} ions in the Lig/CMC domain with rationally designed microstructure provide a good prerequisite for the final conversion of the Fe-Lig/CMC to the Fe_3O_4@LigC/CC nanocomposite.

Figure 1a shows the XRD pattern of the Fe_3O_4@LigC/CC nanocomposite. The sharp XRD diffraction peaks are strictly indexed by the crystalline Fe_3O_4 phase (PDF#19-0629). In addition, an obvious broad peak located near 26 degrees two theta could be ascribed to the LigC/CC carbonaceous supporting matrix. Further characterization information about Fe_3O_4@LigC/CC was obtained from the Raman spectrum in Figure 1b. These peaks, located at 1333 and 1594 cm^{-1}, may be related to the D-band induced by the defective and disordered sp3 carbon and the G-band induced by the ordered sp2 carbon, respectively [20–22]. The calculated I_D/I_G value for the Fe_3O_4@LigC/CC sample is 0.93, which also means that the LigC/CC carbon skeleton is successfully transformed from the natural Lig and CMC products during further calcination.

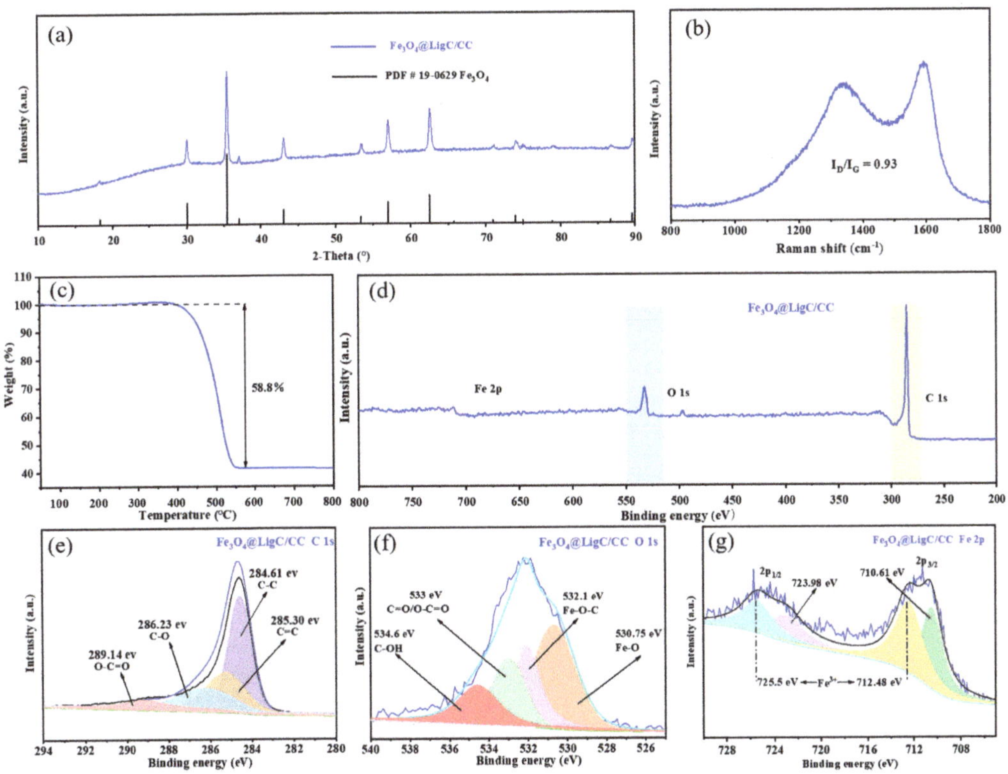

Figure 1. XRD pattern (**a**), Raman spectrum (**b**), TGA curve (**c**), survey XPS spectrum (**d**), high-resolution C 1s (**e**), O 1s (**f**), and Fe 2p (**g**), XPS spectra of the Fe_3O_4@LigC/CC sample.

Figure 1c displays the TGA curve of the Fe_3O_4@LigC/CC sample examined in air to determine the Fe_3O_4 and carbon content in the nanocomposite. The first negligible weight drop is usually explained by the elimination of the physically absorbed water, while the following slight weight increase from 250 to 350 °C can be ascribed to the further oxidation of Fe^{2+} to Fe^{3+} from the Fe_3O_4 crystals in air. In the temperature range of 400 to 550 °C, the rapid weight decrease is apparent. This result is caused by the removal of the LigC/CC carbonaceous supporting matrix. Moreover, it is also revealed that the overall mass drop is about 58.8% for the Fe_3O_4@LigC/CC sample, leaving about 41.2% of the original weight for the residual in Fe_2O_3 form. According to the above result, it can be calculated that the Fe_3O_4 crystals and LigC/CC matrix occupy about 40.4% and 59.6% of the Fe_3O_4@LigC/CC sample in weight, respectively.

According to the overall scanning XPS spectrum for the Fe_3O_4@LigC/CC nanocomposite in Figure 1d, the existence of the three elements C, O, and Fe is validated. Figure 1e reveals the magnified C 1s spectrum, where the divided major peaks at 284.61 and 285.30 eV match with the graphitic C-C and C=C bonds, respectively [23], but the weak ones are highly related to the residual oxygen-contained radicals in the LigC/CC matrix. According to the O 1s spectrum in Figure 1f, the Fe-O peak located at 530.75 eV can be ascribed to lattice oxygen atoms from the Fe_3O_4 crystals, while the one at 532.1 eV can be assigned to the Fe-O-C bond connecting the Fe_3O_4 crystals and the carbon skeleton, implying good accommodation for the Fe_3O_4 crystals within the LigC/CC matrix [24]. The C=O/O-C=O peak at 533.0 eV and the C-OH peak at 534.6 eV show quite weakened strength, which means that the elimination of radicals in the LigC/CC matrix was not thorough [25]. According to the magnified Fe 2p spectrum in Figure 1g, the binding energies of 712.48 and 725.5 eV correspond to Fe 2p3/2 and Fe 2p1/2, respectively, showing an energy gap (ΔE) of 13.02 eV, and these are in good accordance with previously reported spectra of Fe^{2+} in Fe_3O_4 [26].

The microstructural and morphological information of the samples were first analyzed using the FESEM images. For the Fe-Lig/CMC intermediate sample in Figure 2a,b, nano-sized primary particles are aggregated to form the secondary micro-sized powder sample, and the individual Fe-Lig/CMC primary particles have similar sizes with clear space between them. This result means that the Lig and CMC molecules of different sizes have integrated with homogeneous dispersion of the Fe^{3+} ions, which is quite beneficial to the final conversion in the following thermal treatment. For the Fe_3O_4@LigC/CC nanocomposite in Figure 2c,d, the micro-sized powder outlines are successfully inherited from the intermediate sample, while the Fe_3O_4 with similar-sized nanoparticles can be clearly seen in the carbon domains. In particular, no obvious excessive aggregation or over-exposure of these Fe_3O_4 nanocrystals can be verified, which is favorable for the elevation of the electrochemical performance of the Fe_3O_4@LigC/CC anode material. By contrast, the controlled Fe_3O_4@LigC sample in Figure 2e,f exhibits a quite different structure and morphology. Although the Fe_3O_4@LigC sample exhibits a micro-sized particle outline, the Fe_3O_4 nanocrystals with a much larger size and a reduced quantity can be found in the intact pyrolytic carbon. The above structural differences of the Fe_3O_4@LigC/CC and the controlled Fe_3O_4@LigC samples verify that the joint utilization of Lig and CMC of different sizes as starting materials contributes to the optimization of the Fe_3O_4 nanocrystals and the carbon skeleton.

According to the TEM characterization of the Fe_3O_4@LigC/CC nanocomposite in Figure 3a, Fe_3O_4 nanocrystals below 100 nm can be clearly observed in the carbon domain with good dispersion, which matches quite well with the FESEM characterization. The corresponding HRTEM characterization result for Fe_3O_4@LigC/CC in Figure 3b demonstrates that the individual nanocrystal has an average lattice spacing of around 0.253 nm, which belongs to the (3 1 1) direction for Fe_3O_4 crystals. The Fe_3O_4 nanocrystals are well decorated on the carbon-supporting domain. According to the EDS analysis result for Fe_3O_4@LigC/CC in Figure 3c, the Fe, O, and C elements are in good agreement from the mapping with the corresponding atomic percentages of 4.59, 9.59, and 85.23, respectively. In addition, the trace amount of S element probably originates from the Lig product.

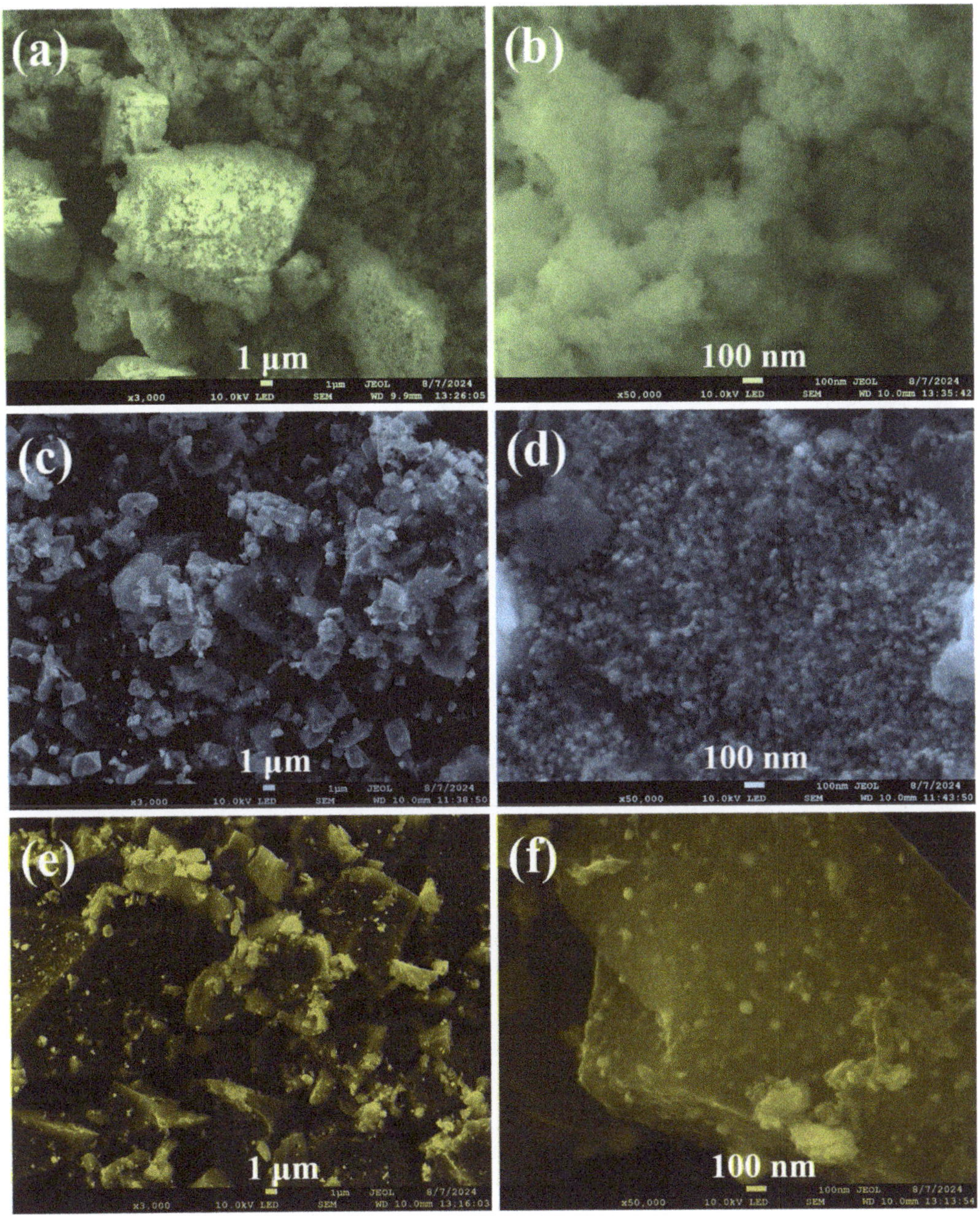

Figure 2. FESEM images of the Fe-Lig/CMC (**a**,**b**) and Fe_3O_4@LigC/CC (**c**,**d**) samples and the controlled Fe_3O_4@LigC sample (**e**,**f**) under corresponding magnifications.

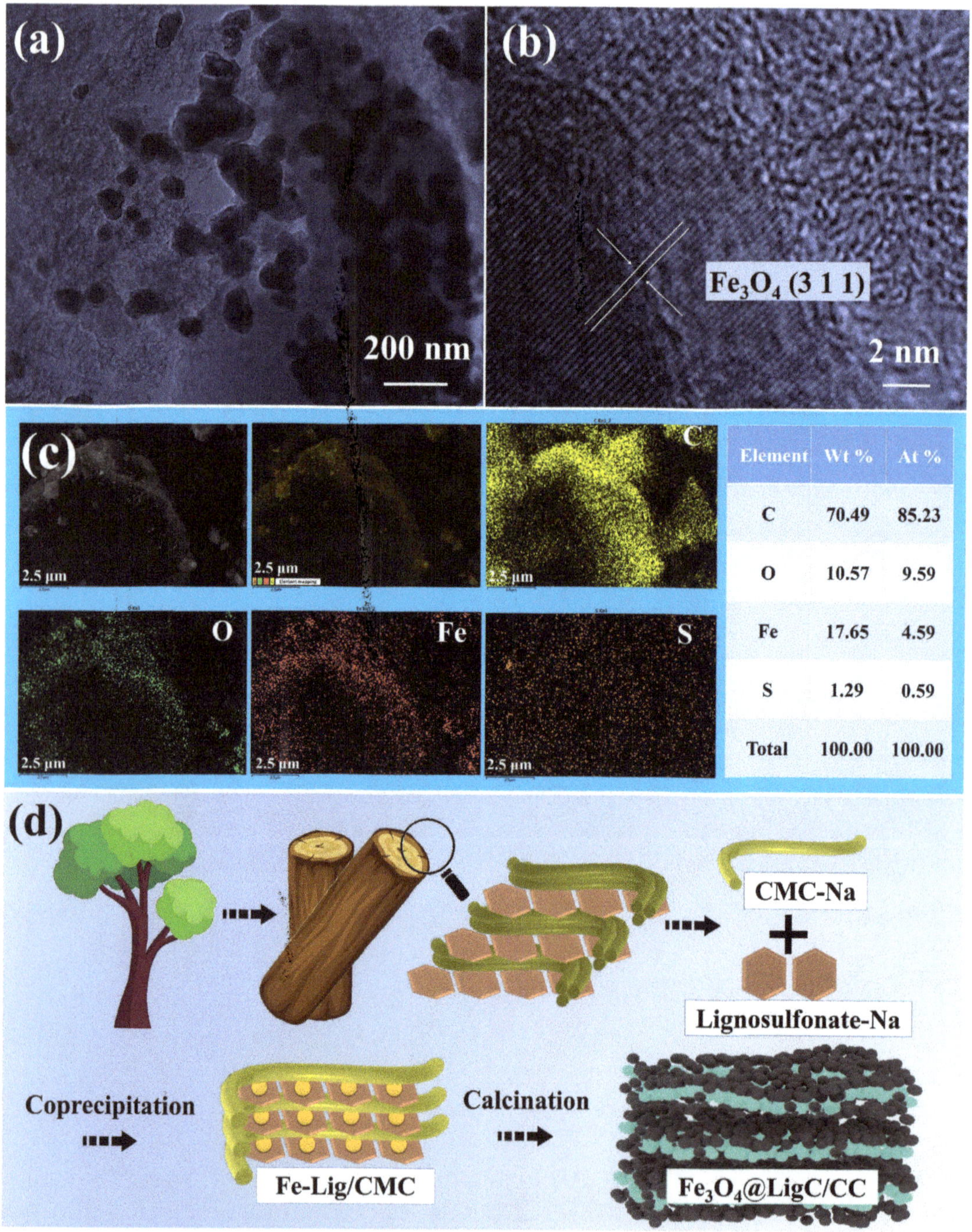

Figure 3. TEM (**a**) and HRTEM (**b**) images and the EDS testing result (**c**) of the Fe_3O_4@LigC/CC sample; illustration of the sustainable and scalable sample design from plant-derived products (**d**).

Summarizing the above sample characterization results, the sample designing and engineering strategy in the work can be illustrated in Figure 3d. Following the principles of green chemistry, the natural plant-derived products Lig and CMC with different sizes are jointly employed as both carbon precursors and coordination organic materials, and water molecules are used as the reaction medium. At an ambient temperature, Fe^{3+} ions can be effectively complexed by the functional groups from both Lig and CMC, while the two organic molecules are reconstructed simultaneously. Therefore, the Fe-Lig/CMC intermediate sample with an optimized microstructure can be directly used to synthesize the final Fe_3O_4@LigC/CC sample via a simple thermal treatment. In detail, the well-dispersed Fe^{3+} ions combine with the oxygen atoms to form iron oxide, which is generated due to the pyrolysis of the Lig/CMC precursor. The reduction capability of LigC/CC carbon atoms may convert the iron oxide phase to Fe_3O_4 crystals in the following stage, for which the crystal size is effectively controlled thanks to the atomic dispersion of the Fe^{3+} ions in the Lig/CMC matrix.

3.2. Lithium-Ion Storage Performances of the Fe_3O_4@LigC/CC Nanocomposite

Figure 4a exhibits the CV testing result of the Fe_3O_4@LigC/CC nanocomposite. In the first cycle, two small cathodic peaks can be seen at about 1.60 and 1.30 V. In addition, the most apparent cathodic one is situated around 0.51 V. The disappearance of these three peaks from the second cycle reveals the related reduction of the Fe_3O_4 species to metallic Fe domains and the accompanying generation of the irreversible solid electrolyte interphase (SEI) on the surface [27–30]. According to corresponding anodic scan curves, the two oxidizing peaks at 1.04 and 1.65 V indicate the reversible conversion of Fe phase to Fe_3O_4. In the resting cathodic curves, the prominent reducing peak can be found at a higher voltage of about 0.97 V. The improved kinetics in the electrode contributes to the above shift, originating from the nano-sized effect of transition metal oxide electrodes for cycling [31]. Therefore, the electrochemical reaction for the Fe_3O_4 nanocrystals in the Fe_3O_4@LigC/CC nanocomposite can be interpreted by the equation: $Fe_3O_4 + 8Li^+ + 8e^- \leftrightarrow 3Fe + 4Li_2O$. The outlines of the CV curves from the second cycle are nearly identical, demonstrating the satisfactory structure stability of the Fe_3O_4@LigC/CC sample [32].

The testing curves in relation to the specific capacities and working voltage for Fe_3O_4@LigC/CC are described in Figure 4b. In the starting discharging stage, the obvious long voltage platform belonging to the initial lithiation process could be identified near 0.7 V. This voltage platform shifts to a higher position at about 0.8 to 1.0 V for the resting ones. On the other hand, the corresponding charging parts demonstrate the main voltage platform between 1.5 and 2.0 V. These platforms significantly support the results of the CV testing. The Fe_3O_4@LigC/CC nanocomposite can deliver the initial specific charging and discharging capacities of 936.6 and 1571.6 mAh g^{-1}, respectively, and the corresponding coulombic efficiency of 59.6% can be obtained. The irreversible capacity can be commonly attributed to the irreversible SEI generation in this cycle [33,34].

Figure 4c displays the low-current cycling performance of the Fe_3O_4@LigC/CC nanocomposite at 200 mAh g^{-1}. It can be found that the specific charging and discharging capacities were 668.5 and 1037.4 mAh g^{-1} in the first cycle, respectively. Therefore, a higher coulombic efficiency of about 64.4% could be obtained compared with that at 1000 mAh g^{-1}. The Fe_3O_4@LigC/CC composite delivers a gradual declining trend about the specific capacities for the first 15 cycles before rising for the following cycles. Stabilized specific capacity is delivered for the same sample and can be observed in the subsequent cycles, for which a high reversible one of 820.6 mAh g^{-1} after 100 cycles could be given. By contrast, the controlled Fe_3O_4@LigC sample delivers a significantly inferior specific capacity at each cycle.

The rate capability for the Fe_3O_4@LigC/CC nanocomposite can be seen in Figure 4d. With staged increases of current rates, the corresponding capacities drop in reverse. When tested by 100, 200, 500, 1000, 2000, and 5000 mA g^{-1}, mean capacities of about 922.5, 869.3, 793.3, 710.8, 602.0, and 381.4 mAh g^{-1} could be given in return, respectively. For the last stage of 100 mAh g^{-1}, it is not surprising that the corresponding specific capacities significantly rise and remain stable, with a high average capacity of 1002.9 mAh g^{-1}. However, the control Fe_3O_4@LigC sample has a much inferior electrochemical behavior in the rate testing.

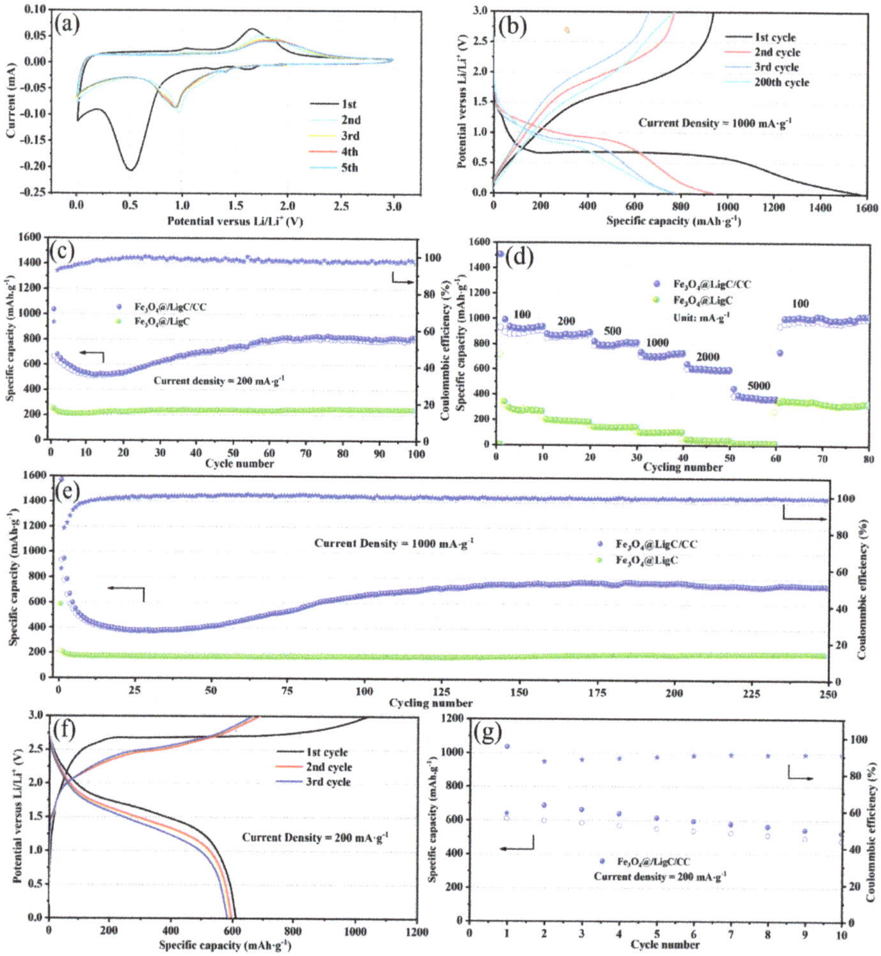

Figure 4. The CV curves for the initial five cycles (**a**) and charge and discharge voltage profile (**b**) of the Fe_3O_4@LigC/CC sample; comparisons of the low-current cycling performances (**c**), rate capabilities (**d**), and high-current cycling performances (**e**) of the Fe_3O_4@LigC/CC and Fe_3O_4@LigC samples; charge/discharge voltage profile (**f**) and cycling performance (**g**) of the Fe_3O_4@LigC/CC sample in Fe_3O_4@LigC/CC//$LiFePO_4$ full cell.

The high rate performances of the above two samples are compared in Figure 4e. Similar to the observations at 200 mAh g^{-1}, the Fe_3O_4@LigC/CC nanocomposite shows a gradual capacity decline during the first 25 cycles before the capacity rises, which remains stable after about 150 cycles. More inspiringly, the Fe_3O_4@LigC/CC sample delivers a specific capacity of 750.5 mAh g^{-1} after the continual 250 cycles, which is much higher than the value of 210.3 mAh g^{-1} for the Fe_3O_4@LigC nanocomposite.

The testing curves in relation to the specific capacities and working voltage for Fe_3O_4@LigC/CC in the Fe_3O_4@LigC/CC//$LiFePO_4$ full cell are illustrated in Figure 4f.

Similar to that in the half cell, this sample can deliver initial specific charging and discharging capacities of 1137.7 and 610.5 mAh g^{-1}, respectively, and a corresponding coulombic efficiency of 58.8% can be obtained. Moreover, a modest discharging voltage platform at about 1.5 V can be observed for the full cell. As can be seen in Figure 4g, a reversible capacity of 483.5 mAh g^{-1} can be maintained after 10 cycles, and the coulombic efficiency gradually rises to 91.7% for this sample.

From the above electrochemical testing results, it is obvious that the coulombic efficiency of the Fe_3O_4@LigC/CC electrode gradually rises in the starting stages and stays stable in the remaining cycles. Moreover, the specific capacity drops at first and rises until it is stable in the subsequent testing. According to our previous research [34], there is additional metallic iron formation in this electrode during the starting cycles, which may lead to a decrease in the specific capacity and the coulombic efficiency. On the other hand, the in situ-formed iron atoms could contribute to the generation of a conductive polymer film at low voltages, which increases the specific capacity via pseudocapacitive behavior. In the last stage, the Fe domains gradually take part in the lithium-ion storage reaction caused by the gradually enhanced reaction environment. In addition, the coulombic efficiency increases accordingly and remains stable.

In Table 1, the key parameters for the lithium-ion storage of the carbon skeleton-supported Fe_3O_4 samples are clearly listed. Compared with the competitors, the Fe_3O_4@LigC/CC nanocomposite could not only provide high and stable reversible capacity but also maintain a long cycling life. Particularly, the raw materials are naturally rich, and the whole sample preparation can be considered to possess the obvious merit of sustainability.

Table 1. Comparison of the lithium-ion storage performance of carbonaceous matrix-supported Fe_3O_4 materials.

Sample Name	Reversible Capacity ($mAh{\cdot}g^{-1}$)	Cycle Number	Current Rate ($mA{\cdot}g^{-1}$)	Year Published
Fe_3O_4@SnO_7/MXene-10 [35]	626.1	900	1000	2024
Fe_2O_3@ Fe_3O_4-5 [36]	707.8	800	1000	2023
H-TiO_2/C/Fe_3O_4@NiO [37]	897.47	200	200	2023
Fe_3O_4/C-500 [38]	718	500	200	2023
Si-QDs/Fe_3O_4/rGO [39]	1367.1	80	100	2023
Fe_3O_4@void@N-Doped C-5 [40]	1222	100	200	2022
Fe_3O_4@C [41]	291.7	300	1000	2022
Fe_3O_4/MWCNT [42]	662	100	50	2022
Fe_3O_4/CNTS@C [43]	612	200	1000	2021
Fe_3O_4@LigC/CC	**820.6** **750.5**	**100** **250**	**200** **1000**	**This work**

The FESEM images of the prepared electrode were further used to examine the morphological change for the Fe_3O_4@LigC/CC electrode. In Figure 5a,b, the micro-sized Fe_3O_4@LigC/CC nanocomposite can be obviously seen in the fresh electrode, which has a slightly rough surface. The Fe_3O_4@LigC/CC nanocomposite shows a similar microstructure to that in Figure 2d. After continual cycling 250 times, in Figure 5c,d, the Fe_3O_4@LigC/CC electrode remains intact except for the slightly smoothened surface caused by SEI generation. In particular, the Fe_3O_4 nanocrystals can be clearly identified in this electrode. This result confirms the good stability of the Fe_3O_4@LigC/CC electrode.

The Nyquist curves with the inset equivalent circuit model are shown in Figure 6a. The Nyquist diagrams are typically constituted by one semicircle as well as a connecting a straight line. The first intercept in the high-frequency region on the Z' axis refers to the inherent resistance in the cell (*Rs*). The next semicircle is located in the high-frequency and mid-frequency region, corresponding to the charge transfer resistor (*Rct*). Lastly, the straight line in the low-frequency region corresponds to the Warburg impedance (*Zw*), representing the lithium-ion diffusion resistance [44–46]. The obviously reduced *Rct* value of the Fe_3O_4@LigC/CC composite electrode after cycling indicates improved charge transfer resistance and enhanced electronic conductivity.

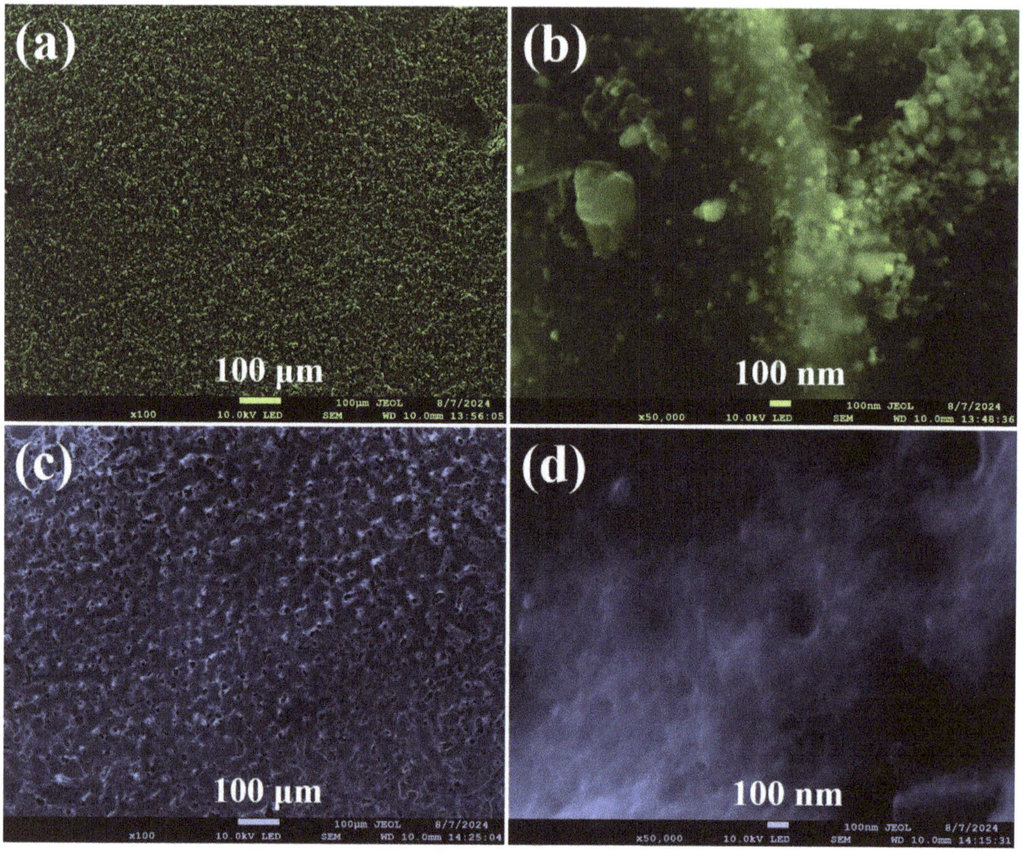

Figure 5. FESEM images of the Fe_3O_4@LigC/CC electrode before cycling (**a**,**b**) and after 250 cycles (**c**,**d**) under the corresponding magnifications.

Cyclic voltammetry was further used to investigate the pseudocapacitive effect for the Fe_3O_4@LigC/CC composite electrode. The series of CV curves in Figure 6b exhibits similar outlines at different current scanning rates. The kinetics of the nanocomposite could be measured on the basis of the equation regarding the peak current (i) and the scan rate (v) using the power-law formula $i = av^b$, in which b is a constant from plotting ln (i) against ln (v) [47]. A b value close to 0.5 implies that the diffusion process possesses the dominating position, while a b value close to 1.0 means that the pseudocapacitive process plays the main role [48]. Figure 6c reveals the plotting of ln (i)-ln (v) in Fe_3O_4@LigC/CC electrode. The b values obtained by the cathodic and anodic peaks are 0.69 and 0.70, accordingly, indicating the co-existence of both pseudocapacitive and diffusion behaviors for the electrode [49]. The pseudocapacitive contributions can be further evaluated by the equation $i = k_1v + k_2v^{1/2}$ [50,51]. Therefore, the pseudocapacitive contribution in Figure 6d increases following the increase in the scanning rate of the Fe_3O_4@LigC/CC electrode. The pseudocapacitive behavior of this electrode is probably triggered by the lithiation of Fe_3O_4 nanoparticles and adsorption on the unique LigC/CC carbon matrix [29,52,53]. Consequently, the accelerated surface lithium-ion storage reaction accompanied by the reduced structural destruction of the nanocomposite lead to the superior electrochemical performance of the Fe_3O_4@LigC/CC electrodes.

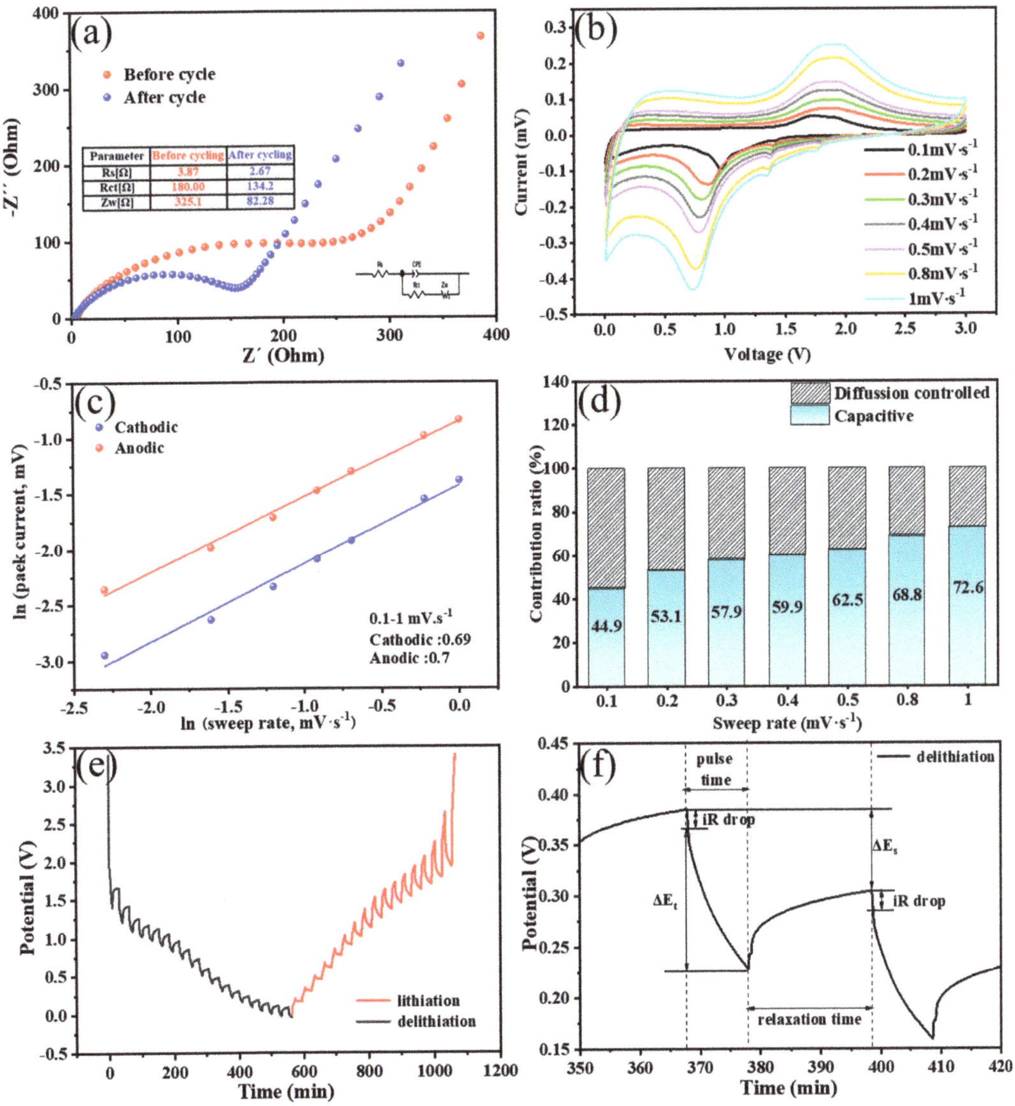

Figure 6. Nyquist plot of the Fe_3O_4@LigC/CC electrode before cycling and after cycling with inset equivalent circuit obtained by fitting the EIS data (**a**); CV curves of the Fe_3O_4@LigC/CC electrode at various scan rates from 0.1 to 1.0 mV s^{-1} (**b**); the ln (i)-ln (v) plot (**c**); capacitive contribution and diffusion ratios at different rates (**d**); GITT curves of the Fe_3O_4@LigC/CC electrode (discharge/charge state) (**e**); voltage (V vs Li^+/Li) versus time curve for one single GITT test (**f**).

The lithium-ion diffusion coefficient (D_{Li^+}) for the Fe_3O_4@LigC/CC nanocomposite electrode is important for understanding charging and discharging behaviors. Hence, the constant current intermittent titration technique (GITT) is employed. The GITT curve of the Fe_3O_4@LigC/CC electrode is illustrated in Figure 6e,f. According to Fick's second law, the D_{Li^+} for Fe_3O_4@LigC/CC can be obtained as follows: $D_{Li^+} = \frac{4L^2}{\pi\tau}\left(\frac{\Delta E_s}{\Delta E_t}\right)^2$, where τ is the pulse time, and ΔE_t and ΔE_s represent the gaps in the equilibrium potential and the current pulse, respectively [54–57]. The calculated results of LgD_{Li^+} show that the average value of D_{Li^+} in the lithiation state (discharge process) is 1.50×10^{-12} cm^2 s^{-1}, and the average value of D_{Li^+} in the delithiation state (charge process) is 1.98×10^{-12} cm^2 s^{-1}.

4. Conclusions

Following the principles of green chemistry and bionics, the natural plant-derived products Lig and CMC with different sizes are jointly employed as both carbon precursors and coordination organic materials to react with Fe^{3+} ions in mild hydrothermal conditions. The Fe-Lig/CMC intermediate sample with optimized microstructure is subsequently converted to the Fe_3O_4@LigC/CC sample after calcination. The Fe_3O_4@LigC/CC electrode shows more attractive specific capacities of 820.6 mAh g^{-1} after 100 cycles under a current rate of 100 mA·g^{-1} and 750.5 mAh g^{-1} after 250 cycles, together with a good rate capability. The robust mechanical strength improves the electric and lithium-ion transferring condition, and the capacitive behaviors result in improved lithium-ion storage performance. This sustainable sample synthesis protocol and engineered Fe_3O_4@LigC/CC nanocomposite can be further developed and have good prospects for application in wider fields

Supplementary Materials: The following supporting information can be downloaded at: https://www.mdpi.com/article/10.3390/batteries10100357/s1.

Author Contributions: B.X. proposed the ideas, steps, and details of the experiment and wrote the article; most of the experiments were performed by H.Z.; and all the authors analyzed the data and discussed the conclusions. All authors have read and agreed to the published version of the manuscript.

Funding: This work was financially supported by the World-Class University and Discipline, the Taishan Scholar's Advantageous and Distinctive Discipline Program, and the World-Class Discipline Program of Shandong Province.

Institutional Review Board Statement: Not applicable.

Informed Consent Statement: Not applicable.

Data Availability Statement: The data supporting the findings of this study are fully included within the main text of the article. Additional information or raw data may be available upon reasonable request from the corresponding author.

Conflicts of Interest: The authors declare no conflicts of interest.

References

1. Goodenough, J.B.; Park, K.S. The Li-ion rechargeable battery: A perspective. *J. Am. Chem. Soc.* **2013**, *135*, 1167–1176. [CrossRef] [PubMed]
2. Mihit, H.P.; Manikandan, P.; Vilas, G.P. Reserve lithium-ion batteries: Deciphering in situ lithiation of lithium-ion free vanadium pentoxide cathode with graphitic anode. *Carbon* **2023**, *203*, 561–570. [CrossRef]
3. Kebede, A.; Kalogiannis, T.; Van Mierlo, J.; Berecibar, M. A comprehensive review of stationary energy storage devices for large scale renewable energy sources grid integration. *Renew. Sustain. Energy Rev.* **2022**, *159*, 112213. [CrossRef]
4. Shia, C.; Wang, T.Y.; Liao, X.B.; Qie, B.Y.; Yang, P.F.; Chen, M.J.; Wang, X.; Srinivasane, A.; Cheng, Q.; Ye, Q.; et al. Accordion-like stretchable Li-ion batteries with high energy density. *Energy Storage Mater.* **2019**, *17*, 136–142. [CrossRef]
5. Cavers, H.; Molaiyan, P.; Abdollahifar, M.; Lassi, U.; Kwade, A. Perspectives on Improving the Safety and Sustainability of High Voltage Lithium-Ion Batteries Through the Electrolyte and Separator Region. *Adv. Energy Mater.* **2022**, *12*, 2200147–2200179. [CrossRef]
6. Chen, M.; Wang, E.; Liu, Q.; Guo, X.; Chen, W.; Chou, S.L.; Dou, S.X. Recent progress on iron- and manganese-based anodes for sodium-ion and potassium-ion batteries. *Energy Storage Mater.* **2019**, *19*, 163–178. [CrossRef]
7. Wang, Y.; Song, Y.; Xia, Y. Electrochemical capacitors: Mechanism, materials, systems, characterization and applications. *Chem. Soc. Rev.* **2016**, *45*, 5925–5950. [CrossRef]

8. Jia, R.X.; Yu, L.B.; Han, Z.Q.; Liu, S.; Shang, P.P.; Deng, S.Q.; Liu, X.H.; Xu, B.H. Synthesis of a MOF-derived magnetite quantum dots on surface modulated reduced graphene oxide composite for high-rate lithium-ion storage. *RSC Appl. Interfaces.* **2024**, *1*, 233–244. [CrossRef]
9. Xu, Z.; Yu, S.Y.; Xie, X.Y.; Li, Q.X.; Ding, L.; Chen, M.L.; Tu, J.; Xing, K.Y.; Cheng, Y.H. Target-triggered Fe_3O_4@NPC-UCNPs assembly for photoactivatable biosensing of Aflatoxin B1. *Chem. Eng. J.* **2023**, *407*, 144028. [CrossRef]
10. Zhang, Y.F.; Qiu, L.G.; Yuan, Y.P.; Zhu, Y.J.; Jiang, X.; Xiao, J.D. Magnetic Fe_3O_4@C/Cu and Fe_3O_4@CuO core–shell composites constructed from MOF-based materials and their photocatalytic properties under visible light. *Appl. Catal. B-Environ.* **2014**, *144*, 863–869. [CrossRef]
11. Purna, K.B.; Gitashree, D.; Priyakshree, B.; Benjamin, L.O.; Manash, R.D. Fe_3O_4 quantum dots anchored on functionalized graphene: A multimodal platform for sensing and remediation of Cr (VI). *Chem. Eng. J.* **2023**, *474*, 145797. [CrossRef]
12. Wang, L.J.; Liu, F.H.; Pal, A.; Ning, Y.S.; Wang, S.; Zhao, B.Y.; Bradley, R.; Wu, W.P. Ultra-small Fe_3O_4 nanoparticles encapsulated in hollow porous carbon nanocapsules for high performance supercapacitors. *Carbon* **2021**, *179*, 327–336. [CrossRef]
13. Dühnen, S.; Betz, J.; Kolek, M.; Schmuch, R.; Winter, M.; Placke, T. Toward Green Battery Cells: Perspective on Materials and Technologies. *Small Methods* **2020**, *4*, 2000039. [CrossRef]
14. Wu, X.J.; Jiang, C.; Wang, J.; Pu, Y.; Ragauskas, A.; Li, S.; Yang, B. Lignin-derived electrochemical energy materials and systems. *Biofuels. Bioprod. Biorefin.* **2020**, *14*, 650–672. [CrossRef]
15. Norgren, M.; Edlund, H. Lignin: Recent advances and emerging applications. *Curr. Opin. Colloid Interface Sci.* **2014**, *19*, 409–416. [CrossRef]
16. Boerjan, W.; Ralph, J.; Baucher, M. Lignin Biosynthesis. *Annu. Rev. Plant Biol.* **2003**, *54*, 519–546. [CrossRef]
17. Feng, X.; Ren, D.; He, X.; Ouyang, M. Mitigating Thermal Runaway of Lithium-Ion Batteries. *Joule* **2020**, *4*, 743–770. [CrossRef]
18. Rinaldi, R.; Jastrzebski, R.; Clough, M.; Ralph, T.J.; Kennema, M.P.C.; Bruijnincx, B.M. Weckhuysen. Paving the Way for Lignin Valorisation: Recent Advances in Bioengineering, Biorefining and Catalysis. *Angew. Chem. Int. Ed. Engl.* **2016**, *55*, 8164. [CrossRef]
19. Lai, X.; Jin, C.; Yi, W.; Han, X.; Feng, X.; Zheng, Y.; Ouyang, M. Mechanism, modeling, detection, and prevention of the internal short circuit in lithium-ion batteries: Recent advances and perspectives. *Energy Storage Mater.* **2021**, *35*, 470–499. [CrossRef]
20. Pachfule, P.; Shinde, D.; Majumder, M.; Xu, Q. Fabrication of carbon nanorods and graphene nanoribbons from a metal–organic framework. *Nat. Chem* **2016**, *8*, 718–724. [CrossRef]
21. Graf, D.; Molitor, F.; Ensslin, K.; Stampfer, C.; Jungen, A.; Hierold, C.; Wirtz, L. Spatially Resolved Raman Spectroscopy of Single- and Few-Layer Graphene. *Nano Lett.* **2007**, *7*, 238–242. [CrossRef] [PubMed]
22. Chen, W.F.; Li, S.R.; Chen, C.H.; Yan, L.F. Self-Assembly and Embedding of Nanoparticles by In Situ Reduced Graphene for Preparation of a 3D Graphene/Nanoparticle Aerogel. *Adv. Mater.* **2011**, *23*, 5679–5683. [CrossRef] [PubMed]
23. Shen, Z.; Xing, H.; Zhu, Y.; Ji, X.; Liu, Z.; Wang, L. Synthesis and enhanced microwave-absorbing properties of SnO_2/α-Fe_2O_3@RGO composites. *J. Mater. Sci. Mater. Electron.* **2017**, *28*, 13896–13904. [CrossRef]
24. Chandra, V.; Park, J.; Chun, Y.; Lee, J.W.; Hwang, I.C.; Kim, K.S. Water-Dispersible Magnetite-Reduced Graphene Oxide Composites for Arsenic Removal. *ACS Nano* **2010**, *4*, 3979–3986. [CrossRef]
25. Li, Y.; Wu, X.; Liu, C.; Wang, S.; Zhou, P.; Zhou, T.; Miao, Z.; Xing, W.; Zhuo, S.; Zhou, J. Fluorinated multi-walled carbon nanotubes as cathode materials of lithium and sodium primary batteries: Effect of graphitization of carbon nanotubes. *J. Mater. Chem. A* **2019**, *7*, 7128–7137. [CrossRef]
26. Li, H.H.; Saini, A.; Xu, R.Y.; Wang, N.; Lv, X.X.; Wang, Y.P.; Yang, T.; Chen, L.; Jiang, H.B. Hierarchical Fe_3O_4@C nanofoams derived from metal–organic frameworks for high-performance lithium storage. *Rare Met.* **2020**, *39*, 1072–1081. [CrossRef]
27. Wei, Z.; Sarwar, S.; Azam, S.; Ahasan, M.R.; Voyda, M.; Zhang, X.; Wang, R. Ultrafast microwave synthesis of $MoTe_2$@graphenecomposites accelerating polysulfide conversion and promoting Li_2S nucleation for high-performance Li-S batteries. *J. Colloid Interface Sci.* **2023**, *635*, 391–405. [CrossRef]
28. Wang, X.; Liu, X.; Wang, G.; Xia, Y.; Wang, H. One-dimensional hybrid nanocomposite of high-density monodispersed Fe_3O_4 nanoparticles and carbon nanotubes for high-capacity storage of lithium and sodium. *J. Mater. Chem. A* **2016**, *47*, 18532–18542. [CrossRef]
29. Cui, Y.; Feng, W.; Liu, W.; Li, J.; Zhang, Y.; Du, Y.; Li, M.; Huang, W.; Wang, H.; Liu, S. Template-assisted loading of Fe_3O_4 nanoparticles inside hollow carbon "rooms" to achieve high volumetric lithium storage. *Nanoscale* **2020**, *12*, 10816–10826. [CrossRef]
30. He, Q.; Chen, H.; Chen, X.; Zheng, J.; Que, L.; Yu, F.; Zhao, J.; Xie, Y.; Huang, M.; Lu, C.; et al. Tea-Derived Sustainable Materials. *Adv. Funct. Mater.* **2024**, *34*, 2310226. [CrossRef]
31. Xu, B.H.; Guan, X.G.; Zhang, L.Y.; Liu, X.W.; Jiao, Z.B.; Liu, X.H.; Hu, X.; Zhao, X.S. A simple route to preparing γ-Fe_2O_3/RGO composite electrode materials for lithium-ion batteries. *J. Mater. Chem. A* **2018**, *6*, 4048–4054. [CrossRef]
32. Ma, C.; Shi, J.; Zhao, Y.; Song, N.J.; Wang, Y. A novel porous reduced microcrystalline graphene oxide supported Fe_3O_4@C nanoparticle composite as anode material with excellent lithium storage performances. *Chem. Eng. J.* **2017**, *326*, 507–517. [CrossRef]
33. Zhang, W.; Li, X.; Liang, J.; Tang, K.; Zhu, Y.; Qian, Y. One-step thermolysis synthesis of two-dimensional ultrafine Fe_3O_4 particles/carbon nanonetworks for high-performance lithium-ion batteries. *Nanoscale* **2016**, *8*, 4733–4741. [CrossRef] [PubMed]

34. Kong, X.L.; Shan, L.; Zhang, R.; Bao, S.C.; Tu, M.Y.; Jia, R.X.; Yu, L.B.; Li, H.; Xu, B.H. Controllable engineering magnetite nanoparticles dispersed in a hierarchical amylose derived carbon and reduced graphene oxide framework for lithium-ion storage. *J. Colloid Interface Sci.* **2022**, *628*, 1–13. [CrossRef] [PubMed]
35. Hu, G.; Zhong, K.; Yu, R.; Liu, Z.; Zhang, Y.; Wu, J.; Zhou, L.; Mai, L. Enveloping SiOx in N-doped carbon for durable lithium storage via an eco-friendly solvent-free approach. *J. Mater. Chem.* **2020**, *8*, 13285–13291. [CrossRef]
36. Duan, X.; Liu, J.Q.; Lv, F.S.; Liu, T.; Cui, W.B.; Wang, J.; Wang, Q.; Yuan, S. 3D porous structure Fe_3O_4@SnO_2/MXene composites with enhanced electrochemical performance for lithium ion battery anode. *J. Energy Storage.* **2024**, *86*, 111308. [CrossRef]
37. Lv, X.X.; Zhang, Y.; Lin, W.; Yang, A.W.; Liang, J. Facile synthesis of Fe_3O_4 ultrathin layer coated Fe_2O_3 composite anode for enhanced lithium-ion storage. *J. Electroanal Chem.* **2023**, *947*, 117758. [CrossRef]
38. Wu, Q.C.; Qiu, S.; Yin, M.; Li, R.; Wang, Y.; Jin, D.Q.; Yang, R.H.; Yong, D.M.; Xie, W. Hydrogenated titanium dioxide modifed core–shell structure Fe3O4@NiO for lithium-ion battery anode material. *Ionics* **2023**, *29*, 2227–2240. [CrossRef]
39. Li, M.; Ma, W.; Tan, F. Fe_3O_4@C-500 anode derived by commercial ammonium ferric citrate for advanced lithium ion batteries. *J. Power Sources* **2023**, *574*, 233146. [CrossRef]
40. Hu, W.; He, K.; Wu, S. Ordered sandwich silicon quantum dot/Fe_3O_4/reduced graphene oxide architectures for high-performance lithium-ion batteries. *J. Alloy Compd.* **2023**, *943*, 168947. [CrossRef]
41. Wang, J.; Hu, Q.; Hu, W. Preparation of hollow core-shell Fe_3O_4/nitrogen-doped carbon nanocomposites for lithium-ion batteries. *Molecules* **2022**, *27*, 396. [CrossRef] [PubMed]
42. Huang, J.; Dai, Q.; Cui, C. Cake-like porous Fe_3O_4@C nanocomposite as high-performance anode for Li-ion battery. *J. Electroanal. Chem.* **2022**, *918*, 116508. [CrossRef]
43. Sumit, R.S.; Jagannatham, M.; Gautam, R.; Rao Rikka, V.; Prakash, R.; Mallikarjunaiah, K.J.; Srinivas Reddy, G. A facile synthesis of raspberry-shaped Fe_3O_4 nanoaggregate and its magnetic and lithium-ion storage properties. *Mater Sci. Eng. B Adv.* **2022**, *282*, 115771. [CrossRef]
44. Xu, J.L.; Zhang, X.; Miao, Y.X.; Wen, M.-X.; Yan, W.-Y.; Lu, P.; Wang, Z.-R.; Sun, Q. In-situ plantation of Fe_3O_4@C nanoparticles on reduced graphene oxide nanosheet as high-performance anode for lithium/sodium-ion batteries. *Appl. Surf. Sci.* **2021**, *546*, 149163. [CrossRef]
45. Ding, R.; Zhang, J.; Qi, J.; Li, Z.; Wang, C.; Chen, M. N-Doped Dual Carbon-Confined 3D Architecture rGO/Fe_3O_4/AC Nanocomposite for High-Performance Lithium-Ion Batteries. *ACS Appl. Mater. Interfaces.* **2018**, *10*, 13470–13478. [CrossRef]
46. Xie, Y.; Qiu, Y.; Tian, L.; Liu, T.; Su, X. Ultrafine hollow Fe_3O_4 anode material modified with reduced graphene oxides for high-power lithium-ion batteries. *J. Alloys Compd.* **2022**, *894*, 162384. [CrossRef]
47. Li, H.; Wang, J.; Li, Y.; Zhao, Y.; Tian, Y.; Kurmanbayeva, I.; Bakenov, Z. Hierarchical sandwiched Fe_3O_4@C/Graphene composite as anode material for lithium-ion batteries. *J. Electroanal. Chem.* **2019**, *847*, 113240. [CrossRef]
48. Yin, L.; Gao, Y.J.; Jeon, I.; Yang, H.; Kim, J.P.; Jeong, S.Y.; Cho, C.R. Rice-panicle-like γ-Fe_2O_3@C nanofibers as high-rate anodes for superior lithium-ion batteries. *Chem. Eng. J.* **2019**, *356*, 60–68. [CrossRef]
49. Yan, Y.; Lu, X.; Li, Y.J.; Song, J.Q.; Tian, Q.; Yang, L.; Sui, Z. Dispersive Fe_3O_4 encapsulated in porous carbon for high capacity and long-life anode of lithium-ion batteries. *J. Alloys Compd.* **2022**, *899*, 163342. [CrossRef]
50. Ma, Q.; Zhao, Z.Q.; Zhao, Y.; Xie, H.W.; Xing, P.F.; Wang, D.H.; Yin, H.Y. A self-driven alloying/dealloying approach to nanostructuring micro-silicon for high-performance lithium-ion battery anodes. *Energy Storage Mater.* **2021**, *34*, 768–777. [CrossRef]
51. Yuan, J.; Gan, Y.; Xu, X.; Mu, M.; He, H.; Li, X.; Zhang, X.; Liu, J. Construction of Fe_7Se_8@Carbon nanotubes with enhanced sodium/potassium storage. *J. Colloid Interface Sci.* **2022**, *626*, 355–363. [CrossRef] [PubMed]
52. Augustyn, V.; Simon, P.; Dunn, B. Pseudocapacitive oxide materials for high-rate electrochemical energy storage. *Energy Environ. Sci.* **2014**, *7*, 1597–1614. [CrossRef]
53. Song, J.; Ji, Y.; Li, Y.; Lu, X.; Ren, W.; Tian, Q.; Chen, J.; Yang, L. Porous carbon assisted carbon nanotubes supporting Fe_3O_4 nanoparticles for improved lithium storage. *Ceram. Int.* **2021**, *47*, 26092–26099. [CrossRef]
54. Wang, F.M.; Alemu, T.; Yeh, N.H.; Wang, X.C.; Lin, Y.W.; Hsu, C.C.; Chang, Y.J.; Liu, C.H.; Chuang, C.I.; Hsiao, L.H.; et al. Interface Interaction Behavior of Self-Terminated Oligomer Electrode Additives for a Ni-Rich Layer Cathode in Lithium-Ion Batteries: Voltage and Temperature Effects. *ACS Appl. Mater. Interfaces* **2019**, *11*, 39827–39840. [CrossRef]
55. Wu, H.B.; Chen, J.S.; Hng, H.H.; Lou, X.W. Nanostructured metal oxide-based materials as advanced anodes for lithium-ion batteries. *Nanoscale* **2012**, *4*, 2526–2542. [CrossRef]
56. Bal, B.; Ozdogru, B.; Nguyen, D.T.; Li, Z.; Murugesan, V.; Çapraz, Ö.Ö. Probing the Formation of Cathode-Electrolyte Interphase on Lithium Iron Phosphate Cathodes via Operando Mechanical Measurements. *ACS Appl. Mater. Interfaces* **2023**, *15*, 42449–42459. [CrossRef]
57. Zhang, H.; Zong, P.; Chen, M.; Jin, H.; Bai, Y.; Li, S.; Ma, F.; Xu, H.; Lian, K. In Situ Synthesis of Multilayer Carbon Matrix Decorated with Copper Particles: Enhancing the Performance of Si as Anode for Li-Ion Batteries. *ACS Nano* **2019**, *13*, 3054–3062. [CrossRef]

batteries

MDPI

Review

Synthesis Methods of Si/C Composite Materials for Lithium-Ion Batteries

Inkyu Park, Hanbyeol Lee and Oh B. Chae *

School of Chemical, Biological and Battery Engineering, Gachon University, Seongnam-si 13120, Republic of Korea; dlsrb778@gachon.ac.kr (I.P.); byulbbang99@gachon.ac.kr (H.L.)
* Correspondence: obchae@gachon.ac.kr; Tel.: +82-31-750-8944

Abstract: Silicon anodes present a high theoretical capacity of 4200 mAh/g, positioning them as strong contenders for improving the performance of lithium-ion batteries. Despite their potential, the practical application of Si anodes is constrained by their significant volumetric expansion (up to 400%) during lithiation/delithiation, which leads to mechanical degradation and loss of electrical contact. This issue contributes to poor cycling stability and hinders their commercial viability, and various silicon–carbon composite fabrication methods have been explored to mitigate these challenges. This review covers key techniques, including ball milling, spray drying, pyrolysis, chemical vapor deposition (CVD), and mechanofusion. Each method has unique benefits; ball milling and spray drying are effective for creating homogeneous composites, whereas pyrolysis and CVD offer high-quality coatings that enhance the mechanical stability of silicon anodes. Mechanofusion has been highlighted for its ability to integrate silicon with carbon materials, showing the potential for further optimization. In light of these advancements, future research should focus on refining these techniques to enhance the stability and performance of Si-based anodes. The optimization of the compounding process has the potential to enhance the performance of silicon anodes by addressing the significant volume change and low conductivity, while simultaneously addressing cost-related concerns.

Keywords: silicon; carbon; Si/C composite; lithium-ion batteries; anode

1. Introduction

Amid growing concerns regarding environmental pollution and global warming, efforts to reduce CO_2 emissions have driven the development of the electric vehicle (EV) market as an alternative to internal combustion engines [1–4]. Currently, lithium-ion batteries are predominantly used in these electric vehicles and are also widely employed in portable electronic devices [5–12]. Lithium-ion batteries are composed of a cathode, an anode, a separator, and an electrolyte. The cathode and anode store electrical energy in the form of chemical energy. When charging a battery, the key considerations include stability, energy density, and cycle life [13–15]. However, lithium-ion batteries currently do not deliver the desired amount of energy, which is increasingly demanded by users [16]. To overcome this challenge, it is necessary to enhance both the energy density and capacity of lithium-ion batteries [17]. Consequently, significant research has been conducted to improve these aspects [3,18–22]. Graphite is currently used as the anode material in lithium-ion batteries [23,24]. The theoretical capacity of this anode material is 372 mAh/g [25–27], which contributes approximately 150 Wh/kg of energy. However, this is insufficient to satisfy the energy demands of electric vehicles [28,29]. Therefore, there is a need to develop alternative anode materials to replace graphite [30–32]. Si-based anode materials are an alternative to graphite anodes [33]. This is because, while graphite can accommodate one lithium ion for every six carbon atoms, silicon can host up to fifteen lithium ions for every four silicon atoms [34,35]. Silicon has a theoretical capacity of approximately 4200 mAh/g, which is about ten times higher than that of graphite, and its operating voltage is approximately 0.4 V versus Li/Li^+ [36–38]. In addition, Si is environment-friendly and abundant, making it a promising candidate for next-generation anode materials [39]. However, the primary reason for the limited application of Si anodes is the significant volumetric expansion (approximately 400%) that occurs during electrochemical reactions, leading to particle cracking [20,40,41]. As a result, direct contact occurs between the electrolyte and silicon particles, leading to undesirable side reactions. Continuous charge and discharge cycles cause repeated breakdown and formation of a solid electrolyte interphase (SEI) layer, which in turn depletes the amount of active lithium [42]. Finally, the low electrical and ionic conductivities of Si are also important issues [43]. This results in slower charge and discharge rates and increased resistance within the electrode [44]. To address these issues, ongoing research has focused on several approaches, including nanostructuring [45–49], composite materials, electrolyte optimization [50–55], binder improvements [56–60], and interface engineering [61–65]. Among these approaches, this study aims to examine silicon–carbon composites. The rationale for this choice is that the use of Si alone has limitations in effectively addressing these issues [66]. Therefore, utilizing Si and carbon composite anode materials is a promising approach [67,68]. The silicon–carbon composites are advantageous because they leverage the high theoretical capacity of silicon while utilizing carbon to provide electrical conductivity and act as a buffer for volumetric expansion. Carbon-based materials possess structural flexibility, which enables them to alleviate the effects of volumetric expansion [69]. This review introduces ball milling, pyrolysis, spray drying, chemical vapor deposition (CVD), and mechanofusion processes for composite formation.

2. Synthesis Methods

2.1. Ball Milling

Figure 1 illustrates the appearance of balls and particles inside the chamber during ball milling. Ball milling is a complex process used to obtain nanostructured materials in powder form with an average particle size of less than 100 nm [70]. In addition, ball milling is a suitable method for large-scale production. This involves the high-speed collision of stainless steel or zirconia balls, which reduces larger particles to finer ones and promotes the formation of new bonds [71]. A ball mill consists of a cylindrical chamber that rotates around the horizontal axis. Generally, ball milling is used by filling the chamber with small balls made of materials such as iron or zirconia. The inside of the chamber is made of materials such as iron and rubber to prevent wear. Ball milling can be performed for approximately 100 to 150 h and can yield sized fine powders. It is characterized by the ability to maintain a vacuum or a specific gas atmosphere within the chamber. In addition, it offers the advantage of enabling ceramic nano-reinforcement [72]. This section presents the formation and analysis of silicon–carbon composites via ball milling, as investigated in a previous study [73–75].

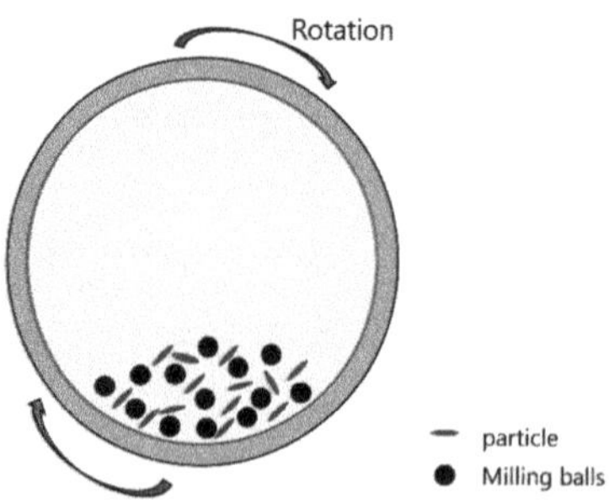

Figure 1. Schematic of the ball milling process.

Korrg et al. [76] synthesized silicon–carbon nanotube (Si/CNT) composites using silicon wafers and carbon nanotubes as precursors through high-energy ball milling

(Figure 2a). The ball milling process was conducted for durations of 10, 20, and 30 h. The silicon particles milled for 10–20 h remained in the micrometer range. However, after 30 h of milling, all particles exhibited sizes of hundreds of nanometers. The performance of the Si30h/CNT composite with particles in the nanometer range was compared with that of a Si electrode subjected to 30 h of ball milling (Figure 2b). The Si30h/CNT electrode cell demonstrated initial coulombic efficiency (ICE) of 98.06% and discharge capacity of 2470 mAh/g. After 100 cycles, the Si30h/CNT electrode showed a capacity retention of 71.28. This is a significant improvement over the Si30h electrode, which showed 52.96%. This performance improvement was able to mitigate the volume change in Si particles due to the strong Si–C bond between CNT and Si. Furthermore, the exceptional mechanical properties of CNTs help maintain particle cohesion, reducing the likelihood of particle disintegration and disruption of the SEI, thereby improving the cycle stability of the Si/CNT electrode. Gu et al. [77] used coarse silicon and citric acid as carbon sources to form a silicon–carbon composite (Si/C) through ball milling. Ball milling was performed for 2, 5, 8, and 10 h to analyze the cycle performance. The discharge capacities of the composites were 232.4, 626.7, 496.6, and 524.4 mAh/g, respectively. Composites ball-milled for 5 h showed the best capacity retention (Figure 2c). Figure 2d–g shows the SEM images of the samples milled for 2, 5, 8, and 10 h, respectively. As shown by the SEM images, the duration of ball milling significantly affected the particle size of the composite. The composite synthesized without ball milling exhibited a greater tendency toward agglomeration than the composite produced from the ball-milled precursors. Ball milling for 5 h resulted in fewer agglomerates and smaller particle sizes. Ball milling for longer than 5 h increased the tendency for agglomeration. The Cabello et al. [78] Si/C composites were prepared by each method using wet and dry ball milling and their particle and electrochemical properties were compared. The pore size distribution of the dry ball-milled material ranges from 6 to 0.5 μm, 0.3 to 0.07 μm, and 0.015 to 0.008 μm (Figure 2h). The broad range of pore size distribution increases exposure to the electrolyte, which may result in continuous SEI formation and subsequent lithium loss. The pore size distribution resulting from wet ball milling (Figure 2i) ranges from 1 to 0.25 μm. This narrow range of pore sizes indicates a more uniform distribution of particles. Figure 2j,k shows SEM images of the composite obtained using the dry and wet ball milling process.

Zhang et al. [79] fabricated Si/graphite/multi-walled carbon nanotube (MWNT) composite anodes using ball milling and evaluated their electrochemical performance. This study compared pure silicon (Si) with silicon/graphite/multiwalled carbon nanotube (SGM) composites. The cycling performance of the SGM composite (Figure 3a) observed a plateau at approximately 0.4 V, indicating the integrity of the SGM electrode. Figure 3b shows a plateau at 0.1–0.2 V during lithiation, which corresponds to the electrochemical reaction of lithium and silicon. During delithiation, a plateau is observed at 0.4 V. Also, the initial discharge capacity was found to be 3343 mAh/g. However, after just four charge/discharge cycles, the discharge capacity dropped significantly to 256 mAh/g. This seems to be due to a large volume change in the Si. On the other hand, SGM exhibited an initial discharge capacity of about 2270 mAh/g, and after four charge/discharge cycles, the capacity decreased to about 1750 mAh/g with capacity retention of about 77.1%. Han et al. [80] prepared nitrogen-doped carbon and Si composites by ball milling using a carbon source containing nitrogen-doped carbon, specifically polyvinylpyrrolidone (PVP). PVP, a polymeric surfactant, is widely used in the chemical industry as a dispersant, emulsifier, and antiprecipitant. In addition to acting as a nitrogen-doped carbon source, PVP promotes the dispersion of Si particles during ball milling, thereby facilitating the formation of the Si/PVP precursor. The sample with a PVP content of 1.0 g was labeled as Si@C-2. The cycling performance of the Si and Si@C-2 electrodes was analyzed at a current density of 1000 mA/g (Figure 3c). A significant decrease in capacity was observed for the Si electrode, indicating poor cycling performance. In contrast, Si@C-2 exhibited improved charging/discharging cycle safety with a capacity of 797.7 mAh/g at the 1000th cycle, thus demonstrating superior long-term performance. The Si@C-2 composite showed advantages in lithium capture due to their Si-O-C chemical bonding and the presence of N-doped carbon. In addition, the ball milling process further tightened the bond between the silicon particles and PVP.

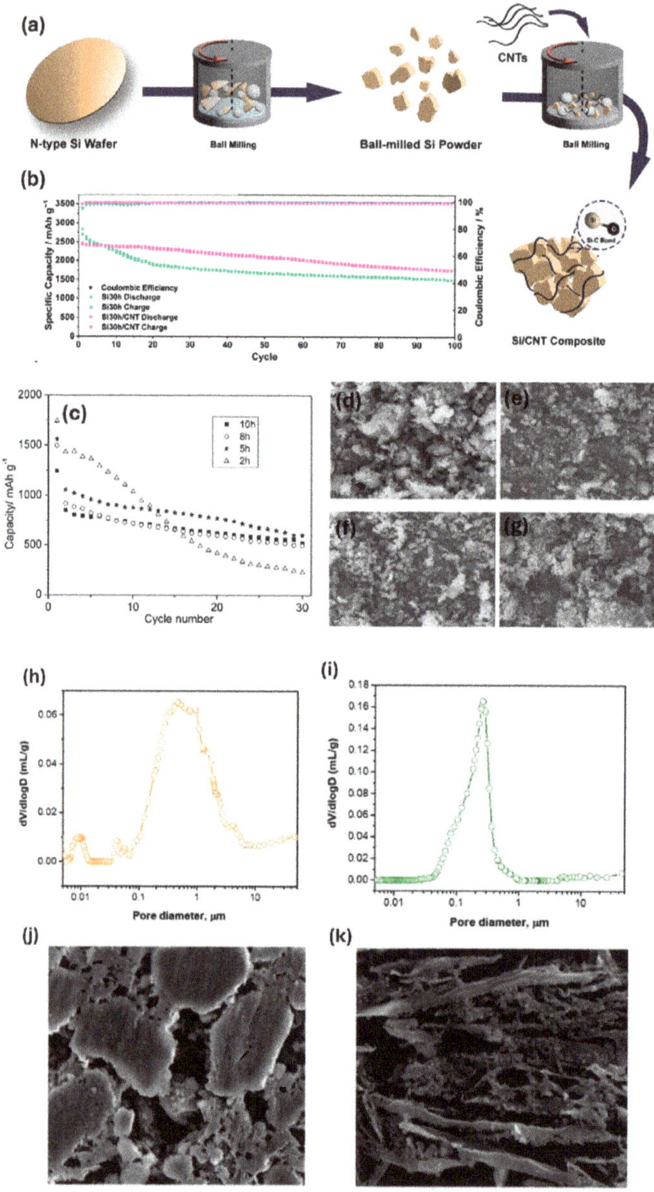

Figure 2. (**a**) Synthesis of Si/CNT; (**b**) cycle performance of Si/CNT [76] Reprinted with permission from Ref. [76]. Copyright © 2022 MDPI). (**c**) Ball milling of Si/C for (**d**) 2 h, (**e**) 5 h, (**f**) 8 h, and (**g**) 10 h [77] Reprinted with permission from Ref. [77]. Copyright © 2010 Elsevier. Pore distribution in (**h**) dry ball milling and (**i**) wet ball milling and SEM images of Si/C produced by (**j**) dry ball milling and (**k**) wet ball milling [78]. Reprinted with permission from Ref. [78]. Copyright © 2020 MDPI.

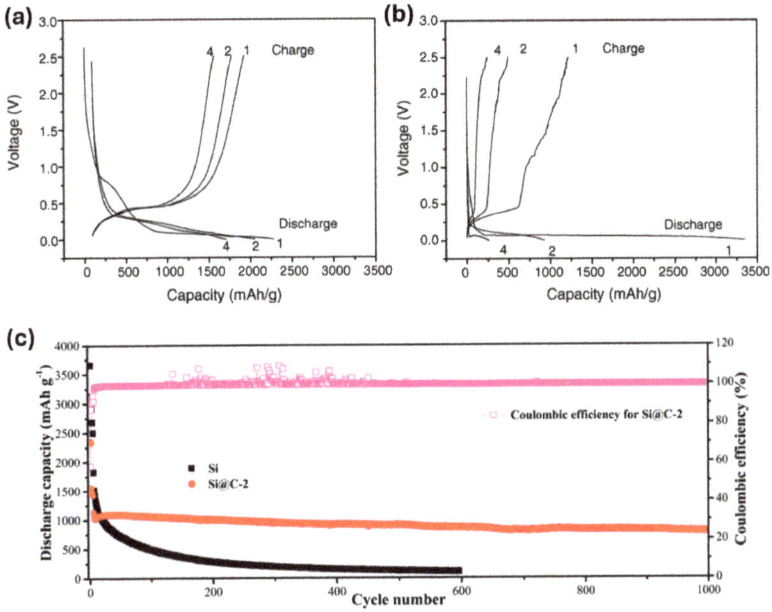

Figure 3. Voltage profiles of (**a**) Silicon/graphite/multi-walled carbon nanotubes and (**b**) pure Si [79] Reprinted with permission from Ref. [79]. Copyright © 2006 Elsevier. (**c**) Cycle performance of silicon and silicon–carbon composites [80] Reprinted with permission from Ref. [80]. Copyright © 2023 Elsevier.

2.2. Pyrolysis

Figure 4 shows the pyrolysis process [81]. Pyrolysis is a fundamental chemical reaction that occurs when heat is applied to a material in the absence of oxygen and serves as an essential step prior to the vaporization and combustion of solid fuels. Pyrolysis can be classified into several types based on the process conditions, parameters (such as residence time, temperature, and heating rate), and products. The major pyrolysis methods include carbonization, conventional pyrolysis, rapid pyrolysis, flash pyrolysis (both liquid and gas), ultrapyrolysis, vacuum pyrolysis, sequential pyrolysis, and methanation pyrolysis. The reactor configuration includes a fluidized bed, rotary kiln, auger, vacuum, and a controlled-atmosphere pyrolysis reactor [82]. The gaseous pyrolysis products are removed from the tube by nitrogen flow, after which they are condensed to collect the polymeric organics. Low-molecular-weight organic gases are processed using a gas treatment system [83]. This section discusses the production of silicon–carbon composites through pyrolysis and analyzes the composites.

Chen et al. [84] produced carbon nanofibers (CNFs) through pyrolysis using polyacrylonitrile (PAN). Silicon and carbon nanofibers (CNFs) were used to produce Si/CNFs, which were heat-treated with sucrose to produce Si/C-CNFs (Figure 5a). The samples were designated as Si/C-CNFs-10, 20, and 30 based on the amount of sucrose used. Figure 5b shows the cycling performances of CNFs, Si/CNFs, Si/C-CNFs-10, 20, and 30 at a current density of 600 mA/g. The discharge capacity of the Si/CNF composite rapidly decreased to 293.8 mAh/g after 50 cycles. This decrease was due to the volume change in the Si particles, which crack and fracture and loses electrical contact with the CNF. Consequently, the capacity decreased as many Si particles became unavailable for lithium-ion interactions during the lithiation and delithiation. In contrast, the Si/C-CNFs composite exhibited enhanced cycling stability compared to the Si/CNFs composite. Specifically, Si/C-CNFs-20 maintained a discharge capacity of 1215.2 mAh/g even after 50 cycles, thereby minimizing the capacity loss. This improvement is attributed to the carbon coating, which mitigates the volume expansion of Si and maintains the structural stability of the electrode. The Si/C-CNFs-30 composite showed stable performance in the second cycle, but the overall capacity was relatively low at 500 mAh/g compared to the other composites. The reason for the decreased capacity is that the excessively thick coating layer allows fewer silicon particles to react with the lithium during the lithium-silicon alloying/dealloying process. Nevertheless, the carbon coating helps establish pathways between the nanofibers for lithium-ion transport, thereby enhancing the rate of performance. Consequently, the Si/C-CNFs-20 electrode demonstrates markedly improved rate capability compared to the Si/CNFs electrode (Figure 5c). Si/C-CNFs-20 had a capacity of approximately 350 mAh/g at a current density of 5000 mA/g. When the current density was decreased to 600 mA/g, the capacity was recovered to approximately 1200 mAh/g. The cycling performance of the CNF anode at 600 mA/g revealed a capacity of below 300 mAh/g. Yang et al. [85] utilized nano-Si, graphite, and coal tar pitch as precursors to form Si/C composites in two stages: heat treatment and pyrolysis. A comparative analysis was conducted based on the primary pyrolysis temperature. Heat treatment was conducted at temperatures of 0 °C (no treatment), 120 °C, 180 °C, and 200 °C. The S180 composite, treated at 180 °C, displayed an irregular morphology with silicon and carbon elements evenly distributed across the particle surfaces, suggesting that all raw materials were well-dispersed within the composite carbon layers, graphite, and silicon, all of which were observed as amorphous. This was because the pitch filled these voids at temperatures exceeding the glass transition temperature (T_g). Hu et al. [86] used nanosilicon and PAN to synthesize silicon–carbon composites via oxidative pyrolysis. The process was carried out at temperatures of 250, 300, 350, and 400 °C for 30 min. It is important to note that temperatures above 350 °C pose a risk of oxidation of the copper current collector, potentially leading to the formation of Cu_2O, which would contribute to the capacity but is not desirable for accurate electrochemical analysis. The electrode treated at 250 °C showed a discharge capacity of 630 mAh/g at the 10th cycle, which dropped to nearly 0 after the 60th cycle. Increasing the processing temperature to 300 °C significantly enhanced performance, yielding a discharge capacity of 1415 mAh/g at the 10th cycle and maintaining a capacity of 1132 mAh/g after 100 cycles. This represents a marked improvement compared to the electrode processed at 250 °C. The Si/C composite thermally treated at 400 °C had a capacity of more than 2000 mAh/g up to 60 cycles and a capacity of 1555 mAh/g up to 100 cycles. It has a retention rate of approximately 78%. This improved performance is due to the high amount of silicon and high current density. The Si/C composite fabricated at high temperatures exhibited superior rate capability. The composite annealed at 250 °C performed the worst among the other composites, while the composite exhibited at 400 °C performed the best. Under 3C cycling conditions, the sample that was treated at 400 °C showed an impressive discharge capacity of approximately 500 mAh/g, whereas the other samples exhibited negligible capacity. When the C-rate was returned to C/10, most of the composites achieved a capacity of 3000 mAh/g, demonstrating that the electrodes could recover their original capacity after cycling at a high C-rate. Su et al. [87] synthesized Si/C composites by pyrolyzing nanosilicon and flake graphite phenolic resins. Figure 5d shows that the synthesized Si/C composite particles were notably small. 1, 2 and 3 in Figure 5d are shown in Figures 5e, 5f and 5g, respectively. Figure 5d–g presents a detailed examination of the particle structure of the Si/C composite. Nanosilicon was uniformly distributed within the graphite matrix, whereas the carbon derived from phenolic resin pyrolysis coated the surface of the Si/graphite structure. This carbon coating effectively alleviated the volumetric expansion of Si. The electronic conductivity of the anode was also improved by coating it with carbon. Figure 5g shows a Fourier transformation of Region 4 from a High-Resolution Transmission Electron Microscopy (HRTEM) image, which indicates that lattice distances in both directions are well in agreement with interlayer distances of Si and lamellar graphite. This showed that the graphite and nanoscale Si powders were thoroughly dispersed. Figure 5h shows the capacity according to the cycle of the Si/C composite at 100 mA/g. The initial capacities were 867.6 and 640.5 mAh/g, respectively, with an ICE of 73.82%. In the early stages of insertion, the observed plateaus are mainly due to Li-alloying of the composite and Li-ion insertion into the carbon matrix. This behavior was evident for the Si/C composites in the voltage area of approximately 0–0.20 V. Reflecting their superior cycling performance, the Si/C composites exhibited capacity retention rates of 96.72%, 100.94% and 110.64% after 20 cycles, as shown in Figure 5i. Even after 40 cycles, all samples exhibited capacity retention rates above these values. This improvement is due to the partial consumption of the composite during the initial cycles, which remains inactive until the subsequent charging and discharging processes. As the active material was activated later, it facilitated a wider diffusion path for Li^+ ions, thereby improving the overall capacity of the material. This outstanding cycle performance was primarily due to the superior structuring characteristics of the composite. The uniform electronic conduction network created by amorphous carbon, along with the enhanced adhesion between the binder and current collector, significantly contributes to the long cycle stability and effectiveness of the material.

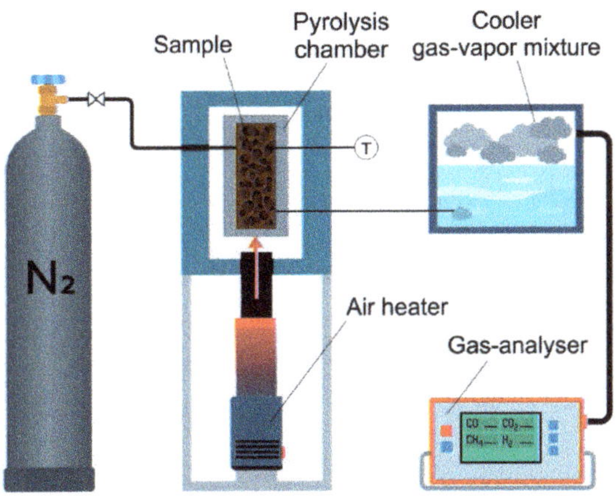

Figure 4. Schematic of pyrolysis [81] Reprinted with permission from Ref. [75]. Copyright © 2021 MDPI.

Wang et al. [88] employed nanosilicon and phenolic resin to fabricate Si/C composites via pyrolysis. Due to the desirable electrochemical properties of pyrolyzed carbon, phenolic resin was selected as the carbon source, which exhibits a reversible capacity of approximately 400 mAh/g and exhibits outstanding cycle performance. Cyclic voltammetry (CV) curves of the synthesized composites were analyzed to assess their electrochemical behavior. The cyclic voltammetry (CV) curves of pure Si and the Si/C composite are shown in Figure 6a,b, respectively. In the early phase cathodic scans of both materials, a distinct reduction peak was observed in the range of 0 to 0.2 V, which likely indicates the transformation of silicon from a crystalline to an amorphous phase. Additionally, the formation of an SEI layer on the electrode surface was indicated by an irreversible cathodic peak observed in the range of 0.5 to 0.75 V. This peak disappeared after the first cycle. For Si nanoparticles, two peaks were observed at approximately 0.32 V and 0.54 V during the first anodic scan, which are associated with the dealloying of Li-Si. Conversely, for the Si/C nanocomposite, the observed anodic peak at approximately 0.25 V suggests that a significant portion of the silicon transitioned to an amorphous Li_xSi phase. Additionally, the Si/C nanocomposite exhibited significantly larger current peaks than the Si nanoparticles, indicating that the lithium ions migrated more effectively into the silicon within the Si/C composite than in the pure Si. Figure 6c,d shows the voltage profiles of pure Si and the Si/C composite, which align with the CV results. In the delithiation process of pure Si, distinct plateaus are observed at 0.32 V and 0.54 V (Figure 6c). In contrast, the Si/C composite exhibits a voltage slope at approximately 0.25 V (Figure 6d), indicating that a greater portion of Si has transitioned to the amorphous Li_xSi phase compared to pure Si. Additionally, while pure Si shows a dramatic capacity decrease with a large overpotential, the Si/C composite demonstrates better capacity retention and similar voltage profiles over 50 cycles.

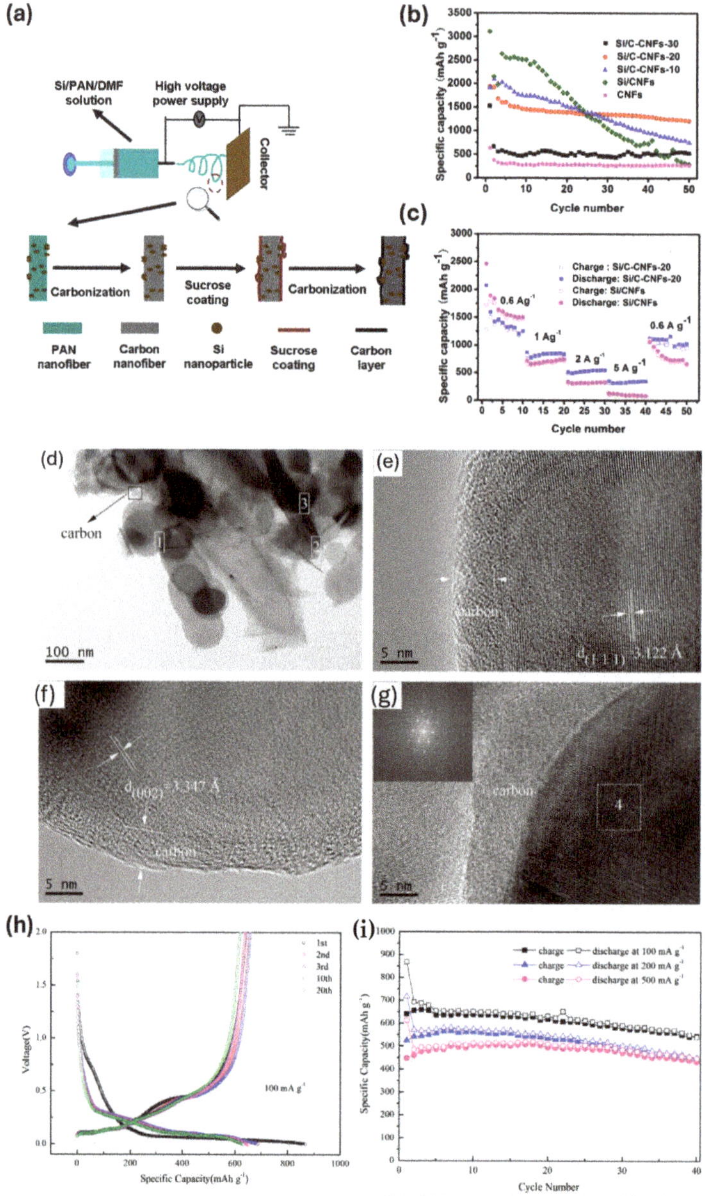

Figure 5. (**a**) Schematic of the manufacturing process of Si/C-CNF by pyrolysis. (**b**) Cycle performance of Si/C-CNF and CNF. (**c**) C-rate capability of Si/CNFs and Si/C-CNFs [84] Reprinted with permission from Ref. [84]. Copyright © 2015 Elsevier. (**d**) TEM of Si/C. (**e–g**) HRTEM of Si/C. (**h**) Voltage profile at a current density of 100 mA/g. (**i**) Cycle performance of the Si/C composite [87] Reprinted with permission from Ref. [87]. Copyright © 2013 Elsevier.

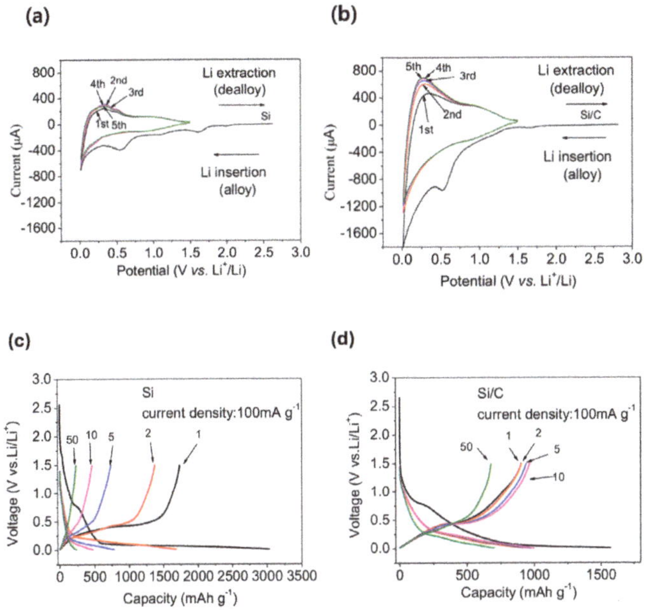

Figure 6. Cyclic voltammetry curves of (**a**) pure Si and (**b**) Si/C composite. Voltage profiles of (**c**) pure Si and (**d**) Si/C composite in 1, 2, 5, 10 and 50 cycles. [88] Reprinted with permission from Ref. [88]. Copyright © 2013 Elsevier.

2.3. Spray Drying

Spray drying produces a dry powder by rapidly evaporating a liquid solution using a high-temperature gas. This method includes a heat treatment method and forms uniform and spherical particles. The heated drying medium is typically air; however, nitrogen is employed when the liquid contains flammable solvents, such as ethanol, or when the product is sensitive to oxygen. During the spray drying process, spherical particles are formed because of the surface tension acting on the liquid or slurry during the spraying stage [89]. Figure 7a shows the spray drying equipment. The spray process is transmitted to the chamber to dry through the nozzle. Types of nozzles include nozzles that rotate and spray and two-fluid nozzles that are supplied to the pump. The atomization of a fluid is achieved through centrifugation, pressure, or kinetic energy, depending on the type of nozzle used. Droplets formed by injection are separated by a cyclone that releases gas to be dried. This is to quickly evaporate the solvent and accumulate the dried droplets in the glass collector. The resulting solid product exhibits superior chemical and physical stability compared to the liquid. Spray drying is broadly used in various industries, such as pharmaceutical, chemical, materials, cosmetics, and food. Generally, this technique is efficient owing to its ability to rapidly and continuously produce results in a single step, with minimal modifications required for scaling up [90]. The electrochemical performance and particle shape of Si/C composite materials manufactured using this method were analyzed. Through spray drying. Pan et al. [91] synthesized a Si, carbon, and graphene oxide composite, Si@C@RGO, to achieve both composite formation and spheronization. The TEM images in Figure 7b show the carbon surrounding silicon nanoparticles (SiNPs) and also show the (111) plane corresponding to the lattice of Si. Synthesis of additional graphene from Si/C composite did not change the carbon coating on SiNPs. Figure 7c shows that carbon surrounds the nanosilicon particle and graphene is adhered to it. Although the first cycle of pure Si showed a capacity of 2368 mAh/g. After 50 cycles, it decreased significantly to 500 mAh/g, and the capacity retention rate was only 21%. On the other hand, Si@C shows a capacity of 1041 mAh/g even after 50 cycles, which shows better performance than pure silicon due to improved effective carbon coating and conductive networks. In addition, based on 100 cycles, the CE of Si@C was about 15%, and Si@C@RGO composites were 40%, which seems that the additional synthesis of graphene alleviated the volume expansion of Si (Figure 7d). The rate capability characteristics of Si@C@RGO composites achieved a capacity retention rate of 94.9% up to 100 cycles and showed excellent cycle performance. On the other hand, Si@C composites continuously appear to have lower specific capacity than Si@C@RGO composites. This can be seen as proving the effectiveness of the graphene addition synthesis mentioned in Figure 7d (Figure 7e).

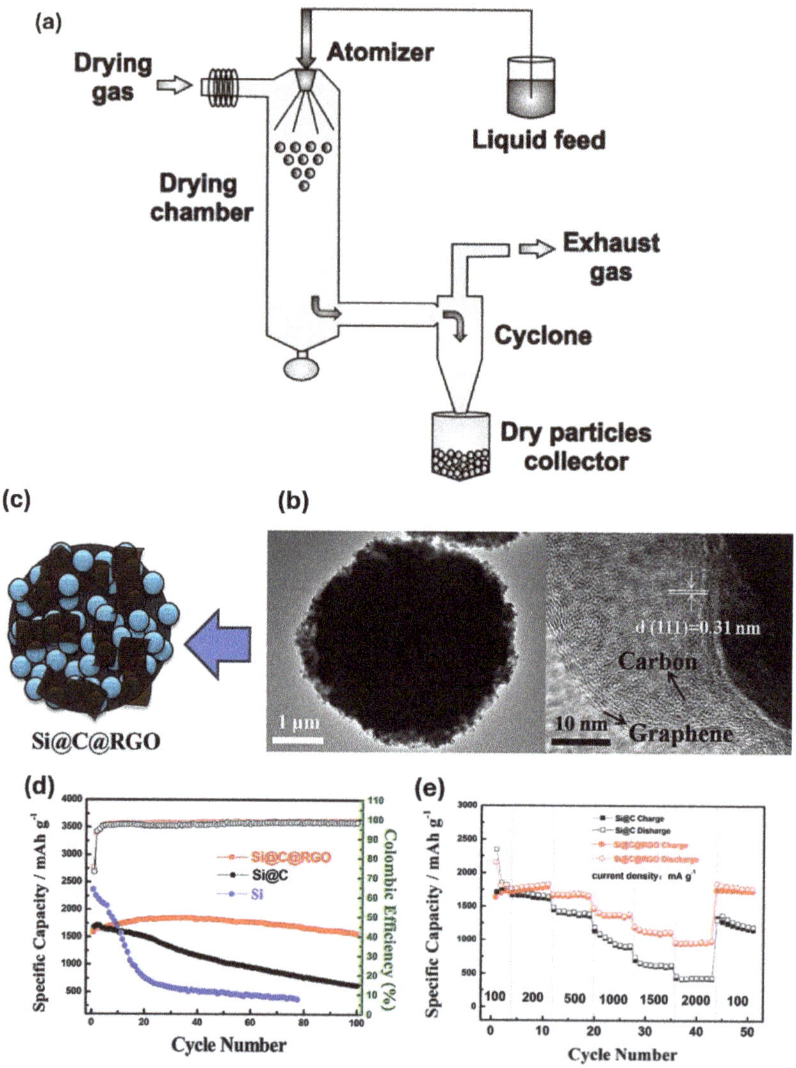

Figure 7. (**a**) Schematic of the spray drying process [90] Reprinted with permission from Ref. [90]. Copyright © 2015 Elsevier. (**b**) TEM of the Si@C@RGO composite. (**c**) Schematic of the Si@C@RGO composite, (**d**) cycle performance of the composites and pure Si, and (**e**) C-rate capability of the composites [91] Reprinted with permission from Ref. [91]. Copyright © 2017 Elsevier.

Lai et al. [92] adopted graphite and glucose as the carbon sources of silicon–carbon composites. They employed spray drying to produce spherical Si/C composites. Figure 8a,b illustrates the electrochemical performance according to the cycle of the previously prepared Si/C composite. The characteristics of this composite material showed a high initial CE of 69.71%. As the SEI layer is formed in the initial cycle, irreversible capacity occurs at about 0.8 V due to lithium consumption. After that, it showed a CE of 96% and an excellent performance with a capacity of 552 mAh/g in the 20th cycle. Su et al. [93] used spray drying of silicon and graphene to synthesize Si/C composites. The composites prepared with silicon and graphene in ratios of 1:2, 1:4, and 1:6 were labeled G1, G2, and G3, respectively. Figure 8c,d shows the performance of composite materials synthesized with different ratios of silicon and graphene. Figure 8c presents the voltage profile of the initial cycles of the Si/C composites with different proportions at a current density of 100 mA/g. Like the previous group, this group's study also showed a significant slope due to the formation of the SEI layer at 0.6~0.8 V. The Si/C composite showed a plateau of about 0.1 V. This is because the SEI layer was formed in the initial discharge stage, a stable film was formed and operated. The plateau of about 0.45 V on the charging curve is the release of some silicon. The charge–discharge curves of different proportions of silicon–carbon composites show similar curves. In the initial cycle, the capacity values of each Si/C composite were 2662.4, 2869.9, and 1671.5 mAh/g. This means that the inclusion of graphene improves the capacity. Figure 8d illustrates the cycle performance of Si/C composites at a current density of 100 mA/g. The capacity of the Si/C composite cell decreased as the cycle progressed. Among the composite materials, the capacities of the composite material with a graphene ratio of 2 times and 6 times greater than silicon were 323.2 mAh/g and 298.6 mAh/g, respectively, and the composite material with a capacity of 4 times greater than silicon showed the best performance with a capacity of more than 600 mAh/g. Carbon with a relatively low carbon content does not seem to be able to improve the conductivity and volume of silicon; on the contrary, the composite material with a high carbon content does not seem to be able to use a high capacity of silicon due to a lot of carbon but rather to increase the resistance. Paireau et al. [94] used silicon and polyvinyl alcohol (PVA) to produce Si/C composites via spray drying. PVA was chosen as the carbon source because of its water solubility, which makes it an environmentally friendly option. After manufacturing the Si/C composites with the selected materials, different Si/C ratios were used for the charging–discharging cycle tests. Figure 8e shows the results for pristine silicon and Si/carbon composites with Si/C ratios of 98:2, 77:23, and 63:37. The low irreversible capacity, an initial problem with silicon–carbon composites, decreased as the carbon content increased. As a result, ICE increased. Although the carbon addition was improved over pure Si, no significant cycle performance improvement was observed. Each Si/C composite material decreased dramatically in capacity during the initial cycle and then stabilized. This decrease in capacity became less pronounced as the carbon content increased. However, even when the carbon content increased to 37 wt%, a significant capacity loss still occurred. The carbon content increased to 37 wt%, but as a result, it performed worse than expected. This suggests that the excess carbon did not improve the low conductivity of the silicon particles, but rather served as a resistive layer. Wang et al. [95] manufactured Si/C composites by spray drying, and citric acid was adopted as the carbon source. Citric acid was used at concentrations of 1.25, 2.5, and 3.5% by weight to produce different Si/C composites. Figure 8f displays the voltage profiles for pristine Si and Si/C composites with different amounts of carbon. Pristine Si has an ICE of 80%, and the initial reversible capacity is 3960 mAh/g and 3180 mAh/g. Si/C composites ranged from 2210 to 2710 mAh/g, but a composite with a carbon content of 1.25% among the prepared Si/C composite showed 84% with higher ICE values than pristine Si. This is because carbon is dispersed between Si particles, improving the low conductivity of silicon and increasing the stable SEI layer.

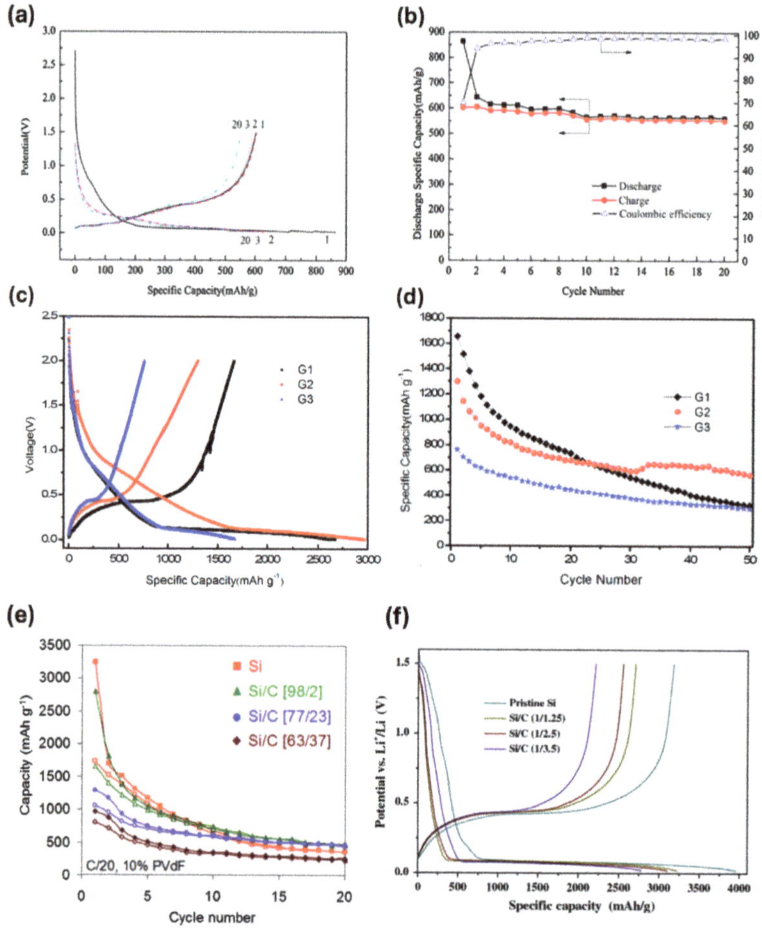

Figure 8. (**a**) Voltage profile and (**b**) cycle performance of graphite@silicon@carbon [92] Reprinted with permission from Ref. [92]. Copyright © 2012 Elsevier. (**c**) Voltage profile and (**d**) cycle performance of silicon@graphene [93] Reprinted with permission from Ref. [93]. Copyright © 2019 Elsevier. (**e**) Cycle test of pure-Si and Si/C [94] Reprinted with permission from Ref. [94]. Copyright © 2015 Elsevier. (**f**) Voltage profile of Si/C [95] Reprinted with permission from Ref. [95]. Copyright © 2014 Elsevier.

2.4. Chemical Vapor Deposition

Chemical vapor deposition (CVD) equipment is shown in Figure 9. CVD is a technique that is widely used for the growth of materials that are used in a variety of electronic and optoelectronic devices, chemical sensors, nanocomposites, and energy storage systems. In CVD, gaseous reactants are transported through a tube furnace where they decompose to generate atomic or molecular species. These species are then deposited on the surface of the substrate. CVD is primarily employed to produce graphite carbon-coated silicon and carbon nanotubes. Moreover, it can be used to deposit silicon onto substrates using precursors such as SiH_4 or $SiHCl_3$ [96]. The composite structure produced by CVD varies with changes in temperature and gas flow rate. As the temperature increases in CVD, the deposition yield also rises. This occurs because higher deposition temperatures enhance chemical kinetics and lead to a more structured atomic arrangement during crystal growth.

As the gas flow rate increases, the amount of deposition also rises. However, as deposition increases, resistance also escalates, leading to a deterioration in performance. Therefore, it is essential to adjust the gas flow rate to an appropriate level [97,98]. Based on this principle, the CVD method was used to synthesize Si/C composites, followed by a comprehensive evaluation of their performance.

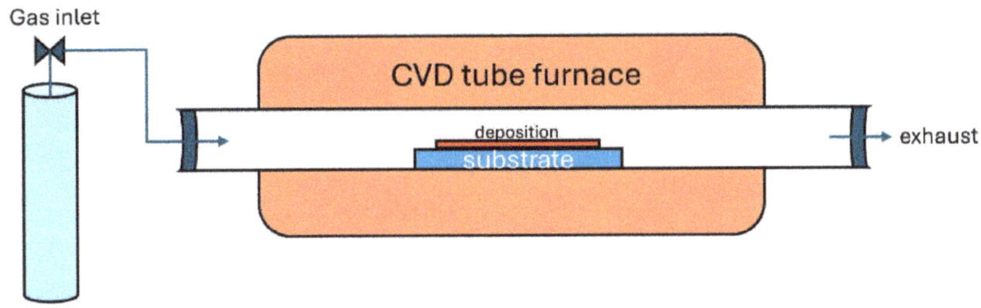

Figure 9. Schematic of the chemical vapor deposition process.

Wang et al. [99] developed nanoscale Si/C composites utilizing a dual-step CVD process. Following the procedure depicted in Figure 10a, xylene (C_8H_{10}) was employed as the source of hydrocarbons and iron produced by decomposition at high temperatures using ferrocene ($Fe(C_5H_5)_2$) as a catalyst to promote the production of multi-walled carbon nanotubes (MWNTs). Subsequently, silane gas was introduced to deposit Si on the MWNTs, completing the fabrication process. The Si/C ratio was determined after calculating by measuring the substrate by adjusting the parameters. The cycling performance of Si/CNT is shown in Figure 10b. The initial reversible capacities were 2552 and 2049 mAh/g, showing an ICE of 80.3%. It shows a capacity loss of 19.7%. In the subsequent cycle, the reversible capacity increased due to the activation of additional silicon atoms, and the CE was improved to 99%. It is shown that the electrode is stabilized by the formation of a stable SEI. Figure 10c shows the charge/discharge capacity at various current densities. As can be seen in the figure, this composite material showed higher capacity and stable performance with a capacity of about 1000 mAh/g at a C-rate of 2.5C than graphite, which is a commonly used cathode material. This performance is a result of the effect of the silicon/CNT nanostructure to reduce particle aggregation and manage volume change. In addition, the improved electronic conductivity and mechanical strength provided by the CNTs contributed to these results. Fu et al. [100] used nanostructured Si and carbon nanofibers (CNFs) to form Si/C composites. PAN was employed as the carbon precursor, and CVD was employed to deposit a carbon coating on the silicon surface, resulting in a Si/CNF-C composite. During the process, acetylene (C_2H_2) was used as the carbon source for deposition. In order to prevent detachment between the nanofiber surface and silicon, Si@CNF composite material manufacturing adopts CVD technology. The Si@CNF-C composites were analyzed by varying the CVD coating time. Figure 10d–f shows high-magnification TEM images of composites with different CVD treatment times. In the figure, a red arrow indicates the carbon coating layer formed during the CVD. The SEI layer depends on the type of carbon coating of the particles. Si/CNF-C composites were effective in forming an SEI layer with a stable carbon coating layer. As a result, ICE was improved positively. In addition, the electrochemical performance of the composite material coated with CVD for 90 min was remarkably improved. Si/CNF composites and Si/CNF-C composites were compared to confirm the effect with or without additional carbon coating. Si/CNF without carbon coating showed a capacity retention rate of 65% after the 30th cycle and the charging capacity decreased to 570 mAh/g. In contrast, all Si/CNF-C composites demonstrated improved performance, retaining approximately 76% of their capacity and achieving approximately 700 mAh/g after 30 cycles. This result is the exfoliation of Si attached to carbon nanofibers due to the volume change in Si in the Si/CNF composite. In contrast, the CVD-coated Si/CNF-C composite effectively mitigated the volumetric changes in Si, thereby improving the structural integrity and performance of the composite. Jin et al. [101] adopted porous silicon (pSi) and carbon nanofibers (CNFs) and synthesized Si/C composites through CVD using Fe-Ni as a catalyst. Consequently, the Si/C composite particle size of the composite ranged from 5 to 20 μm, and the CNFs were uniformly coated. Figure 10g shows the voltage profile curve for each cycle of the composite. The porous silicon–carbon nanofiber (pSi-CNF) electrode showed capacities of 1895 mAh/g and 1282 mAh/g in the first cycle, indicating an ICE of 67.7%. The use of silicon-based anode materials has been observed to result in an almost low ICE. This is primarily due to its high reactivity with Li^+, which results in an electrochemical side reaction, such as the formation of Li_2O or Li_4SiO_4. In this process, the electrolyte is decomposed, and Li^+ is consumed. This phenomenon is commonly observed during the initial cycle. In the initial lithiation of pSi-CNF, a voltage plateau was observed at 0.1 V, indicating that lithium had alloyed with silicon. During the initial charging phase, a plateau was observed at 0.5 V. The conversion of the material into amorphous Si is achieved through dealloying of the Si component, which occurs subsequent to its alloying with Li ions. In the subsequent cycle, a capacity of approximately 1300–1400 mAh/g was observed, accompanied by stable performance. Figure 10h shows the electrochemical behavior of Si/C composites with varying carbon contents. The initial capacities of each pSi-CNF-C composite material with CNF contents of 10, 22, and 45% were 1971, 1321, and 1270 mAh/g, respectively. The ICE values were 67.1, 65.0, and 60.3%. Furthermore, the capacity based on 100 cycles exhibited a tendency to decrease, reaching 1441, 848, and 648 mAh/g. This is attributed to the fact that an elevated CNF content results in a greater resistance layer, leading to an increased consumption of Li^+ during the SEI formation process. This contrasts with the desired outcome of establishing a conductive network.

Liu et al. [102] synthesized a silicon nanowire (SiNW)/spherical graphite (GM) composite (C@SiNW@GM) in which silicon was deposited on spherical microsized graphite using CVD and an additional carbon layer was formed using acetylene. Figure 11a illustrates the process of gas-phase Si deposition onto micro-sized spherical graphite, which subsequently transforms into SiNWs. As the CVD processing time increased, the silicon transitioned into silicon nanowires. Nucleation of silicon was formed by flowing SiH_4 or $SiHCl_3$ into the gas phase. According to the C@SiNW@GM composite electrochemical behavior reported at a current density of 0.2C, the section decreasing from 1.0 V to 0.5 V during the first discharge shows a reduced slope due to the formation of SEI. Subsequently, a stable plateau was observed during the subsequent cycle, which can be attributed to the electrochemical reaction of graphite and Si. Jinglu et al. [103] carbonized toluene at 800 °C to coat the surface of silicon particles with carbonized carbon through a CVD process. Toluene was chosen because of its ability to decompose at this temperature and form a coating on the surface of the silicon particles. The composites were prepared by performing CVD treatments at 800 °C for durations of 1, 2, and 3 h. These composites were labeled Si@C-1, 2, and 3, respectively. Figure 11b shows the structure of the Si@C-2 composite, which exhibited the most stable performance among the various composites. The particle distribution of the pristine Si was primarily in the range of 50–200 nm. The particle sizes of Si@C-1, 2, and 3 exhibited an increase of 7 nm, 12 nm, and 15 nm, respectively, which can be attributed to the carbon deposition process through CVD, resulting in an enlargement of the composite particle size. Additionally, the results of electrochemical behavior for pristine Si, Si@C-1, 2, and 3 indicate the polarization increased with higher carbon content. Consequently, the resistance rises in conjunction with the expansion of the carbon layer, which functions as a resistance layer in itself. Figure 11c shows the cycling performance analysis of each composite electrode, along with that of the pristine Si electrode. Initially, Si@C-1 exhibited a rapid decrease in capacity. This was likely due to its lower carbon coating content, which provided insufficient mitigation of the volumetric expansion. In contrast, composites with high carbon coating contents had relatively low cycle performance due to increased voltage polarization during charging/discharging. The ICE values of the Si@C-1, 2, and 3 composites were found to be 83.6%, 80.1% and 78.9%, as indicated in the results. This indicates that the carbon content affects the CE, with a higher carbon content leading to a lower efficiency. This is because the amorphous carbon structure tends to trap the inserted lithium, thereby consuming more lithium and reducing the overall CE.

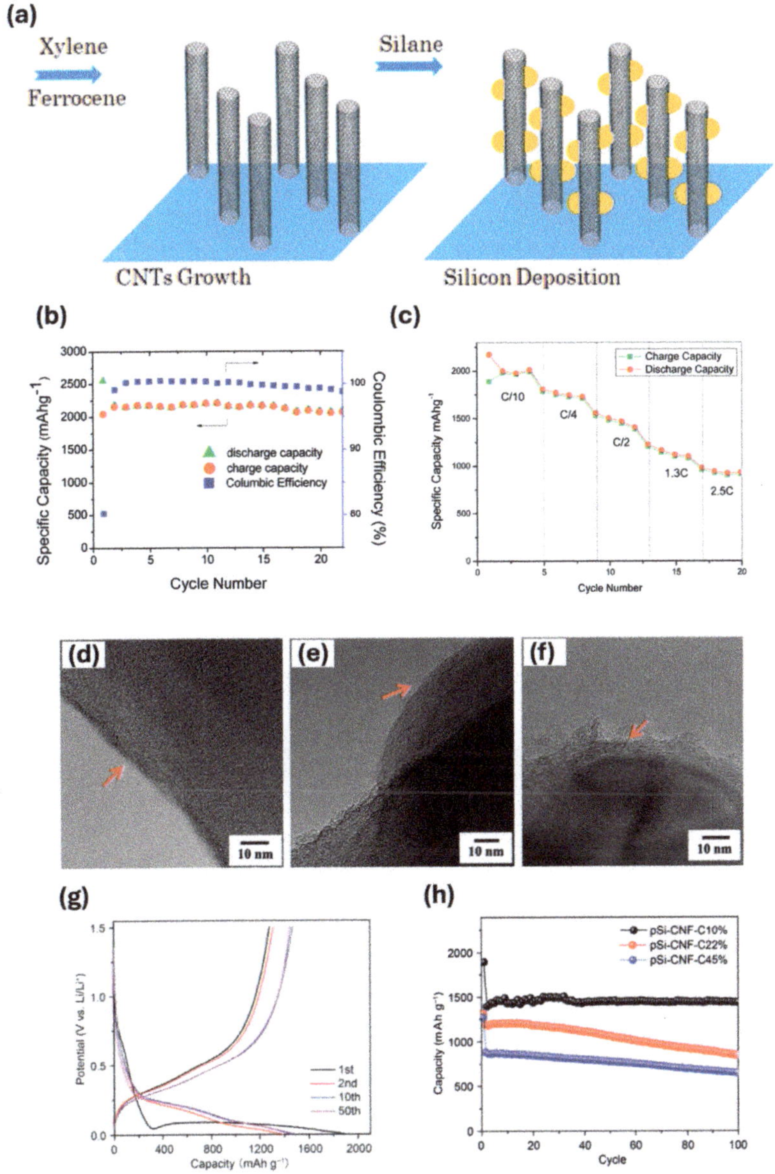

Figure 10. (**a**) Schematic of Si/C composite produced using CVD. (**b**) Cycle performance of Si/CNT, (**c**) C-rate capability of Si/CNT [99] Reprinted with permission from Ref. [99]. Copyright © 2010 ACS Publications. (**d–f**) TEM images of composites CVD-treated for 30, 60, and 90 min [100] Reprinted with permission from Ref. [100]. Copyright © 2013 Elsevier. (**g**) Voltage profile of the pSi-CNF composite and its (**h**) cycle performance [101] Reprinted with permission from Ref. [101]. Copyright © 2021 Elsevier.

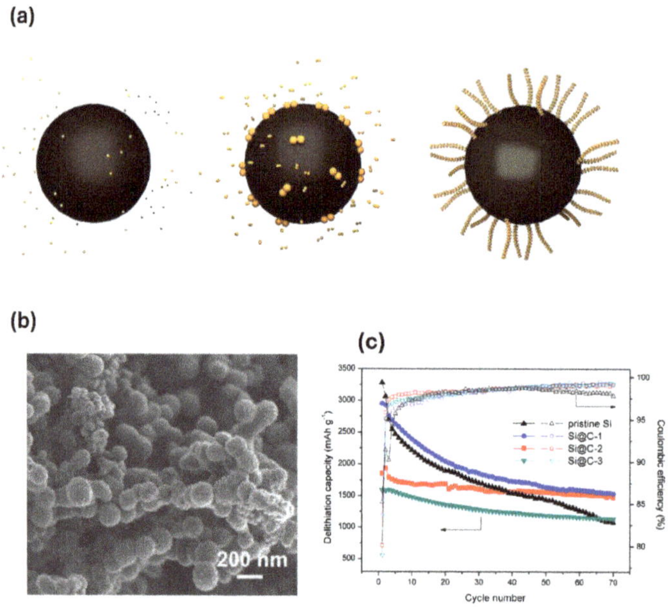

Figure 11. (**a**) CVD deposition process of SiNW@GM [102] Reprinted with permission from Ref. [102]. Copyright © 2021 ACS Publications. (**b**) SEM image of Si@C-2 composite and (**c**) cycle performance and CE of Si@C composite and pure Si [103] Reprinted with permission from Ref. [103]. Copyright © 2014 ACS Publications.

2.5. Mechanofusion

Figure 12 illustrates the configuration and interparticle interactions involved in the mechanofusion process. The configuration includes a press head, scraper, and chamber. In a mechanofusion system, the chamber rotates to generate a centrifugal force [104]. This centrifugal force causes the powders to converge and move between the press head and the inner wall of the chamber. Through this process, the powders experience friction and collisions between particles, causing small particles to stick to the surface of the large particles and disperse them [105]. To produce core-shell structured particles using mechanofusion, it is essential to use particles of different sizes: larger particles for the core and smaller particles for the shell. Owing to the continuous application of the centrifugal force, the powders adhere to the chamber walls. The scraper is responsible for removing the powder attached to the inner wall of the chamber due to centrifugal force. The mechanical composite formation was achieved by continuously repeating this process. Mechanofusion offers the advantage of coating particles without the need for solvents or other substances [106].

Kim et al. [107] fabricated SiO_x@C composites via mechanofusion with silicon oxide and pitch, as shown in Figure 13a. In this process, the silicon oxide (SiO_x) that has relatively large particles was structured as a core, while the pitch, which is a smaller particle, forms a shell structure. During the manufacturing process, the rotation of the chamber generated high temperatures that surpassed the softening point of the pitch, allowing for an effective coating. In order to confirm the effect of carbon coating, a performance evaluation was conducted, comparing the SiO_x@C composite material with pure SiO_x. The SiO_x@C composite material demonstrated superior performance compared to the SiO_x electrode, which was attributed to the carbon coating. Although the SiO_x@C composite initially exhibited a CE of 60%, the SiO_x electrode achieved a CE of only 12.6% after the 100th cycle. In addition, while the SiO_x electrode reached a CE of 99% by the 10th cycle, the SiO_x@C composite achieved the same efficiency by the 4th cycle. As a result of further evaluating the rate capability, the SiO_x@C composite material showed a capacity retention rate of

77.2% at 0.5C and a lower capacity retention rate of 47.8% for the SiO_x electrode. It was found that the carbon coating at the SiO_x@C composite site contributed to the mitigation against volume change in Si and the formation of a stable SEI layer. Cao et al. [108] used mechanofusion to fabricate Si-alloy/graphite composites. The composite is distinguished by the silicon alloy embedded within the layered structure of graphite. The capacity exhibited at a high C-rate was approximately 900 mAh/g to 950 mAh/g. Consequently, it was established that the graphite layer facilitated the mitigation of the volume change in the silicon alloy. Jiang et al. [109] used mechanofusion to incorporate nanosilicon into graphite pores, followed by carbon coating to create a nanosilicon/graphite composite (n-Si/graphite). In this structure, Si served as the guest particle, whereas graphite acted as the host matrix. Figure 13c–f shows the SEM images of the mechanofusion process conducted on the composite for 1, 2, 3, and 5 h, respectively. These images illustrate the morphological changes in the particles over time. After 1 h (Figure 13c), the large particle was graphite and the small particle was n-Si. It has been established that the smaller particles are capable of adhering to the surface of the larger particles. After 2 h (Figure 13d), the n-Si particles became less discernible, and the graphite particles began to adopt a more spherical shape. After 3 h (Figure 13e), the graphite surface became smoother. Finally, after 5 h (Figure 13f), the bright spots in the image represent n-Si particles, and no further changes were observed in graphite particles. Wutthiprom et al. [110] used SiO_2 and carbon nanospheres to create a SiO_2@C composite through mechanofusion. Subsequently, a reduction process was employed to introduce porosity into the composite, thereby obtaining Si@C composite. The choice of SiO_2 as the precursor for silicon was motivated by its cost-effectiveness, being approximately 3000 times cheaper than nanosilicon. This renders the process economically advantageous. The electrochemical behavior of the Si@C composite material prepared by this method was confirmed. Regarding the C-rate performance (Figure 13g) of the Si@C composite material, the ICE was 33.8% and the discharge capacity was 4308 mAh/g. Subsequent increases in the C-rate resulted in a decreasing trend in performance. However, the capacity recovered to 1259 mAh/g at 0.05C. This was initially unstable due to the formation of the SEI layer at the beginning. In comparison, pure Si demonstrated 25.5% ICE. These low ICE lack a stable carbon layer, resulting in the formation of an unstable SEI layer. The pre-lithiated Si@C composite electrode (Figure 13h) exhibited 98.8% ICE due to the stable SEI layer.

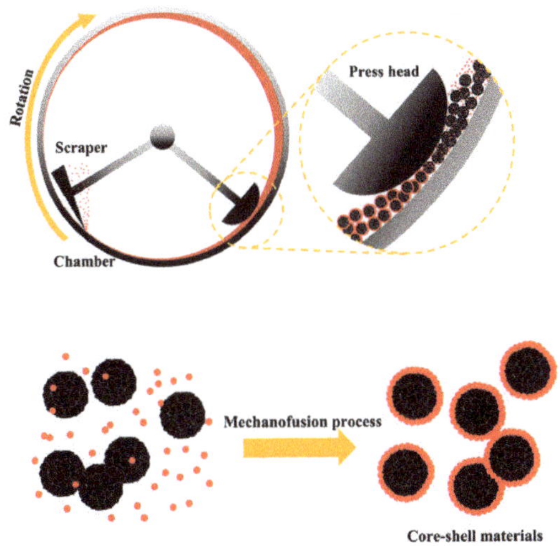

Figure 12. Configuration and process of mechanofusion [106] Reprinted with permission from Ref. [106]. Copyright © 2023 Wiley-VCH.

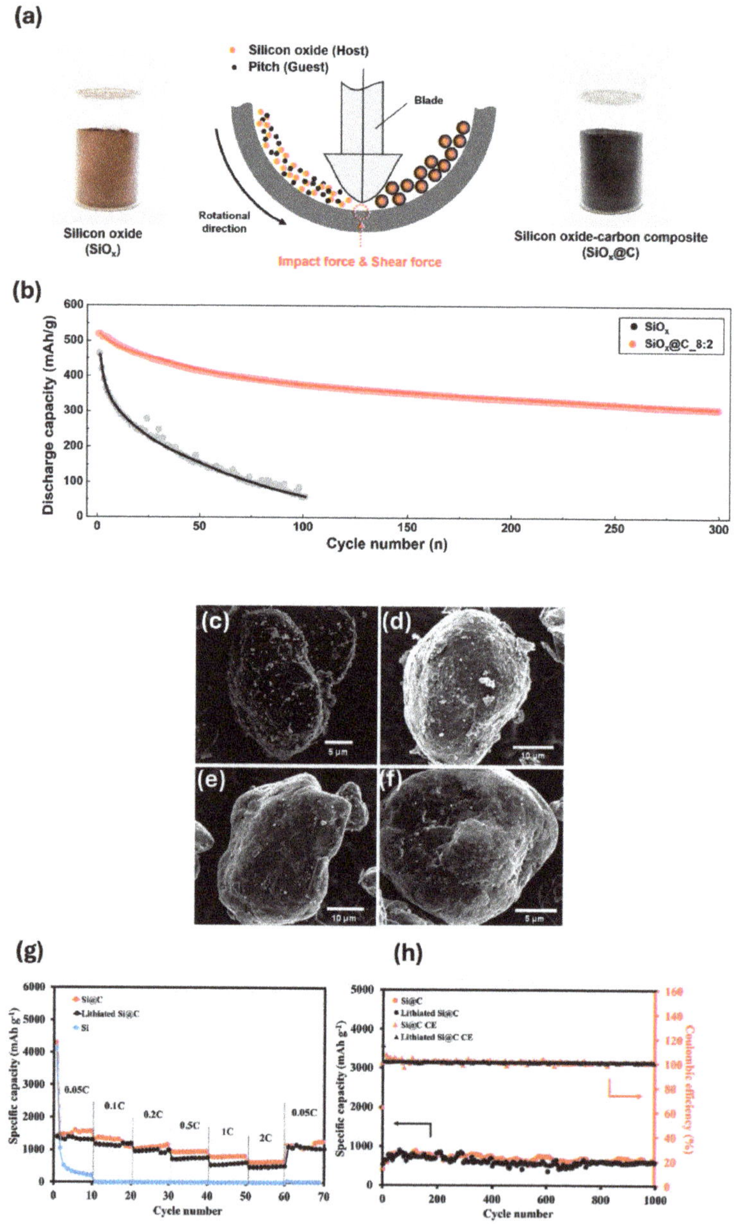

Figure 13. (**a**) Si@C manufacturing process by mechanofusion, (**b**) cycle performance of SiO_x@C composite [107] Reprinted with permission from Ref. [107]. Copyright © 2023 MDPI. (**c–f**) SEM images of composite treated with CVD for 1, 2, 3, and 5 h [109] Reprinted with permission from Ref. [109]. Copyright © 2023 iopscience. (**g**) C-rate performance of Si@C and Si, (**h**) cycle performance and CE [110] Reprinted with permission from Ref. [110]. Copyright © 2020 Elsevier.

3. Conclusions

Si anode material is regarded as a promising candidate for next-generation lithium-ion batteries due to its high theoretical capacity and natural abundance. Si is one of the most plentiful and environmentally friendly resources on Earth. However, its practical application is hindered by significant challenges, including substantial volume expansion and low electrical conductivity during lithium insertion and extraction processes. These issues lead to rapid capacity fading and mechanical degradation, preventing silicon from reaching the stage of widespread commercialization. This review systematically explores various methods for manufacturing silicon–carbon composite anodes, including ball milling, spray drying, pyrolysis, CVD, and mechanofusion. These methods were employed to enhance the structural stability and electrochemical performance of the silicon–carbon composite, providing a pathway toward overcoming the limitations of pure silicon anodes. Ball milling is an effective method for large-scale production. Nevertheless, it faces the limitation of producing non-uniform particle sizes. Pyrolysis offers high yields, but it requires substantial high-temperature energy inputs. Spray drying is a simple method that promotes the formation of spherical particles; still, the size of the equipment requires considerable factory space. CVD stands out for its ability to create diverse particle designs and uniform particles yet, the high costs associated with this method have hindered its commercialization. Lastly, mechanofusion can produce spherical particles. However, its high cost has posed challenges for commercialization. The advantages, disadvantages, and ICE of each synthesis method, along with a comparative analysis, are summarized in Table 1.

Table 1. Comparison of the advantages, disadvantages, and ICE of different processes.

Synthesis Method	Advantages	Disadvantages	Materials	Capacity	Ref.
Ball milling	Large-scale production	Non-uniform particles	Si/CNT	1760 mAh/g after 100 cycles	[76]
Pyrolysis	High yield	High energy	Si/C	1555 mAh/g after 100 cycles	[86]
Spray Drying	Spherical particle, simple process	Bulky equipment	Si/C	1041 mAh/g after 50 cycles	[91]
CVD	Uniform particle, various particle design	Expensive process	pSi-CNF	848 mAh/g after 100 cycles	[101]
Mechanofusion	Spherical particle	Expensive process	Si/C	600 mAh/g after 1000 cycles	[109]

Both cycle performance and coulombic efficiency (CE) remain insufficient, presenting challenges for applications in battery-dependent devices, such as electric vehicles (EVs) and IT systems. Future research should prioritize addressing the limitations of the presented processes and enhancing material performance. Process optimization is critical to minimize costs and ensure uniform product output. Rather than adopting a single process, employing a combination of approaches may further improve overall performance. A key step toward commercialization is the establishment of a pilot-scale production facility to validate these processes and support further investigation. By resolving key challenges, such as volume expansion and conductivity limitations, it will become feasible to develop batteries with enhanced capabilities. For example, a single charge could enable EVs to achieve significantly longer driving ranges, while emerging technologies like robotics and urban air mobility (UAM) systems would benefit from more efficient power storage. Overcoming these technological hurdles will be essential in accelerating the path toward successful commercialization.

Author Contributions: Conceptualization, I.P., H.L. and O.B.C.; writing—original draft preparation, I.P.; writing—review and editing, I.P., H.L. and O.B.C.; supervision, O.B.C. All authors have read and agreed to the published version of the manuscript.

Funding: This work was supported by the Gachon University research fund of 2023 (GCU-202400930001).

Conflicts of Interest: The authors declare no conflicts of interest.

References

1. Dou, F.; Shi, L.Y.; Chen, G.R.; Zhang, D.S. Silicon/Carbon Composite Anode Materials for Lithium-Ion Batteries. *Electrochem. Energy Rev.* **2019**, *2*, 149–198. [CrossRef]
2. Li, M.; Lu, J.; Chen, Z.W.; Amine, K. 30 Years of Lithium-Ion Batteries. *Adv. Mater.* **2018**, *30*, 1800561. [CrossRef] [PubMed]
3. Lee, D.; Lee, S.; Jung, D.S.; Roh, K.C.; Seo, J.; Kim, J.; Kim, K.; Kim, P.J.; Choi, J. Synergistically enhanced LiF-rich protective layer for highly stable silicon anodes. *Appl. Surf. Sci.* **2024**, *661*, 160023. [CrossRef]
4. Lee, H.; Yoon, T.; Chae, O. Strategies for Enhancing the Stability of Lithium Metal Anodes in Solid-State Electrolytes. *Micromachines* **2024**, *15*, 453. [CrossRef]
5. Su, H.P.; Barragan, A.A.; Geng, L.X.; Long, D.H.; Ling, L.C.; Bozhilov, K.N.; Mangolini, L.; Guo, J.C. Colloidal Synthesis of Silicon-Carbon Composite Material for Lithium-Ion Batteries. *Angew. Chem. Int. Ed.* **2017**, *56*, 10780–10785. [CrossRef]
6. Nzereogu, P.U.; Omah, A.D.; Ezema, F.I.; Iwuoha, E.I.; Nwanya, A.C. Anode materials for lithium-ion batteries: A review. *Appl. Surf. Sci. Adv.* **2022**, *9*, 100233. [CrossRef]
7. Ko, J.; So, S.; Kim, M.; Kim, I.; Ahn, Y.N.; Hur, J. Promoting Zn^{2+} migration through polar perovskite dielectric layer on Zn metal anode for the enhanced aqueous Zn-ion batteries. *Chem. Eng. J.* **2023**, *462*, 142308. [CrossRef]
8. Nam, G.; Hwang, J.; Kang, D.H.; Oh, S.; Chae, S.; Yoon, M.; Ko, M.S. Mechanical densification synthesis of single-crystalline Ni-rich cathode for high-energy lithium-ion batteries. *J. Energy Chem.* **2023**, *79*, 562–568. [CrossRef]
9. Song, W.; Chae, O.B.; Ryu, J.H. Surface Nitridation of Nano-sized Anatase TiO_2 using Urea and Thiourea for Enhanced Electrochemical Performance in Lithium-ion Batteries. *J. Electrochem. Sci. Technol.* **2024**, *15*, 512–520. [CrossRef]
10. Kim, J.; Yoon, T.; Chae, O.B. Behavior of NO^{3-}-Based Electrolyte Additive in Lithium Metal Batteries. *Batteries* **2024**, *10*, 135. [CrossRef]
11. Phan Nguyen, T.; Thi Giang, T.; Tae Kim, I. Restructuring NiO to $LiNiO_2$: Ultrastable and reversible anodes for lithium-ion batteries. *Chem. Eng. J.* **2022**, *437*, 135292. [CrossRef]
12. Sui, D.; Liu, J. Constriction-susceptible lithium support for fast cycling of solid-state lithium metal battery. *Chin. Chem. Lett.* **2024**, 110417. [CrossRef]
13. Goodenough, J.B.; Park, K.S. The Li-Ion Rechargeable Battery: A Perspective. *J. Am. Chem. Soc.* **2013**, *135*, 1167–1176. [CrossRef] [PubMed]
14. Kim, J.Y.; Chae, O.B.; Kim, G.; Peterson, A.A.; Wu, M.; Jung, H.T. Long-Range Uniform Deposition of Ag Nanoseed on Cu Current Collector for High-Performance Lithium Metal Batteries. *Small* **2024**, *20*, e2307200. [CrossRef] [PubMed]
15. Park, K.; Myeong, S.; Lee, D.; Yoo, H.E.; Kim, J.; Kim, C.; Kim, J.; Sun, S.; Kwon, J.; Kim, S.C.; et al. Improved Li-ion kinetics of the anode by kneading process of binder for lithium-ion batteries with high energy density. *Electrochim. Acta* **2023**, *464*, 142900. [CrossRef]
16. de las Casas, C.; Li, W.Z. A review of application of carbon nanotubes for lithium ion battery anode material. *J. Power Sources* **2012**, *208*, 74–85. [CrossRef]
17. Chan, C.K.; Peng, H.L.; Liu, G.; McIlwrath, K.; Zhang, X.F.; Huggins, R.A.; Cui, Y. High-performance lithium battery anodes using silicon nanowires. *Nat. Nanotechnol.* **2008**, *3*, 31–35. [CrossRef]
18. Chae, O.B.; Rynearson, L.; Lucht, B.L. Distance-Dependent Solid Electrolyte Interphase Control by Electrochemical Pretreatment. *ACS Energy Lett.* **2022**, *7*, 3087–3094. [CrossRef]
19. Nguyen, T.P.; Kim, I.T. Self-Assembled Few-Layered MoS on SnO Anode for Enhancing Lithium-Ion Storage. *Nanomaterials* **2020**, *10*, 2558. [CrossRef]
20. Lee, G.; Kim, I.; Hur, J. Highly conductive and robust telluride-carbon hybrid matrix for enhanced copper diphosphide anode in Li-ion batteries. *J. Alloys Compd.* **2023**, *950*, 169914. [CrossRef]
21. So, S.; Ko, J.; Ahn, Y.N.; Kim, I.; Hur, J. Unraveling improved electrochemical kinetics of In_2Te_3-based anodes embedded in hybrid matrix for Li-ion batteries. *Chem. Eng. J.* **2022**, *429*, 132395. [CrossRef]
22. Feng, K.; Li, M.; Liu, W.W.; Kashkooli, A.G.; Xiao, X.C.; Cai, M.; Chen, Z.W. Silicon-Based Anodes for Lithium-Ion Batteries: From Fundamentals to Practical Applications. *Small* **2018**, *14*, 1702737. [CrossRef] [PubMed]
23. Kim, B.C.; Uono, H.; Satou, T.; Fuse, T.; Ishihara, T.; Ue, M.; Senna, M. Cyclic properties of Si-Cu/carbon nanocomposite anodes for Li-ion secondary batteries. *J. Electrochem. Soc.* **2005**, *152*, A523–A526. [CrossRef]
24. Zhang, Y.X.; Wu, B.R.; Mu, G.; Ma, C.W.; Mu, D.B.; Wu, F. Recent progress and perspectives on silicon anode: Synthesis and prelithiation for LIBs energy storage. *J. Energy Chem.* **2022**, *64*, 615–650. [CrossRef]
25. Quilty, C.D.; Wu, D.R.; Li, W.Z.; Bock, D.C.; Wang, L.; Housel, L.M.; Abraham, A.; Takeuchi, K.J.; Marschilok, A.C.; Takeuchi, E.S. Electron and Ion Transport in Lithium and Lithium-Ion Battery Negative and Positive Composite Electrodes. *Chem. Rev.* **2023**, *123*, 1327–1363. [CrossRef]
26. Choi, W.S.; Kim, M.; Kim, I.T. Te-rP-C Anodes Prepared Using a Scalable Milling Process for High-Performance Lithium-Ion Batteries. *Micromachines* **2023**, *14*, 2156. [CrossRef]
27. Kidanu, W.G.; Hur, J.; Kim, I.T. Gallium-Indium-Tin Eutectic as a Self-Healing Room-Temperature Liquid Metal Anode for High-Capacity Lithium-Ion Batteries. *Materials* **2022**, *15*, 168. [CrossRef]
28. Luo, F.; Liu, B.N.; Zheng, J.Y.; Chu, G.; Zhong, K.F.; Li, H.; Huang, X.J.; Chen, L.Q. Review-Nano-Silicon/Carbon Composite Anode Materials Towards Practical Application for Next Generation Li-Ion Batteries. *J. Electrochem. Soc.* **2015**, *162*, A2509–A2528. [CrossRef]

29. Roy, P.; Srivastava, S.K. Nanostructured anode materials for lithium ion batteries. *J. Mater. Chem. A* **2015**, *3*, 2454–2484. [CrossRef]
30. Bärmann, P.; Diehl, M.; Göbel, L.; Ruttert, M.; Nowak, S.; Winter, M.; Placke, T. Impact of the silicon particle size on the pre-lithiation behavior of silicon/carbon composite materials for lithium ion batteries. *J. Power Sources* **2020**, *464*, 228224. [CrossRef]
31. Shi, C.; Wang, T.; Liao, X.; Qie, B.; Yang, P.; Chen, M.; Wang, X.; Srinivasan, A.; Cheng, Q.; Ye, Q.; et al. Accordion-like stretchable Li-ion batteries with high energy density. *Energy Storage Mater.* **2019**, *17*, 136–142. [CrossRef]
32. Zhang, H.; Yang, Y.; Ren, D.; Wang, L.; He, X. Graphite as anode materials: Fundamental mechanism, recent progress and advances. *Energy Storage Mater.* **2021**, *36*, 147–170. [CrossRef]
33. Li, P.; Kim, H.; Myung, S.T.; Sun, Y.K. Diverting Exploration of Silicon Anode into Practical Way: A Review Focused on Silicon-Graphite Composite for Lithium Ion Batteries. *Energy Storage Mater.* **2021**, *35*, 550–576. [CrossRef]
34. Gonzalez, A.F.; Yang, N.H.; Liu, R.S. Silicon Anode Design for Lithium-Ion Batteries: Progress and Perspectives. *J. Phys. Chem. C* **2017**, *121*, 27775–27787. [CrossRef]
35. Yoshio, M.; Wang, H.Y.; Fukuda, K.; Umeno, T.; Dimov, N.; Ogumi, Z. Carbon-coated Si as a lithium-ion battery anode material. *J. Electrochem. Soc.* **2002**, *149*, A1598–A1603. [CrossRef]
36. Sun, L.; Liu, Y.X.; Shao, R.; Wu, J.; Jiang, R.Y.; Jin, Z. Recent progress and future perspective on practical silicon anode-based lithium ion batteries. *Energy Storage Mater.* **2022**, *46*, 482–502. [CrossRef]
37. Preman, A.N.; Vo, T.N.; Choi, S.; Lee, H.; Lim, Y.E.; Kim, I.; Ahn, S.K. Self-Healable Poly(Acrylic Acid) Binder toward Optimized Electrochemical Performance for Silicon Anodes: Importance of Balanced Properties. *ACS Appl. Energy Mater.* **2024**, *7*, 749–759. [CrossRef]
38. Gu, M.; He, Y.; Zheng, J.M.; Wang, C.M. Nanoscale silicon as anode for Li-ion batteries: The fundamentals, promises, and challenges. *Nano Energy* **2015**, *17*, 366–383. [CrossRef]
39. Chen, X.; Li, H.X.; Yan, Z.H.; Cheng, F.Y.; Chen, J. Structure design and mechanism analysis of silicon anode for lithium-ion batteries. *Sci. China Mater.* **2019**, *62*, 1515–1536. [CrossRef]
40. Liu, X.Y.; Zhu, X.J.; Pan, D. Solutions for the problems of silicon-carbon anode materials for lithium-ion batteries. *R. Soc. Open Sci.* **2018**, *5*, 172370. [CrossRef]
41. Park, E.; So, S.; Hur, J. Carbon-free hydrated cobalt vanadium oxide as a promising anode for lithium-ion batteries. *Appl. Surf. Sci.* **2022**, *579*, 152182. [CrossRef]
42. Li, P.; Zhao, G.Q.; Zheng, X.B.; Xu, X.; Yao, C.H.; Sun, W.P.; Dou, S.X. Recent progress on silicon-based anode materials for practical lithium-ion battery applications. *Energy Storage Mater.* **2018**, *15*, 422–446. [CrossRef]
43. Ozanam, F.; Rosso, M. Silicon as anode material for Li-ion batteries. *Mater. Sci. Eng. B-Adv.* **2016**, *213*, 2–11. [CrossRef]
44. Jin, Y.; Zhu, B.; Lu, Z.D.; Liu, N.; Zhu, J. Challenges and Recent Progress in the Development of Si Anodes for Lithium-Ion Battery. *Adv. Energy Mater.* **2017**, *7*, 1700715. [CrossRef]
45. Shen, T.; Yao, Z.J.; Xia, X.H.; Wang, X.L.; Gu, C.D.; Tu, J.P. Rationally Designed Silicon Nanostructures as Anode Material for Lithium-Ion Batteries. *Adv. Eng. Mater.* **2018**, *20*, 1700591. [CrossRef]
46. Szczech, J.R.; Jin, S. Nanostructured silicon for high capacity lithium battery anodes. *Energy Environ. Sci.* **2011**, *4*, 56–72. [CrossRef]
47. Wu, H.; Cui, Y. Designing nanostructured Si anodes for high energy lithium ion batteries. *Nano Today* **2012**, *7*, 414–429. [CrossRef]
48. Rahman, M.A.; Song, G.S.; Bhatt, A.I.; Wong, Y.C.; Wen, C.E. Nanostructured Silicon Anodes for High-Performance Lithium-Ion Batteries. *Adv. Funct. Mater.* **2016**, *26*, 647–678. [CrossRef]
49. Teki, R.; Datta, M.K.; Krishnan, R.; Parker, T.C.; Lu, T.M.; Kumta, P.N.; Koratkar, N. Nanostructured Silicon Anodes for Lithium Ion Rechargeable Batteries. *Small* **2009**, *5*, 2236–2242. [CrossRef]
50. Cao, Z.; Zheng, X.Y.; Qu, Q.T.; Huang, Y.H.; Zheng, H.H. Electrolyte Design Enabling a High-Safety and High-Performance Si Anode with a Tailored Electrode-Electrolyte Interphase. *Adv. Mater.* **2021**, *33*, 2103178. [CrossRef]
51. Zhang, S.; He, M.N.; Su, C.C.; Zhang, Z.C. Advanced electrolyte/additive for lithium-ion batteries with silicon anode. *Curr. Opin. Chem. Eng.* **2016**, *13*, 24–35. [CrossRef]
52. Xu, C.; Lindgren, F.; Philippe, B.; Gorgoi, M.; Björefors, F.; Edström, K.; Gustafsson, T. Improved Performance of the Silicon Anode for Li-Ion Batteries: Understanding the Surface Modification Mechanism of Fluoroethylene Carbonate as an Effective Electrolyte Additive. *Chem. Mat.* **2015**, *27*, 2591–2599. [CrossRef]
53. Liu, H.B.; Sun, Q.; Zhang, H.Q.; Cheng, J.; Li, Y.Y.; Zeng, Z.; Zhang, S.; Xu, X.; Ji, F.J.; Li, D.P.; et al. The application road of silicon-based anode in lithium-ion batteries: From liquid electrolyte to solid-state electrolyte. *Energy Storage Mater.* **2023**, *55*, 244–263. [CrossRef]
54. He, Y.; Jiang, L.; Chen, T.W.; Xu, Y.B.; Jia, H.P.; Yi, R.; Xue, D.C.; Song, M.; Genc, A.; Bouchet-Marquis, C.; et al. Progressive growth of the solid-electrolyte interphase towards the Si anode interior causes capacity fading. *Nat. Nanotechnol.* **2021**, *16*, 1113–1120. [CrossRef]
55. Xu, Z.X.; Yang, J.; Li, H.P.; Nuli, Y.N.; Wang, J.L. Electrolytes for advanced lithium ion batteries using silicon-based anodes. *J. Mater. Chem. A* **2019**, *7*, 9432–9446. [CrossRef]
56. Li, Z.H.; Zhang, Y.P.; Liu, T.F.; Gao, X.H.; Li, S.Y.; Ling, M.; Liang, C.D.; Zheng, J.C.; Lin, Z. Silicon Anode with High Initial Coulombic Efficiency by Modulated Trifunctional Binder for High-Areal-Capacity Lithium-Ion Batteries. *Adv. Energy Mater.* **2020**, *10*, 1903110. [CrossRef]

57. Gao, S.L.; Sun, F.Y.; Brady, A.; Pan, Y.Y.; Erwin, A.; Yang, D.D.; Tsukruk, V.; Stack, A.G.; Saito, T.; Yang, H.B.; et al. Ultra-efficient polymer binder for silicon anode in high-capacity lithium-ion batteries. *Nano Energy* **2020**, *73*, 104804. [CrossRef]
58. Magasinski, A.; Zdyrko, B.; Kovalenko, I.; Hertzberg, B.; Burtovyy, R.; Huebner, C.F.; Fuller, T.F.; Luzinov, I.; Yushin, G. Toward Efficient Binders for Li-Ion Battery Si-Based Anodes: Polyacrylic Acid. *ACS Appl. Mater. Interfaces* **2010**, *2*, 3004–3010. [CrossRef]
59. Liu, D.; Zhao, Y.; Tan, R.; Tian, L.L.; Liu, Y.D.; Chen, H.B.; Pan, F. Novel conductive binder for high-performance silicon anodes in lithium ion batteries. *Nano Energy* **2017**, *36*, 206–212. [CrossRef]
60. Preman, A.N.; Lee, H.; Yoo, J.; Kim, I.; Saito, T.; Ahn, S.K. Progress of 3D network binders in silicon anodes for lithium ion batteries. *J. Mater. Chem. A* **2020**, *8*, 25548–25570. [CrossRef]
61. Luo, W.; Chen, X.Q.; Xia, Y.; Chen, M.; Wang, L.J.; Wang, Q.Q.; Li, W.; Yang, J.P. Surface and Interface Engineering of Silicon-Based Anode Materials for Lithium-Ion Batteries. *Adv. Energy Mater.* **2017**, *7*, 1701083. [CrossRef]
62. Jiang, C.L.; Xiang, L.; Miao, S.J.; Shi, L.; Xie, D.H.; Yan, J.X.; Zheng, Z.J.; Zhang, X.M.; Tang, Y.B. Flexible Interface Design for Stress Regulation of a Silicon Anode toward Highly Stable Dual-Ion Batteries. *Adv. Mater.* **2020**, *32*, e1908470. [CrossRef]
63. Lv, Y.Y.; Han, Z.B.; Jia, R.R.; Shi, L.Y.; Yuan, S. Porous interface for fast charging silicon anode. *Battery Energy* **2022**, *1*, 20220009. [CrossRef]
64. Wang, L.; Lu, J.J.; Li, S.Y.; Xi, F.S.; Tong, Z.Q.; Chen, X.H.; Wei, K.X.; Ma, W.H. Controllable Interface Engineering for the Preparation of High Rate Silicon Anode. *Adv. Funct. Mater.* **2024**, *34*, 2403574. [CrossRef]
65. Wang, L.; Yu, J.; Li, S.Y.; Xi, F.S.; Ma, W.H.; Wei, K.X.; Lu, J.J.; Tong, Z.Q.; Liu, B.; Luo, B. Recent advances in interface engineering of silicon anodes for enhanced lithium-ion battery performance. *Energy Storage Mater.* **2024**, *66*, 103243. [CrossRef]
66. Zuo, X.X.; Zhu, J.; Müller-Buschbaum, P.; Cheng, Y.J. Silicon based lithium-ion battery anodes: A chronicle perspective review. *Nano Energy* **2017**, *31*, 113–143. [CrossRef]
67. Andersen, H.F.; Foss, C.E.L.; Voje, J.; Tronstad, R.; Mokkelbost, T.; ErikVullum, P.; Ulvestad, A.; Kirkengen, M.; Mæhlen, J.P. Silicon-Carbon composite anodes from industrial battery grade silicon. *Sci. Rep.* **2019**, *9*, 14814. [CrossRef]
68. Liu, Y.; Hanai, K.; Yang, J.; Imanishi, N.; Hirano, A.; Takeda, Y. Silicon/carbon composites as anode materials for Li-ion batteries. *Electrochem. Solid-State Lett.* **2004**, *7*, A369–A372. [CrossRef]
69. Wang, J.Z.; Zhong, C.; Chou, S.L.; Liu, H.K. Flexible free-standing graphene-silicon composite film for lithium-ion batteries. *Electrochem. Commun.* **2010**, *12*, 1467–1470. [CrossRef]
70. Mhadhbi, M. Modelling of the High-Energy Ball Milling Process. *Adv. Mat. Phys. Chem.* **2021**, *11*, 31–44. [CrossRef]
71. Yu, S.X.; Guo, B.B.; Zeng, T.B.; Qu, H.Q.; Yang, J.L.; Bai, J.M. Graphene-based lithium-ion battery anode materials manufactured by mechanochemical ball milling process: A review and perspective. *Compos. Pt. B-Eng.* **2022**, *246*, 110232. [CrossRef]
72. Bhagyaraj, S.M.; Oluwafemi, O.S.; Kalarikkal, N.; Thomas, S. *Synthesis of Inorganic Nanomaterials: Advances and Key Technologies*; Woodhead Publishing: Duxford, UK; Cambridge, MA, USA, 2018; p. 12. 300p.
73. Zhang, Y.; Cheng, Y.; Song, J.; Zhang, Y.; Shi, Q.; Wang, J.; Tian, F.; Yuan, S.; Su, Z.; Zhou, C.; et al. Functionalization-assistant ball milling towards Si/graphene anodes in high performance Li-ion batteries. *Carbon* **2021**, *181*, 300–309. [CrossRef]
74. Wang, D.; Gao, M.; Pan, H.; Wang, J.; Liu, Y. High performance amorphous-Si@SiO_x/C composite anode materials for Li-ion batteries derived from ball-milling and in situ carbonization. *J. Power Sources* **2014**, *256*, 190–199. [CrossRef]
75. Qian, L.; Lan, J.-L.; Xue, M.; Yu, Y.; Yang, X. Two-step ball-milling synthesis of a Si/SiO_x/C composite electrode for lithium ion batteries with excellent long-term cycling stability. *RSC Adv.* **2017**, *7*, 36697–36704. [CrossRef]
76. Koraag, P.Y.E.; Firdaus, A.M.; Hawari, N.H.; Refino, A.D.; Dempwolf, W.; Iskandar, F.; Peiner, E.; Wasisto, H.S.; Sumboja, A. Covalently Bonded Ball-Milled Silicon/CNT Nanocomposite as Lithium-Ion Battery Anode Material. *Batteries* **2022**, *8*, 165. [CrossRef]
77. Cu, P.; Cai, R.; Zhou, Y.K.; Shao, Z.P. Si/C composite lithium-ion battery anodes synthesized from coarse silicon and citric acid through combined ball milling and thermal pyrolysis. *Electrochim. Acta* **2010**, *55*, 3876–3883.
78. Cabello, M.; Gucciardi, E.; Herrán, A.; Carriazo, D.; Villaverde, A.; Rojo, T. Towards a High-Power Si@graphite Anode for Lithium Ion Batteries through a Wet Ball Milling Process. *Molecules* **2020**, *25*, 2494. [CrossRef]
79. Zhang, Y.; Zhang, X.G.; Zhang, H.L.; Zhao, Z.G.; Li, F.; Liu, C.; Cheng, H.M. Composite anode material of silicon/graphite/carbon nanotubes for Li-ion batteries. *Electrochim. Acta* **2006**, *51*, 4994–5000. [CrossRef]
80. Han, J.; Zhao, C.C.; Wang, L.; Song, J.; Yang, D.; Tian, Q.H. Simple ball milling-assisted method enabling N-doped carbon embedded Si for high performance lithium-ion battery anode. *J. Alloys Compd.* **2023**, *966*, 171668. [CrossRef]
81. Gorshkov, A.; Berezikov, N.; Kaltaev, A.; Yankovsky, S.; Slyusarsky, K.; Tabakaev, R.; Larionov, K. Analysis of the Physicochemical Characteristics of Biochar Obtained by Slow Pyrolysis of Nut Shells in a Nitrogen Atmosphere. *Energies* **2021**, *14*, 8075. [CrossRef]
82. Kaur, R.; Gera, P.; Jha, M.K.; Bhaskar, T. Chapter 8—Thermochemical Route for Biohydrogen Production. In *Biohydrogen*, 2nd ed.; Pandey, A., Mohan, S.V., Chang, J.-S., Hallenbeck, P.C., Larroche, C., Eds.; Elsevier: Amsterdam, The Netherlands, 2019; pp. 187–218.
83. Zhang, G.W.; He, Y.Q.; Feng, Y.; Wang, H.F.; Zhang, T.; Xie, W.N.; Zhu, X.N. Enhancement in liberation of electrode materials derived from spent lithium-ion battery by pyrolysis. *J. Clean. Prod.* **2018**, *199*, 62–68. [CrossRef]
84. Chen, Y.L.; Hu, Y.; Shao, J.Z.; Shen, Z.; Chen, R.Z.; Zhang, X.W.; He, X.; Song, Y.Z.; Xing, X.L. Pyrolytic carbon-coated silicon/carbon nanofiber composite anodes for high-performance lithium-ion batteries. *J. Power Sources* **2015**, *298*, 130–137. [CrossRef]

85. Yang, Z.W.; Yang, Y.; Guo, H.J.; Wang, Z.X.; Li, X.H.; Zhou, Y.; Wang, J.X. Compact structured silicon/carbon composites as high-performance anodes for lithium ion batteries. *Ionics* **2018**, *24*, 3405–3411. [CrossRef]
86. Hu, J.Z.; Wang, M.; Meyer, A.W.; Huang, X.S.; Cheng, Y.T. Oxidative Pyrolysis of Si/Polyacrylonitrile Composites as an Unconventional Approach to Fabricate High Performance Lithium Ion Battery Negative Electrodes. *J. Electrochem. Soc.* **2019**, *166*, A3716–A3722. [CrossRef]
87. Su, M.R.; Wang, Z.X.; Guo, H.J.; Li, X.H.; Huang, S.L.; Gan, L. Silicon, flake graphite and phenolic resin-pyrolyzed carbon based Si/C composites as anode material for lithium-ion batteries. *Adv. Powder Technol.* **2013**, *24*, 921–925. [CrossRef]
88. Wang, M.S.; Fan, L.Z. Silicon/carbon nanocomposite pyrolyzed from phenolic resin as anode materials for lithium-ion batteries. *J. Power Sources* **2013**, *244*, 570–574. [CrossRef]
89. Campbell, H.R.; Alsharif, F.M.; Marsac, P.J.; Lodder, R.A. The Development of a Novel Pharmaceutical Formulation of D-Tagatose for Spray-Drying. *J. Pharm. Innov.* **2022**, *17*, 194–206. [CrossRef]
90. Sosnik, A.; Seremeta, K.P. Advantages and challenges of the spray-drying technology for the production of pure drug particles and drug-loaded polymeric carriers. *Adv. Colloid Interface Sci.* **2015**, *223*, 40–54. [CrossRef]
91. Pan, Q.R.; Zuo, P.J.; Lou, S.F.; Mu, T.S.; Du, C.Y.; Cheng, X.Q.; Ma, Y.L.; Gao, Y.Z.; Yin, G.P. Micro-sized spherical silicon@carbon@graphene prepared by spray drying as anode material for lithium-ion batteries. *J. Alloys Compd.* **2017**, *723*, 434–440. [CrossRef]
92. Lai, J.; Guo, H.J.; Wang, Z.X.; Li, X.H.; Zhang, X.P.; Wu, F.X.; Yue, P. Preparation and characterization of flake graphite/silicon/carbon spherical composite as anode materials for lithium-ion batteries. *J. Alloys Compd.* **2012**, *530*, 30–35. [CrossRef]
93. Su, M.R.; Liu, S.; Tao, L.; Tang, Y.P.; Dou, A.C.; Lv, J.; Liu, Y.J. Silicon@graphene composite prepared by spray-drying method as anode for lithium ion batteries. *J. Electroanal. Chem.* **2019**, *844*, 86–90. [CrossRef]
94. Paireau, C.; Jouanneau, S.; Ammar, M.R.; Simon, P.; Béguin, F.; Raymundo-Piñero, E. Si/C composites prepared by spray drying from cross-linked polyvinyl alcohol as Li-ion batteries anodes. *Electrochim. Acta* **2015**, *174*, 361–368. [CrossRef]
95. Wang, D.S.; Gao, M.X.; Pan, H.G.; Liu, Y.F.; Wang, J.H.; Li, S.Q.; Ge, H.W. Enhanced cycle stability of micro-sized Si/C anode material with low carbon content fabricated via spray drying and in situ carbonization. *J. Alloys Compd.* **2014**, *604*, 130–136. [CrossRef]
96. Shi, Q.T.; Zhou, J.H.; Ullah, S.; Yang, X.Q.; Tokarska, K.; Trzebicka, B.; Ta, H.Q.; Rümmeli, M.H. A review of recent developments in Si/C composite materials for Li-ion batteries. *Energy Storage Mater.* **2021**, *34*, 735–754. [CrossRef]
97. Ren, J.; Zhang, Y.; Li, J.; Tian, S.; Fei, T.; Li, H. Effects of deposition temperature and time on HfC nanowires synthesized by CVD on SiC-coated C/C composites. *Ceram. Int.* **2016**, *42*, 5623–5628. [CrossRef]
98. Shukrullah, S.; Mohamed, N.; Shaharun, M.; Saheed, M.; Irshad, M. Effect of CVD process temperature on activation energy and structural growth of MWCNTs. *Metall. Mater. Trans. A* **2016**, *47*, 1413–1424. [CrossRef]
99. Wang, W.; Kumta, P.N. Nanostructured Hybrid Silicon/Carbon Nanotube Heterostructures: Reversible High-Capacity Lithium-Ion Anodes. *ACS Nano* **2010**, *4*, 2233–2241. [CrossRef]
100. Fu, K.; Xue, L.G.; Yildiz, O.; Li, S.L.; Lee, H.; Li, Y.; Xu, G.J.; Zhou, L.; Bradford, P.D.; Zhang, X.W. Effect of CVD carbon coatings on Si@CNF composite as anode for lithium-ion batteries. *Nano Energy* **2013**, *2*, 976–986. [CrossRef]
101. Jin, H.C.; Sun, Q.; Wang, J.T.; Ma, C.; Ling, L.C.; Qiao, W.M. Preparation and electrochemical properties of novel silicon-carbon composite anode materials with a core-shell structure. *New Carbon Mater.* **2021**, *36*, 390–398. [CrossRef]
102. Liu, B.; Huang, P.; Xie, Z.Y.; Huang, Q.Z. Large-Scale Production of a Silicon Nanowire/Graphite Composites Anode via the CVD Method for High-Performance Lithium-Ion Batteries. *Energy Fuels* **2021**, *35*, 2758–2765. [CrossRef]
103. Yu, J.L.; Yang, J.; Feng, X.J.; Jia, H.; Wang, J.L.; Lu, W. Uniform Carbon Coating on Silicon Nanoparticles by Dynamic CVD Process for Electrochemical Lithium Storage. *Ind. Eng. Chem. Res.* **2014**, *53*, 12697–12704. [CrossRef]
104. Chen, W.L.; Dave, R.N.; Pfeffer, R.; Walton, O. Numerical simulation of Mechanofusion system. *Powder Technol.* **2004**, *146*, 121–136. [CrossRef]
105. Alonso, M.; Satoh, M.; Miyanami, K. Mechanism of the combined coating-mechanofusion processing of powders. *Powder Technol.* **1989**, *59*, 45–52. [CrossRef]
106. Tubtimkuna, S.; Danilov, D.L.; Sawangphruk, M.; Notten, P.H.L. Review of the Scalable Core-Shell Synthesis Methods: The Improvements of Li-Ion Battery Electrochemistry and Cycling Stability. *Small Methods* **2023**, *7*, e2300345. [CrossRef]
107. Kim, S.J.; Ha, S.J.; Lee, J.U.; Jeon, Y.P.; Hong, J.Y.; Bedia, J.; Vedyagin, A.A. Preparation of Silicon Oxide-Carbon Composite with Tailored Electrochemical Properties for Anode in Lithium-Ion Batteries. *C-J. Carbon Res.* **2023**, *9*, 114. [CrossRef]
108. Cao, Y.D.; Hatchard, T.D.; Dunlap, R.A.; Obrovac, M.N. Mechanofusion-derived Si-alloy/graphite composite electrode materials for Li-ion batteries. *J. Mater. Chem. A* **2019**, *7*, 8335–8343. [CrossRef]
109. Jiang, H.R.; Salehabadi, M.; Yasmin, S.; Wang, J.; Obrovac, M.N. Nano-Si Filled Graphite Anode Particles by Mechanofusion. *J. Electrochem. Soc.* **2023**, *170*, 120511. [CrossRef]
110. Wutthiprom, J.; Phattharasupakun, N.; Tomon, C.; Sawangphruk, M. Scalable solvent-free mechanofusion and magnesiothermic reduction processes for obtaining carbon nanospheres-encapsulated crystalline silicon anode for Li-ion batteries. *Electrochim. Acta* **2020**, *352*, 136457. [CrossRef]

batteries

Review

Research Progress and Challenges of Carbon/MXene Composites for Supercapacitors

Li Sun [1], Yu Dong [1], Hangyu Li [1], Hanqi Meng [1], Jianfei Liu [1], Qigao Cao [1] and Chunxu Pan [2,*]

1 Northwest Institute for Non-Ferrous Metal Research, Xi'an 710016, China
2 MOE Key Laboratory of Artificial Micro- and Nanostructures, School of Physics and Technology, Wuhan University, Wuhan 430072, China
* Correspondence: cxpan@whu.edu.cn

Abstract: Carbon materials/MXenes composite materials have gained widespread attention in the field of supercapacitors due to their excellent electrochemical performance. Carbon materials are considered ideal electrode materials for supercapacitors due to their high specific surface area, good conductivity, and outstanding electrochemical stability. MXenes, as a novel two-dimensional material, exhibit prominent conductivity, mechanical properties, and ionic conductivity, thereby showing great potential for applications in energy storage devices. The combination of carbon materials and MXenes is expected to fully leverage the advantages of both, optimizing electrode conductivity, enhancing the energy density and power density, and improving the charge–discharge performance. This article reviews the key research progress of carbon/MXenes composite materials in supercapacitors in recent years, including their synthesis methods, structural tuning, and improvements in their electrochemical performance. Finally, the article looks forward to future research directions and proposes potential strategies to enhance the overall performance of the composite materials and achieve large-scale applications. By addressing the existing challenges, carbon/MXenes composite materials are anticipated to achieve higher energy and power outputs for the supercapacitor field in the future, providing strong support for the development of new energy storage technologies such as electric vehicles and wearable devices.

Keywords: supercapacitors; carbon materials/MXenes; energy storage and conversion; composites

1. Introduction

With the rapid development of the global economy and the rapid advancement of technology, the utilization and storage technology of renewable energy has become an important research field [1–4]. Faced with the increasingly severe energy crisis and environmental pollution, it is particularly urgent to develop efficient energy storage devices. Researchers have conducted numerous studies into the development of sustainable and clean energy conversion and storage systems, such as supercapacitors, fuel cells, and metal ion batteries [5–8]. Supercapacitors, as a new type of energy storage device, are excellent electrochemical devices for energy storage in devices with high-quality energy batteries [8,9]. Supercapacitors have the advantages of fast charge–discharge, a long cycle life, and high power density, and are widely used in fields such as electric vehicles, renewable energy storage, large-scale power regulation, and other fields [2,10–12].

According to the charge storage mechanism, supercapacitors can be divided into double-layer capacitors and Faradaic-type capacitors [12–15]. Double-layer capacitors typically have a high power density due to the physical store of charges at the electrode–electrolyte interface through electrostatic attraction [2]. Carbon materials with a high specific surface area, excellent electronic conductivity, and stability are typical electrode materials for double-layer capacitors [16–18]. However, the low-mass density of carbon-based electrode materials can lead to a lower energy density and volumetric capacitance of double-layer capacitors [19]. Faradaic-type capacitors store charges through rapid oxidation-reduction reactions, and have a high energy density but low cycling stability and rate performance, such as transition metal oxides/carbides and conductive polymers [17,20–22]. In view of this, scientists have conducted extensive research in an attempt to solve the dilemma faced by supercapacitors by developing novel electrode materials, including (1) enhancing the energy density and holding a high power density, (2) increasing the cycle stability to extend the service life, and (3) decreasing the expense and complexity of electrode materials [23–25].

MXenes, with a layered structure similar to graphene, are a novel two-dimensional (2D) material [26,27]. Since the first report in 2011, the distinctive properties and structure of MXenes have led to their widespread research in the field of energy storage and conversion [28]. MXenes are constructed alternately with carbon layers and transition metal layers, exhibiting outstanding conductivity ($Ti_3C_2T_x$ up to 10,000 S cm^{-1}) and a rapid electron transfer [29]. In addition, due to the highly ion accessibility of the 2D layered structure and functionalized surface, MXenes can provide double-layer capacitance, intercalation pseudocapacitance, and redox pseudocapacitance [30]. Due to the magnitude of the electrostatic potential difference at the electrode–electrolyte interface and the types of electrolyte ions and surface terminations, the charge storage mechanism in MXene is a non-singular and interchangeable process [1,2,24]. MXenes may exhibit a double-layer capacitance mechanism in environments lacking H+, such as K_2SO_4 and KOH electrolytes. However, the charge storage is mainly dominated by pseudocapacitive behavior in acidic electrolytes (such as 3 M H_2SO_4 electrolyte) [6,19]. The H+ ions undergo highly reversible protonation/deprotonation at the -O terminal of MXenes in the electrolyte [6,10]. In the neutral and alkaline electrolytes, the electrolyte ions act as a dielectric medium to prevent the hybrid orbital between the ions and the MXenes surface terminal, forming a double electric layer [6,10]. On the contrary, MXenes exhibit the combination of double-layer capacitance, intercalation pseudocapacitance, and REDOX pseudocapacitance due to the polarization voltage and the surface terminal composition in the acidic electrolyte [15]. When MXenes are heavily modified by other functional groups, the double-layer capacitance dominates due to the protonation reaction triggered only by the -O terminal [6,23]. When a sufficient negative voltage is applied, the adsorption potential of the intercalated positive ions can be overcome, and hybrid orbitals allow hydrogen ions to directly adsorb with single-bond O terminals, generating intercalation pseudocapacitance [6,28]. As the negative potential increases, the single-bond O terminal continuously adsorbs ions and converts into a single-bond -OH terminal, expanding the MXenes intermediate layer under electrostatic repulsion [6]. As the interlayer space continues to expand, more and more MXenes surfaces are exposed, and ions undergo rapid surface redox reactions, producing redox pseudocapacitance [10]. Modified MXenes have more ion transport channels, which will reduce ion diffusion resistance and improve the electrochemical performance [31]. However, there are strong van der Waals interactions and hydrogen bonding between the MXene nanosheets, which can easily lead to re-stacking and self-oxidation in practical applications, resulting in a decrease in the accessible surface area and a decline in the capacitance performance [32,33]. Secondly, MXenes, such as $Ti_3C_2T_x$, undergo a reaction when in contact with air or moisture. Non-conductive TiO_2 will form due to the oxidation of surface Ti atoms through the interaction of oxygen or H_2O, which reduces electronic conductivity and capacitance active sites, leading to a deterioration in the electrochemical performance [18]. In addition, MXenes-based supercapacitor devices face challenges such as severe self-discharge behavior and limited potential windows.

To overcome the above problems, researchers have found that introducing conductive spacers with stable structures (such as molybdenum carbide, carbon materials, polymers, metal oxides, etc.) into the intermediate layer of MXene nanosheets is an effective and feasible strategy [34–36]. The phenomenon of MXene nanosheets restacking and aggregation can be suppressed, and consecutive electron transport channels and extra capacity can also be provided. Carbon materials have an ultra-high specific surface area, excellent electrical conductivity, good chemical stability, and an outstanding electrochemical performance [37]. Carbon materials are mainly based on adsorbing ions to construct double-layer capacitors for energy storage. Therefore, the contact area between the electrode and the electrolyte is increased by increasing the specific surface area, so as to raise the double-electric-layer capacitance [37]. With the maturity and expansion of functionalization technology for carbon materials, carbon materials with a Faradaic-type capacitors mechanism have attracted widespread attention from researchers, especially heteroatom-doped carbon materials. The difference in electronegativity between heteroatoms and carbon atoms will modify the electronic structure, promote electron migration, increase the surface activity of the material, and thus improve the electrochemical performance [37]. The introduction of heteroatoms into carbon materials can transfer high electrochemical active sites and change the reactivity and kinetics of the material through unconventional charge polarization [37]. Moreover, the abundant active sites enable Faraday reactions to store more charges, significantly improving the specific capacitance and energy density. Moreover, carbon materials with various nanostructures can be introduced into the MXenes structure to prevent restacking, including zero-dimensional (0D) (carbon quantum dots (CQDs), carbon spheres), one-dimensional (1D) (carbon nanotubes (CNT), carbon nanofibers (CNF)), 2D (graphene nanosheets, carbon nanosheets), three-dimensional (3D) (graphene foam, activated carbon) [17,19,23,25]. In addition, carbon materials not only protect MXenes from oxidation to a certain extent, but also reduce the oxidation reaction rate of MXene [17]. Carbon materials coating the surface of MXenes can inhibit oxidation reactions, as oxidants such as oxygen and water are isolated [19]. The introduction of MXenes into carbon materials can increase the interlayer spacing of nanosheets, thus overcoming the structural defects of MXenes [23]. The electron transfer ability and electrochemical performance of carbon/MXenes can be effectively improved through the synergistic effect of carbon materials and MXenes [25]. In recent years, some researchers have confirmed and summarized that the application of carbon/MXene composite materials in supercapacitors has a significant impact on improving the electrochemical performance. Therefore, the introduction of two-dimensional MXenes into carbon materials to construct and improve the microstructure to fully exploit the electrochemical properties has become a widely sought after solution. Zhang et al. summarized the emerging technologies of MXenes and carbon hybridization, and their applications in lithium storage, sodium storage, lithium sulfur batteries, supercapacitors, and electrocatalysis [18]. However, the new preparation strategies of carbon-MXene composites (such as 3D printing technology) and the effects of the introduction of carbon with different structures into MXenes on the performance of supercapacitors require a more in-depth and detailed summary. Cai et al. introduced the definition of MXenes, common strategies for synthesizing $Ti_3C_2T_x$, and the latest progress of $Ti_3C_2T_x$/carbon composite materials as electrode materials for supercapacitors [23]. Nevertheless, further discussion is needed on the preparation strategy and system classification of carbon MXene composite materials, as well as the influence of the existing states of carbon materials with different structures on the performance of supercapacitors. Siddu N. K. et al. briefly reviewed the structure, surface termination, and synthesis of MXenes and MXene/carbon composite materials, as well as their applications in the field of supercapacitors [17]. Further review is needed on the preparation strategies and clear classification of carbon MXene composite materials. Although the above article provides a review of the application of carbon/MXene composite materials in the field of supercapacitors, with the continuous research and exploration of new preparation technologies and construction strategies, the systematic classification and in-depth analysis of preparation techniques and the introduction of carbon materials with different structures into MXenes to achieve their full performance still have important significance.

This article comprehensively summarizes the research progress of traditional and emerging carbon materials/MXenes composites in the field of supercapacitors. Furthermore, we have reclassified and summarized the synthesis methods of carbon/MXene composites and focused on discussing the role of different structured carbon materials in regulating the structure/properties of MXenes and improving the electrochemical performance, as shown in Figure 1, as well as how to achieve the effect of "1 + 1 > 2". Classifying carbon/MXene composites based on their carbon size and analyzing the main challenges and future research directions of carbon/MXene composites as electrode materials for supercapacitors can provide useful guidance for researchers in related fields.

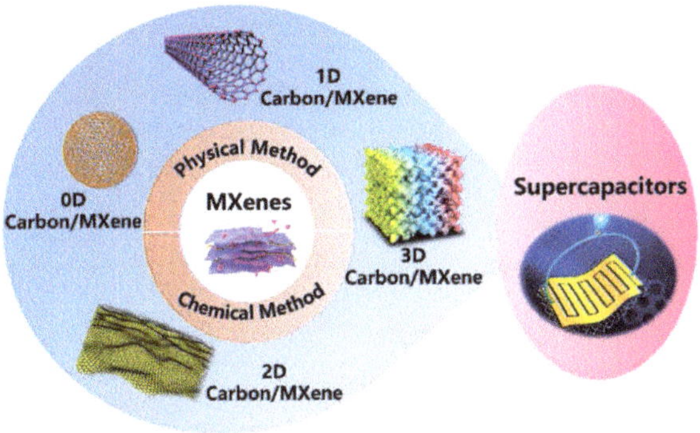

Figure 1. Schematic illustration of carbon/MXene composites for supercapacitors.

2. Synthesis of Carbon Materials/MXene Composites

2.1. Physical Method

Physical methods are the simplest way to prepare composite materials, and no new substances are formed during this process [38]. Moreover, this method enables the uniform mixing of various components of the material, thereby avoiding the complex steps and parameters that require strict control in the chemical reaction process [34]. At present, the physical synthesis strategies for carbon/$Ti_3C_2T_x$ composite materials include the mechanical mixing strategy, self-assembly strategy, vacuum-assisted filtration strategy, and some other preparation strategies.

2.1.1. Mechanical Mixing Strategy

The mechanical mixing strategy utilizes techniques such as stirring, ball milling, and impregnation to uniformly crush or mix different components to form composite materials, which has the advantages of dependable and uncomplicated manipulation [18]. For carbon/MXenes composite materials, during the mechanical mixing process, MXenes can easily establish connections with other components due to the presence of abundant oxygen-containing functional groups [38]. Therefore, the strong interaction as a binding force between MXenes and the carbon structure forms a structurally stable carbon/MXenes composite material [38]. The carbon material as a conductive spacer is introduced into the intermediate layer of MXene nanosheets to form a mixed or sandwich structure, which will improve the performance by increasing the interlayer spacing, ion sites, and providing rapid ion transport [17]. Qi et al. prepared a flexible electrode of CC@MnO_2@MXene by dipping methods so that MXene was completely coated on the CC@MnO_2 fiber surface [39]. When MXene is wrapped around the surface of CC@MnO_2, it not only enhances the conductivity, but also generates more active sites of composite materials [39]. Accordingly, the composite materials provide a high charge storage performance, excellent cycling stability, and excellent energy density. Shi et al. designed a unique solvent-free ball-milling approach to prepare MOF/MXene composites, as shown in Figure 2a [40]. After selenization, metal selenides@carbon/MXene composite materials are prepared by transforming MOF/MXene. During this process, MXene is used as an electronic conductive network and form a mesoporous and macroporous structure of the obtained metal selenides@carbon/MXene composites [40].

2.1.2. Self-Assembly Strategy

The self-assembly process is the spontaneous aggregation of molecules, atoms, or nanomaterials into stable structures with certain regular shapes driven by electrostatic interactions, without any chemical bonds [19]. This method is simple and stable, and has been successfully applied to the synthesis of various composite materials. Due to the preparation process of etching and delamination, most MXenes have abundant surface functional groups (such as -O and -OH) [25]. Therefore, MXenes can easily and spontaneously aggregate with carbon materials to form stable and complete composites driven by electrostatic interactions, hydrogen bonds, and/or covalent bonds [18]. Fu et al. constructed graphene oxide/MXene (GO-M) composite materials using laser-induced electrostatic self-assembly technology, as shown in Figure 2b [41]. The GO-M in a reduced state prepared by the synergistic effect of a laser matter interaction and electrostatic self-assembly exhibits good conductivity and a layered micro/nano structure, which is significantly better than the reduced-state graphene oxide [41]. In addition, the introduction of MXenes makes the composite material rich in active sites, so the capacitance value achieved is five times that of the reduced graphene-based supercapacitors. Depijan et al. prepared $Ti_3C_2T_x$/pg-C_3N_4 composite materials through self-assembly technology and the annealing process [42]. When pg-C_3N_4 is combined with $Ti_3C_2T_x$, the interlayer spacing and conductivity are effectively improved, and ion accessibility is enhanced. Therefore, the composite material exhibits excellent capacitance and cycling stability.

2.1.3. Vacuum-Assisted Filtration Strategy

The vacuum-assisted filtration strategy is used to prepare carbon/MXene composite materials through suction filtration and a pressure difference [34]. This method is simple, efficient, and widely used in the construction of carbon/MXene flexible electrodes. Due to the triple characteristics of metal-like conductivity, high mechanical properties, and surface functionality, MXenes can tightly bond with carbon materials to form flexible substrates during vacuum-assisted filtration processes without the need for additional polymer adhesives [19]. Hong et al. designed a DRCT-like CNT/Ti_3C_2 composite through an alternate filtration process to integrate SACNT membranes with Ti_3C_2 [43]. Benefiting from this biomimetic structure, the DRCT-like CNT/Ti_3C_2 film could endure a mechanical strain of up to 8.01% and withstand 1000 cycles of extensive bending at angles of 90° and 180°. At room temperature, this anode delivered a high specific capacity and maintained a considerable capacity even after 3300 stable cycles. When subjected to a severe temperature of −40 °C, the DRCT-like CNT/Ti_3C_2 anode still retained a substantial specific capacity. Yu et al. employed two-dimensional $Ti_3C_2T_x$ MXene as a flexible, conductive, and electrochemically active binder for one-step vacuum-assisted filtration to prepare MXene-bonded activated carbon as a flexible electrode for supercapacitors, as shown in Figure 2c [44]. Yu et al. demonstrated a MXene-bonded activated carbon composite membrane using one-step vacuum-assisted filtration technology. The MXene layers encapsulate activated carbon particles to form a 3D conductive network, and 2D $Ti_3C_2T_x$ MXene serves as a flexible, conductive, and electrochemically active adhesive. The synergistic effect of MXene and activated carbon expands the interlayer spacing, effectively improving the capacitance and rate capability of the composites, thus exhibiting a high capacitance of 126 F g^{-1} and a capacitance retention rate of 57.9% (0.1–100 A g^{-1}).

2.1.4. Other Strategies

Spray drying is a method of systematic technology applied to material drying [18]. During this process, when the solution is in contact with hot air, the solvent will vaporize rapidly to obtain dry powder or granular products. The particle size distribution and morphological structure can be adjusted flexibly by changing the process parameters such as the feed rate and viscosity of the solution [18]. Wang et al. presented a 3D lightweight porous microsphere composite (3D rGO/MXene/TiO_2/Fe_2C) with a 2D/2D/0D/0D intercalation heterostructure by spray freeze-drying and microwave radiation, as shown in Figure 2d [45]. During the preparation process, defects, porous frameworks, multi-layer assemblies, and multi-component combinations are formed, effectively generating high-density polarized charges and abundant polarization sites, thus enhancing interfacial polarization [45].

Electrospinning is a simple and effective new processing technology for producing nanofibers and nanofiber-based materials [25]. In the electrospinning process, polymer solutions or melts are sprayed in a strong electric field. Due to the repulsive effect of charges, fibers will unfold and orient, while the solvent evaporates or solidifies, ultimately forming a fiber network [38]. This method has a high fiber orientation, controllable structure, and high production efficiency, making it easy to construct polymer solutions containing nanocomposites into fibers with adjustable properties [31]. Lei et al. prepared a $Ti_3C_2T_x$ MXene/carbon nanofiber film using electrospinning technology, as shown in Figure 3a [46]. The number of active sites, electron/ion transport, and diffusion kinetics of the composite material have been effectively improved, resulting in a high desalination capacity, rapid desalination rate, and longer cycle life, which are superior to other carbon-based and MXene-based electrode materials [46].

The 3D printing strategy offers an advanced manufacturing technique that can easily create 3D structures by converting materials into ink, which can then be extruded through a nozzle and deposited layer by layer onto the substrate [47]. This technology can be directly printed into intersecting patterns of 3D microstructures [48]. By adjusting the process parameters, various desired morphological structures can be generated to meet more demands and achieve high efficiency. Dai et al. reported an MXene/RGO conductive scaffold using 3D printing technology, as shown in Figure 3b [48]. The addition of graphene oxide effectively improves the rheological and printability of MXene ink, ensuring the stability and structural integrity of conductive support connections. The improved ink has the flexibility of structural design and the adjustability of conformation, and the printed conductive scaffold has excellent reversible compressibility and excellent resistance to cyclic fatigue, which is significantly better than the reported MXene [48].

2.2. Chemical Method

2.2.1. Hydrothermal Strategy

The hydrothermal strategy, as one of the most common methods for preparing composites, refers to using an aqueous solution as the reaction medium in a specially designed closed-reaction vessel (autoclave), creating a high-temperature (100–1000 °C) and high-pressure (1–100 MPa) reaction environment through heating, so that usually slightly soluble or insoluble substances dissolve and recrystallize [38]. Zhu et al. developed a graphene-reinforced MXene hydrogel with 3D spherical macroporous structures (SGMH) by a hydrothermal reaction, as shown in Figure 3c [49]. In this configuration, graphene not only conductively assembles MXene sheets, but also improves the hydrophobicity of the composite hydrogel skeleton [49]. The stiffness and pore radius of SGMH are improved by the 3D spherical macroporous structures and prefreezing treatment, allowing it to be naturally dried and transition into graphene/MXene aerogel (SGMA) of arbitrary size scales [49]. Additionally, the 3D spherical structure inside the SGMA helps to homogenize the stress distribution and improve the electrical response during compression. Sangili et al. demonstrated a carbon-coated $Ti_3C_2T_x$ MXene ($Ti_3C_2T_x$@C) by a hydrothermal method [50]. Surface oxidation and aggregation can be effectively inhibited by the in situ wrapping of $Ti_3C_2T_x$ in carbon layers, enabling the composite to be stored in water and air for 100 days. Moreover, the composite material exhibits high capacitance and long-term cycling stability.

2.2.2. Heat Treatment Strategy

In addition to the hydrothermal method, the heat treatment strategy is another commonly used method for preparing composite materials. In the heat treatment process, the microstructure and phase state of the materials are changed by controlling the heating and cooling process, so as to improve the crystal structure, eliminate defects, and improve the performance of the materials [38]. Allah et al. reported a $Ti_3C_2T_x$-NOMC heterostructure (composed of $Ti_3C_2T_x$ and nitrogen-doped ordered mesoporous carbon) through heat treatment techniques [51]. The introduction of abundant ordered mesoporous NOMC on the surface of $Ti_3C_2T_x$ nanosheets with high electronic conductivity can suppress stacking and aggregation phenomena, prevent oxidation, increase ion transport pathways, and efficiency. Therefore, the material exhibits an excellent capacitance value, rate performance, and cycle life. Zhang et al. proposed a heat treatment technique for directly preparing 3D carbon-coated $Ti_3C_2T_x$ composite materials (T-MXene@C) in Figure 3d [52]. The composite material presents a 3D silver ear-like structure, and a thin carbon coating completely covering the surface of $Ti_3C_2T_x$ nanosheets, thus inhibiting oxidation and aggregation. This structure endows the composite material with a stable and rich active surface, effectively improving the charge transfer rate, and achieving an ultra-high capacity and an excellent rate performance and cycle life.

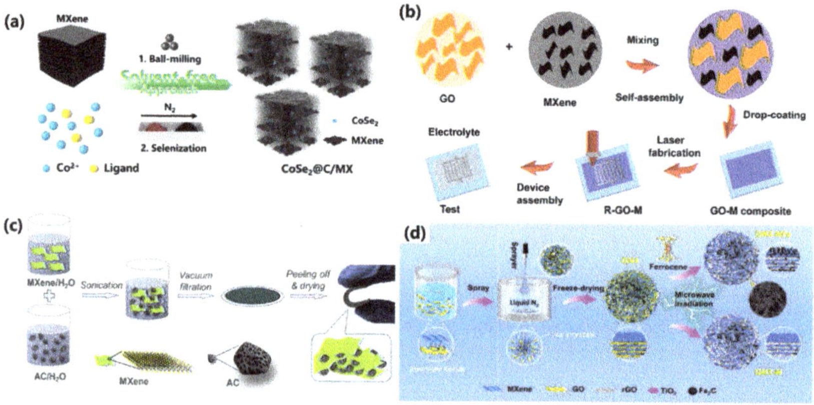

Figure 2. Physical method for the fabrication of carbon/MXene composites: Schematic diagram of (**a**) selenides@carbon/MX [40], copyright 2023, Elsevier, (**b**) GO-M [41], copyright 2023, American Chemical Society, (**c**) AC/MXene [44], copyright 2018, American Chemical Society, and (**d**) rGO/MXene/TiO_2/Fe_2C [45], copyright 2023, Springer.

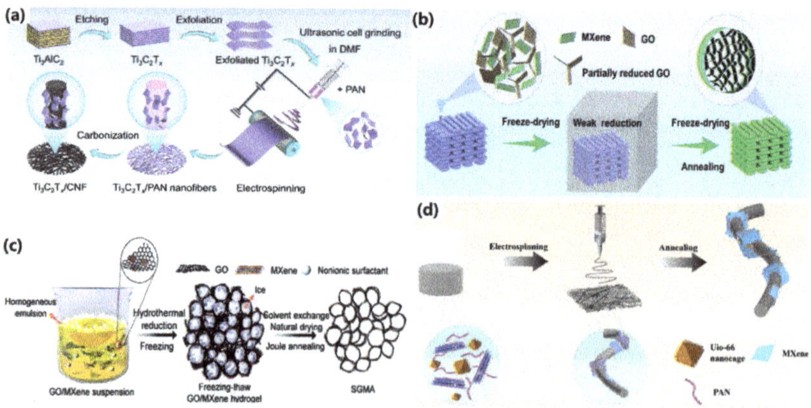

Figure 3. Physical method for the fabrication of carbon/MXene composites: (**a**) Schematic diagram of $Ti_3C_2T_x$/CNF [46], copyright 2023, Royal Society of Chemistry, and (**b**) MXene/RGO [48], copyright 2022, Royal Society of Chemistry. Chemical method for the fabrication of carbon/MXene composites: (**c**) Schematic diagram of SGMA [49], copyright 2024, Wiley-VCH GmbH, and (**d**) T-MXene@C [52], copyright 2024, Elsevier.

3. Carbon Materials/MXenes Composites for Supercapacitors

Combining carbon materials and MXenes to build a composite material can take full advantage of the advantages of both materials to achieve "1 + 1 > 2", thus significantly improving the performance of supercapacitors [19,53]. Firstly, the excellent electrical conductivity of MXenes can effectively improve the electrical conductivity of porous carbon materials, thereby improving the overall electrical conductivity of the electrode [23,54]. The introduction of carbon materials can increase the interlayer spacing of MXene nanosheets, effectively alleviating the occurrence of stacking and re-aggregation phenomena [25,55]. In addition, carbon materials can contribute more storage sites, thereby improving their energy storage capacity [17,56]. Currently, MXenes have been used to construct composite materials with carbon materials of various sizes (0–3D), and have significantly improved the electrochemical performance.

3.1. Zero-Dimensional Carbon/MXenes Composites

Zero-dimensional carbon materials mainly refer to carbon quantum dots (CQDs), which are a promising low-dimensional nanomaterial [19,57]. CQDs have excellent electronic conductivity and high carrier mobility, which makes them stand out in electronic devices [58]. In addition, CQDs have excellent chemical and environmental stability and are considered environmentally friendly nanomaterials [23]. At the microscale, compared to other quantum dots, CQDs typically exhibit a spherical structure filled with sp2 and sp3 hybridized carbons, possessing an overall smaller size (<10 nm) [25]. CQDs containing abundant surface functional groups (such as carboxyl and hydroxyl groups) can provide satisfactory water solubility, and enable CQDs to assemble with MXene at liquid–liquid interfaces [19]. In recent years, CQDs have become a typical carbon material combined with MXene to construct composite materials. Small-sized CQDs can be dispersed on the plane of MXene to prevent nanosheets re-stacking, expose more active sites, and promote electrolyte diffusion within MXene nanosheets [58]. When MXene/CQDs composite materials are used as electrode materials, they have a high specific capacitance and capacitance retention ability, achieving an increase in the energy density of supercapacitors. Wang et al. reported an MXene/carbon dot (p-MC) composite film with a 3D interconnected porous structure [59]. Carbon dots are inserted between MXene nanosheets to prevent re-stacking and aggregation, achieving the generation of more electrochemical active sites. The macropores in 3D porous structures facilitate efficient ion diffusion, and the interlayer pore structure rapidly transfers ions to active sites to participate in reactions. Therefore, the high weight capacitance (688.9 F g^{-1} at 2 A g^{-1}) and excellent rate performance are achieved for p-MC films. Zhang et al. proposed an MXene/CAC (calcium alginate-derived carbon dots) composite film with a large ion accessible active surface and high density through carbon dot intercalation technology, as shown in Figure 4a–d [60]. Carbon dots embedded into MXene nanosheets increase interlayer spacing and facilitate the efficient diffusion of ions in the electrolyte. Therefore, the composite film exhibits a high-volume capacitance value of 1244.6 F cm^{-3}, excellent rate capacity of 53.2% (1–1000 A g^{-1}), and cycling stability (93.5% after 30,000 cycles).

Besides CQDs, 0D porous carbon microspheres are also expected to become conductive spacers, promoting electrolyte migration kinetics and improving the electrochemical performance. Wei et al. fabricated a porous carbon spheres/MXene composite (CPCM/MXene) with a sandwich-like structure through electrostatic interactions between a $Ti_3C_2T_x$ MXene and chitosan-based porous carbon microsphere [61]. The spherical structure of CPCM is protected by $Ti_3C_2T_x$, while CPCM inhibits the re-aggregation of $Ti_3C_2T_x$ nanosheets, achieving an improvement in electrolyte migration kinetics. Driven by the synergistic effect, PCM/MXene exhibits a specific capacitance value of 362 F g^{-1} and a capacitance retention rate of 93.87% after 10,000 cycles. Yang et al. designed a composite film (MXene/MHCS/CNT) composed of hollow carbon spheres (MHCS)/MXenes heterostructures and carbon nanotubes, as shown in Figure 4e–g [62]. This structure and design effectively increase the specific surface area, pore structure, stability, and conductivity of the composite material, thus achieving the improved penetration efficiency of the electrolyte solution and a shortened ion transport process. Therefore, the excellent specific capacitance value of 395F g^{-1}, rate performance of 70.9% (2–1000 mV s^{-1}), and cycling stability (98.3% after 10,000 cycles) exist.

Zero-dimensional carbon materials can serve as an intercalation agent for MXenes, and carbon materials can also form 0D structures with MXenes to enhance electrochemical performance. Li et al. demonstrated a hollow core–shell microsphere ($NiCo_2S_4$/MXene/NC) composed of $NiCo_2S_4$, MXene, and N-doped carbon through a multi-step process, as shown in Figure 4h–k [63]. The introduction of MXene and N-doped carbon materials effectively improves the conductivity of the composite material, while the special hollow structure can suppress the occurrence of unstable factors during the charge and discharge processes. Therefore, the electrode based on the composite material exhibits excellent capacitance (1786 F g^{-1}) and cycling stability (over 100% after 10,000 cycles). Moreover, the corresponding hybrid supercapacitor devices can achieve a maximum energy density of 67 Wh kg^{-1} and a capacitance retention rate of over 80% after 10,000 cycles.

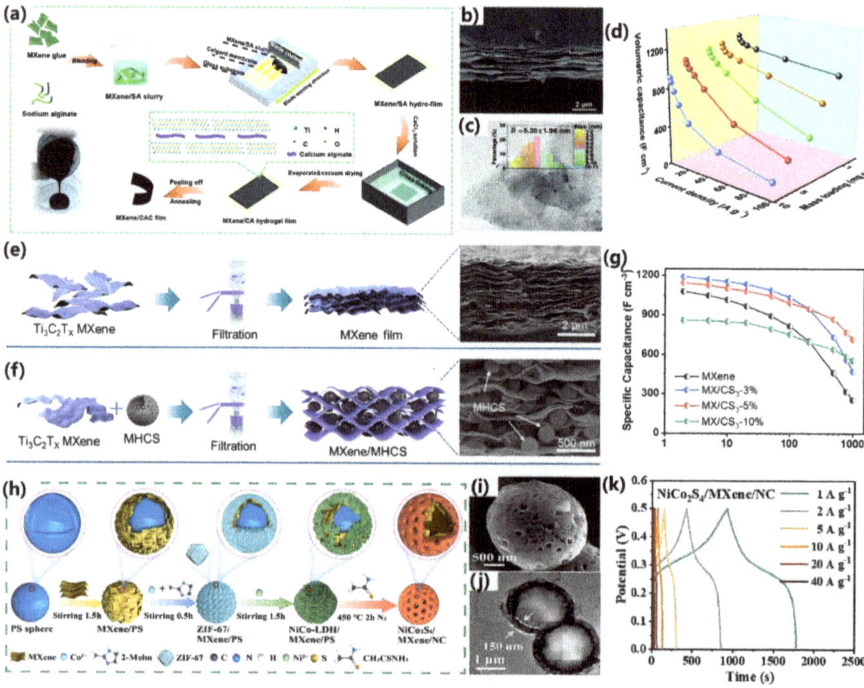

Figure 4. Zero-dimensional carbon/MXene composites: (**a**) Schematic diagram, (**b**) SEM, (**c**) TEM, and (**d**) capacitance value of MXene/CAC films [60], copyright 2022, Wiley-VCH GmbH. Schematic diagram of (**e**) pure MXene film and (**f**) MXene/MHCS film, (**g**) capacitance value [62], copyright 2024, Elsevier. (**h**) Schematic diagram, (**i**) SEM, (**j**) TEM, and (**k**) GCD curves of $NiCo_2S_4$/MXene/NC [63], copyright 2024, Elsevier.

3.2. One-Dimensional Carbon/MXenes Composites

One-dimensional carbon materials typically include nanotubes, nanofibers, or nanoribbons. For example, carbon nanotubes (CNTs) have the advantages of a high mechanical strength, low electrical resistivity, strong conductivity, large specific surface area, good chemical stability, good contact with electrolytes, and their "tubular" structure and higher aspect ratio can provide sufficient high-speed channels for an electron transfer [18,19,64]. Compared to 0D CQDs, the interaction between carbon nanotubes can spontaneously form a highly conductive network, which can be used as a self-supporting material without binders or substrates [23]. More importantly, most nanomaterials with a 1D structure typically have a relatively high mechanical strength to realize a sturdy structure [25]. Therefore, 1D carbon materials are highly suitable as "spacers" to introduce in a 2D MXene nanosheet, stabilize the structure of composite materials, and help further alleviate or avoid self-stacking re-aggregation phenomena [17]. Various structures can be established by coupling 1D carbon-based materials at the nanoscale with 2D MXene through different synthesis or assembly methods.

One-dimensional CNT can grow spontaneously on 2D MXene nanosheets to form a "1D tube on a 2D sheet" pattern through chemical vaper deposition (CVD) or catalytic conversion. Therefore, the interaction between 1D CNT and 2D MXenes is relatively strong, usually resulting in a lower contact impedance [19]. Wang et al. designed MXene/N-CNT composite materials with a hierarchical porous structure that effectively suppressed the self-accumulation of 2D MXene and 1D N-CNT, as shown in Figure 5a–c [65]. This structure increases the interlayer spacing of MXene and the specific surface area of the composite material, obtaining more active sites to participate in reactions and ion diffusion channels. The composite material exhibits a significantly better specific capacitance (167.2 F g^{-1}) than the original MXene electrode. Moreover, the asymmetric supercapacitor device based on this composite material has a high capacitance retention rate of 73.2% after 10,000 cycles, a high Coulombic efficiency of 97.5%, and a maximum energy density of 12.1 Wh kg^{-1}. Li et al. reported an in situ growth of multi-walled carbon nanotubes on MXene nanosheets loaded on a CC (MWCNTs-MXene@CC) composite material, achieving a synergistic combination of a large surface area and excellent electrical conductivity [66]. The growth of MWCNTs on MXene nanosheets can not only effectively suppress the occurrence of re-stacking phenomena, but also serve as charge collectors within and between nanosheets. Based on the 3D interconnected structure and high specific surface area of the composite material, rapid electrolyte penetration and efficient charge collection have been promoted, thus achieving an area-specific capacitance of 114.58 mF cm^{-2} and a high-capacity retention rate of 118%, which is significantly better than early-reported MWCNTs and MXene composite materials.

Hydrogen bonding, as a classic electrostatic interaction, is also considered an important bridge for achieving microscale structural integration. The surface of 2D MXenes contains abundant surface functional groups (-OH, -F, etc.), which can effectively participate in hydrogen bonding to construct composite materials by combining MXenes with 1D tubular carbon nanomaterials with rich surface chemical properties [64]. Hakim et al. grafted MWCNTs and $Mo_2TiC_2T_x$ to construct a composite (MWCNTs@$Mo_2TiC_2T_x$), which achieved a significant improvement in electrochemical properties and reaction kinetics [67]. MWCNT as a conductive medium is introduced into MXene, which effectively improves the specific surface area and conductivity of the composite material, achieving a capacitance value of 1740 F g^{-1}, almost four times that of the original $Mo_2TiC_2T_x$. You et al. reported a hybrid carbon nanofiber membrane (MCNF) derived from MXene/polyacrylonitrile, as shown in Figure 5d–f [68]. As the content of MXene nanosheets increases, the deposition thickness continuously increases and uniformly adheres to the surface of CNFs, which enhances the hydrophilicity of MCNF. Therefore, the electrolyte permeability at the electrode–electrolyte interface is improved, achieving the specific capacitance of 436.5 F g^{-1}, significantly better than that of pure CNFs.

When the size of 1D carbon-based materials increases to a size larger than that of 2D MXene nanosheets, smaller 2D MXene nanosheets will randomly deposit or be located on the surface of 1D carbon-based materials, forming a "2D on 1D" structure [19]. Sun et al. prepared a heterostructure of MXene/biomass-derived carbon fiber (MXene/CF) with a hierarchical porous "skin/skeleton"-like structure, which effectively suppressed the stacking and re-aggregation of MXene nanosheets, as shown in Figure 6a–c [69]. This structure enhances the electrolyte penetration rate, allowing ions to rapidly diffuse/transfer into the electrode due to efficient and stable channels. Therefore, the composite material exhibits an excellent volumetric capacitance of 7.14 F cm^{-3}, rate characteristics of 63.9%

(0.5–100 A g^{-1}), and cycling stability (99.8% after 5000 cycles). The all-solid state symmetrical supercapacitor device based on this composite material not only exhibits an excellent electrochemical performance, but also can withstand 2500 cycles under different bending conditions. Dharmasiri et al. developed a MXene-coated carbon fiber (CF) composite material with a stable structure and excellent mechanical performance, exhibiting a specific capacitance of 157 F g^{-1}, which is approximately 725 times higher than the original non-functionalized CF [70]. When applied to functionalized and coated woven CF pads and used to construct supercapacitor devices, a specific capacitance of 908 mF g^{-1} was achieved, which is approximately 42 times that of the control group.

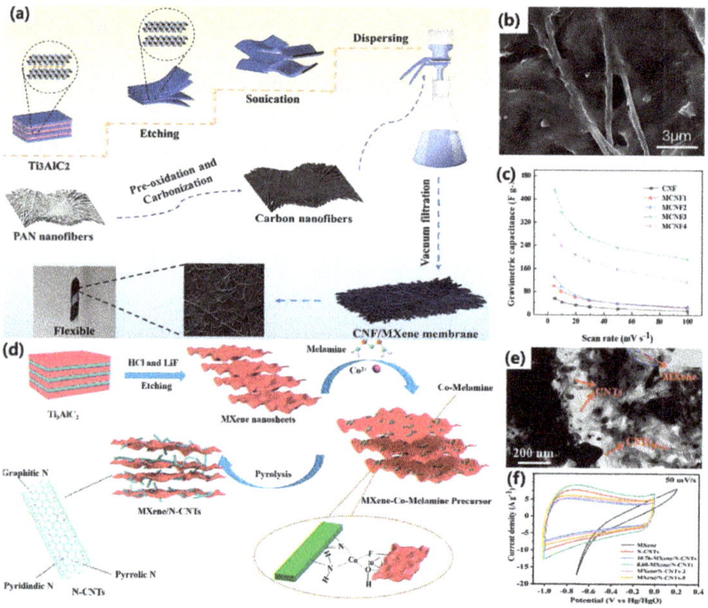

Figure 5. One-dimensional carbon/MXene composites: (**a**) Schematic diagram, (**b**) SEM, and (**c**) capacitance value of MXene/N-CNT [65], copyright 2023 American Chemical Society. (**d**) Schematic diagram and (**e**) TEM of MCNFs, (**f**) CV curves [68], copyright 2024, Springer.

Transforming the hybrid mode of 1D carbon materials/2D MXene into nanotube or nanofiber structures is also a method that can fundamentally avoid the overlap phenomenon of 2D MXene nanosheets, named "2D to 1D" [23]. Song et al. integrated MXene into N-doped carbon nanofibers with cavity-interconnected porous structures to construct a composite carbon film [71]. This design and structure expose more active sites, exhibiting a significantly enhanced electrochemical performance, such as a maximum energy density of 26.2 Wh kg^{-1} and a capacitance retention rate of 96.3% after 10,000 cycles. Zhao et al. developed a multifunctional $Ti_3C_2T_x$ MXene/carbon nanotube (MXene/CNT) hybrid fiber through the wet spinning strategy, as shown in Figure 6d–f [72]. When the CNT content is about 1 wt%, the mixed fiber can achieve a high strength of 61 ± 7 MPa, a conductivity of 1142.08 ± 40.04 S cm^{-1}, and an excellent specific capacitance of about 295 F g^{-1}. As the load increases to 9wt%, a maximum strain of 161 ± 19 MPa and further increased conductivity (1715 ± 22 S cm^{-1}) can be achieved. Therefore, the MXene/CNT fibers can be woven into energy storage textiles due to the excellent mechanical properties and achieve a maximum energy density of approximately 6.08 mW h cm^{-3}.

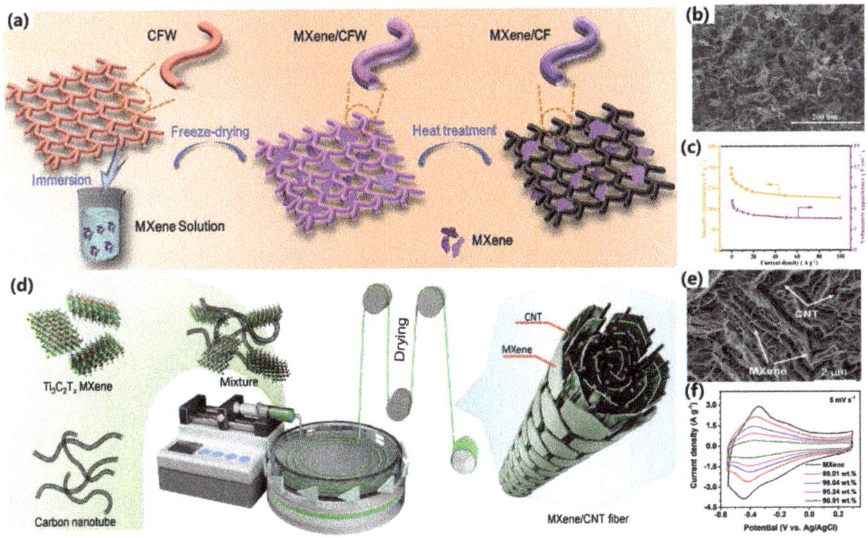

Figure 6. One-dimensional carbon/MXene composites: (**a**) Schematic diagram, (**b**) SEM, and (**c**) Capacitance value of MXene/CF [69], copyright 2020, Elsevier. (**d**) Schematic diagram and (**e**) TEM of MXene/CNT fiber, (**f**) CV curves [72], copyright 2022, Elsevier.

3.3. Two-Dimensional Carbon/MXenes Composites

Two-dimensional carbon materials have the characteristics of a large specific surface area, high electrical conductivity, high stability, and excellent mechanical properties [25]. Different forms of 2D carbon materials' (graphene, g-C_3N_4, carbon nanosheets, etc.) composite materials have been constructed in combination with MXene for applications in supercapacitors and have shown excellent electrochemical properties.

As a typical two-dimensional material, graphene is composed of a single-atom-thick carbon layer, which has unique properties beyond 0D and 1D carbon materials; these include strong in-plane covalent bonds with high conductivity, a suitable layer thickness and a wide transverse size that shortens the ion transport path, and large exposed surfaces that facilitate ion adsorption [19,73]. The surface portion of functionalized graphene serves as an oxidation-reduction center, making a significant contribution to Faradaic-type capacitors [19]. Similar to MXene, graphene nanosheets also exhibit self-aggregation. However, the introduction of graphene into MXene will inhibit the defects of the two materials, and achieve a synergistic effect to enhance the electrochemical properties [19]. Dang et al. constructed a reduced graphene oxide/$Ti_3C_2T_x$/carbon nanotube (rGO/$Ti_3C_2T_x$/CNT) fiber [74]. This composite material effectively suppresses stacking and aggregation phenomena through π–π stacking interactions, hydrogen bonding, and van der Waals forces. Moreover, all solid-state supercapacitor devices based on this material composite show a high-volume capacitance (336.1 F cm^{-3}) after 3000 cycles and a high-volume energy density of 23.11 mWh cm^{-3}. Qin et al. designed and prepared a nanofiber-reinforced MXene/rGO conductive aerogel by the pyrrole-assisted hydrothermal reduction of GO nanosheets, as shown in Figure 7a–d [75]. Pyrroles can not only induce the reduction of GO and self-assemble into an ordered 3D porous structure, but also in situ polymerized polypyrrole (PPy) can achieve strong cross-linking between rGO and MXene nanosheets through covalent and noncovalent bonds [75]. Meanwhile, the introduction of hydroxyl-rich cellulose acetate (CA) nanofibers into the aerogel to construct a "brick-mortar-rebar" structure can endow the material with excellent mechanical properties [75]. The conductive MXene/rGO aerogel served as a supercapacitor electrode and made full use of the electric double-layer capacitance of rGO and the Faradaic-type capacitors of MXene and PPy to store energy. The porous structure can effectively solve the problem of restacking, facilitate the rapid diffusion of ions in the electrode materials, and demonstrate potential applications in energy storage [75]. The all-solid-state supercapacitor based on MXene/rGO aerogel is also fabricated, which shows a high area-specific capacitance of up to 274 mF cm^{-2} at a current density of 1 mA cm^{-2}.

Graphite nitride carbon (g-C_3N_4) is a kind of carbon material with a high nitrogen content and a typical polymer semiconductor, which has a two-dimensional layered structure similar to graphene, high electrochemical reactivity, surface polarity, and wettability [25]. The pyridine type N atom in g-C_3N_4 acts as an active site, making it prone to highly reversible chemical reactions with electrolyte ions [76]. Therefore, the introduction of g-C_3N_4 into MXene can not only achieve the effective inhibition of a nanosheet stack, but also improves the electrochemical properties of the materials. Xu et al. reported an in situ synthesis technique for a g-C_3N_4/$TiVCT_x$ composite material [77]. The coating of g-C3N4 on TiVCTx nanosheets can effectively inhibit the occurrence of oxidation and self-stacks, so that the composite still maintains a high specific capacitance of 508.9 F g^{-1} after 7000 cycles. Zhang et al. constructed a $Ti_3C_2T_x$/g-C_3N_4 heterostructure by the self-assembly method, which avoids stacking, increases layer spacing, and enhances the charge transfer efficiency, as shown in Figure 7e–g [78]. This composite material exhibits a specific capacitance value of 414 F g^{-1}, and the assembled supercapacitor device can not only maintain a stable performance under 180° bending conditions, but can also achieve a maximum energy density of 23.98 Wh kg^{-1}.

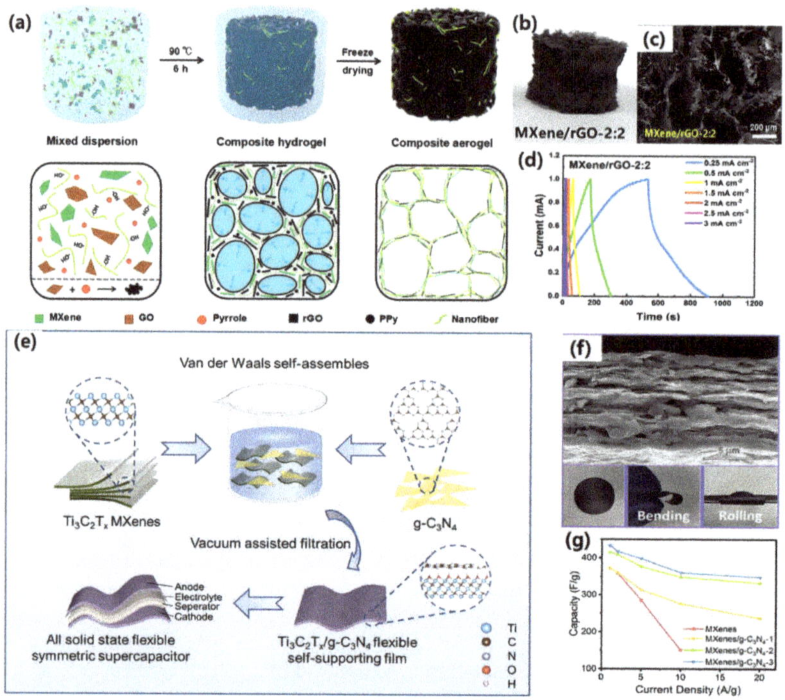

Figure 7. Two-dimensional carbon/MXene composites: (**a**) Schematic diagram, (**b**) Photograph, (**c**) SEM, and (**d**) GCD curves of MXene/rGO aerogel [75], copyright 2024, American Chemical Society. (**e**) Schematic diagram and (**f**) SEM of $Ti_3C_2T_x$/g-C_3N_4 heterostructure, (**g**) capacitance value [78], copyright 2022, Elsevier.

Carbon nanosheets have a two-dimensional sheet-like structure with many unique physical and chemical properties and potential applications, such as a high specific surface area providing more reaction sites, excellent conductivity allowing for very fast electron migration, excellent chemical stability, resisting oxidation and reduction reactions, high mechanical strength, etc., [79]. The 2D/2D composite materials constructed by combining carbon nanosheets with MXenes have large interlayer pores, high conductivity, structural stability, etc., which is conducive to electron transfers and electrode lifetime enhancement, while maintaining the unique crystal structure and characteristics of MXenes [80]. Moreover, the introduction of carbon nanosheets can provide a sufficient porous structure inside the intermediate layer, suppressing the stacking and aggregation of MXene nanosheets [79]. Zhang et al. prepared a composite material (MXene/CCNS) composed of MXene and citrus-derived carbon nanosheets, which can serve as a flexible self-supporting electrode to achieve a high specific capacitance of 1825.6 mF cm^{-2} and a capacity retention rate of 99.82% after 10,000 cycles [79]. Moreover, the composite material can be configured for conductive ink screen printing and large-scale patterned electrodes, respectively achieving a capacitance and energy density of 114.9 mF cm^{-2} and 12.9 μWh cm^{-2} for the application of forked supercapacitors, which constructed a composite material (MCN) by inserting high nitrogen-doped carbon nanosheets (CN) into MXene, as shown in Figure 8a–c [80]. This design effectively suppresses the occurrence of the MXene stacking phenomenon and improves the stability of the structure and ion transport efficiency. The doping of heteroatoms N achieves the control of the electronic structure of the composite materials and contributes to the Faraday capacitance, demonstrating a high specific capacitance of 418.4 F g^{-1}.

Apart from graphene, g-C_3N_4, and carbon nanosheets, graphdiyne (GDY), as an emerging two-dimensional carbon material, has rich carbon chemical bonds, large conjugated systems, wide interplanar spacing, and excellent chemical stability [81]. GDY is a novel all-carbon framework with a unique π-conjugated sp^2 and sp-hybrid backbone and a macroporous structure [82]. When GDY is grown in situ on the surface of MXene, the intercalation capacitance behavior and bulk energy density can be improved by increasing the nanosheet spacing and active sites to effectively promote ion migration dynamics/storage and interfacial electron conduction [83]. Wu et al. reported a hydrogen-rich GDY-$Ti_3C_2T_x$ electrode, which achieved the effective regulation of layer spacing, active sites, and charge storage, as shown in Figure 8d–f [83]. Due to the strong H + electrostatic attraction, low migration resistance, accelerated intercalation pseudocapacitance dynamics, fast electron response, and stable Ti-O-C bond bridge organic–inorganic heterojunction, the composite exhibits a short-range electron transport pathway, high ion diffusion rate, and electrolyte permeation, resulting in a volume capacitance of 2296 F cm^{-3}, 55% rate characteristics (1–50 A cm^{-3}), a volume energy density of 65.6 mWh cm^{-3}, and long-term deformable cycling stability.

3.4. Three-Dimensional Carbon/MXenes Composites

Three-dimensional carbon materials typically have interconnected porous structures that provide large electrochemical active surface areas and excellent electron and mass transfer capabilities [23]. Compared with 0D, 1D, and 2D carbon materials, 3D carbon materials have unique advantages. When combined with two-dimensional MXene layers, the composite materials exhibit excellent physical and chemical properties and demonstrate an outstanding performance in the field of energy storage [17]. Under the premise of a reasonable nanostructure design, carbon materials and MXene exert synergistic effects in 3D/2D composite materials to achieve satisfactory results [19]. In view of the difference in size, 3D carbon materials and MXenes can combine to build composite materials with different structures, such as activated carbon or hierarchical porous carbon embedded in the MXene layer and MXenes deposited on the surface of carbon cloth and carbon foam.

Activated carbon with advantages such as a large specific surface area, good electrical performance, global availability, biodegradability, and low cost is the earliest electrode material used in supercapacitors [17]. The physical or chemical activation process can not only make the specific surface area of activated carbon up to 3000 m^2 g^{-1}, but also build a porous structure, which is conducive to combining with various materials to construct composite materials, thereby improving the electrochemical performance [23]. A group studied a composite material of tin (IV) selenide ($SnSe_2$) growth and inserted activated porous carbon/$Ti_3C_2T_x$ (APC/MXene), as shown in Figure 9a–c [84]. The highly porous APC structure wrapped in $Ti_3C_2T_x$ effectively suppresses the occurrence of re-stacking and exposes more active sites. Therefore, the composite material as an electrode material has a high capacitance value of 815 F g^{-1}, and the assembled supercapacitor device achieves a maximum energy density of 102 Wh kg^{-1}. Yong et al. reported two methods for combining MXene ($Ti_3C_2T_x$) with activated carbon (AC) to construct porous, flexible, and mechanically durable composite films, including multilayer MXene powder deposited on textiles after mechanical mixing with AC pulp, and MXene nanosheets sprayed on the surface of sprayed AC electrodes, achieving the optimal specific capacitance of 148.7 mF cm^{-2} and an energy density of 0.921 mW cm^{-2} [85]. Compared with identical supercapacitors with standard AC electrodes, the composite prepared by combining these two methods has a 220% increase in capacitance.

In addition to activated carbon, hierarchical porous carbon also has unlimited potential to be combined with MXenes to construct composite materials, achieving high chemical stability, an adjustable porous structure, and a specific surface area [23]. Hou et al. reported a porous heterostructure of MXenes/waste PET-derived carbon, as shown in Figure 9d–f [86]. Carbon materials in composite materials inhibit the re-aggregation of $Ti_3C_2T_x$, while $Ti_3C_2T_x$ increases the electrochemical active sites. The large specific surface area (1754.3 m^2 g^{-1}) promotes electrolyte migration kinetics and achieves a high specific capacitance value (404.1 F g^{-1}). The supercapacitor device based on this composite material has a voltage window of 1.8 V, energy density of 31.19 Wh kg^{-1}, and 1% capacitance decay after 15,000 cycles. Seenath et al. demonstrated an MXene-coupled nitrogen-doped porous carbon (MX-MC-N) hybrid material with a high specific surface area (796 m^2 g^{-1}) and hierarchically porous networks, achieving an effective improvement in the electrochemical performance [87]. The composite material has a high capacitance value of 245 F g^{-1}, which is significantly better than porous carbon without MXene, graphene, and nitrogen-doped porous carbon materials. In addition, the symmetrical and asymmetrical supercapacitors based on this composite material exhibit high areal capacitance values of 72 mF cm^{-2} and 228 mF cm^{-2} and an energy density of 13 μWh cm^{-2} and 92.1 μWh cm^{-2}, respectively.

Some organic compounds (glucose, citric acid, lignin, etc.) can be used as carbon sources to construct carbon/MXene composites with 2D MXene nanosheets [19,23,25]. Huang et al. reported a composite material (N, S-MXene/3DPC) composed of N and S co-doped 3D porous carbon and MXene, achieving a specific capacitance of 412.7 F g^{-1} [88]. The 3D conductive network co-doped with N and S can provide multiple transport paths and abundant translation active sites for ions and charges, while avoiding MXene stacking. The introduction of MXene effectively improves the conductivity and mechanical stability. Therefore, the asymmetric supercapacitor based on the material has a capacitance value of 210.5 F g^{-1}, 97.5% capacitance retention after 5000 cycles, and a maximum energy density of 94.7 Wh kg^{-1}. Yang et al. prepared a highly conductive $Ti_3C_2T_x$/CNT/PC composite film by tightly anchoring porous carbon (PC) onto MXene nanosheets using 1D CNT as a bridge, as shown in Figure 9g–i [89]. The composite material exhibits a high specific capacitance of 364.8 mF cm^{-2}, an over 80% rate performance (0.5–50 mA cm^{-2}), and a large area energy density of 10.5 μ Wh cm^{-2}.

When the size of the 3D structural skeleton is larger than that of the 2D MXene nanosheets, the 2D MXene nanosheets are located on the surface or surfaces of the 3D carbonaceous skeleton, forming a "2D in 3D" mode. Highly porous 3D carbonaceous frameworks are typically prepared by the high-temperature carbonization of natural biomass or inherently porous structured polymers. Yao et al. prepared a super-high mechanical strength and self-supporting carbonized wood (CW)/MXene nanocomposite material, which not only prevented the re-aggregation of MXene, but also fully utilized the dual effects of CW (a graded porous structure) and MXene (high conductivity and wettability) [90]. Therefore, the symmetrical supercapacitor based on this composite material exhibits a high capacitance characteristic of 4.24 F cm^{-2} (80.61 F g^{-1}) and a high energy density of 11.2 Wh kg^{-1}. Sun et al. prepared a new type of MXene/N-doped carbon foam (MXene/NCF) compressible composite with a 3D hollow interconnection structure, as shown in Figure 10a–d [91]. NCF can provide additional pseudocapacitance through N atom doping, while supporting MXene nanosheets to form 3D hollow interconnected structures, providing highly stable and efficient channels and more contact sites for ion diffusion/electron transport. MXene enhances the conductivity and hydrophilicity of composite materials. Due to the synergistic effect, the composite material exhibits high capacitance characteristics of 332 F g^{-1} (3162 mF cm^{-3}), a 64% rate performance (0.5–100 A g^{-1}), and 99.2% capacity retention after 10,000 cycles. In addition, the material can withstand multiple 60% strains with a stable morphology and electrochemical properties.

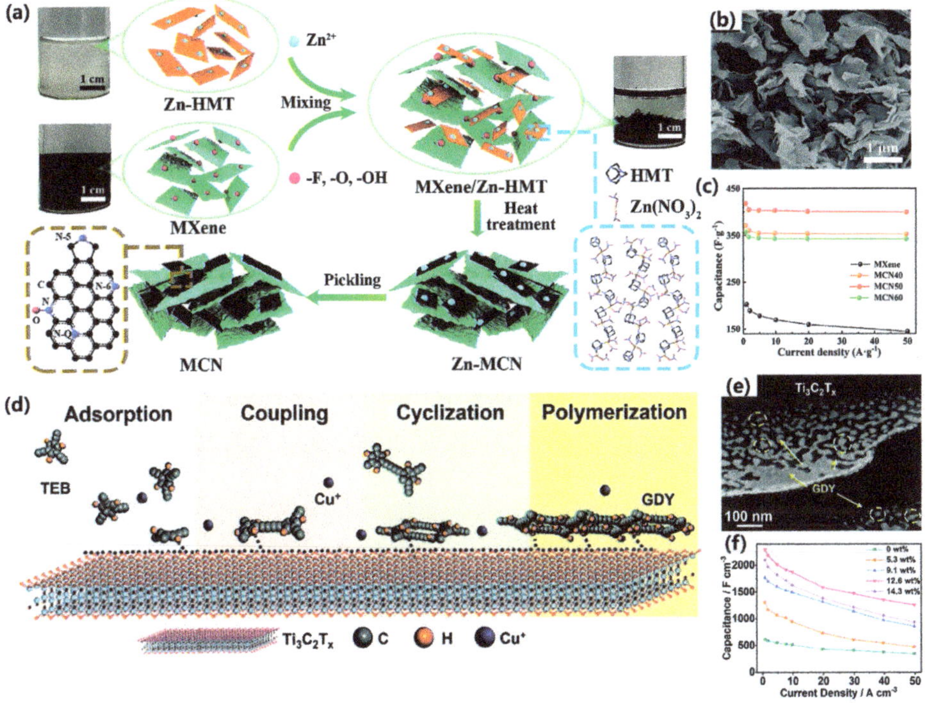

Figure 8. Two-dimensional carbon/MXene composites: (**a**) Schematic diagram and (**b**) SEM of MCN50, (**c**) capacitance value [80], copyright 2024, Wiley-VCH GmbH. (**d**) Schematic diagram and (**e**) SEM of GDY-$Ti_3C_2T_x$ heterostructure, (**f**) capacitance value [83], copyright 2024, Wiley-VCH GmbH.

Apart from a 3D structural skeleton, researchers have also studied the composite material of $Ti_3C_2T_x$ with carbon cloth. Li et al. reported a composite material composed of nitrogen-doped superhydrophilic carbon cloth (ENCC) and $Ti_3C_2T_x$ MXene nanosheets, effectively suppressing the stacking of MXene nanosheets and achieving an area-specific capacitance of 2080.1 mF cm^{-2}, as shown in Figure 10e–g [92]. Moreover, the symmetrical supercapacitor based on this composite material has a wide voltage window of 1.8V and a capacitance retention rate of 91% after 10,000 cycles. Fan et al. demonstrated a 3D porous composite material (MXene/CMDF) derived from MXene/needle punched denim felt, which achieved the removal of specific harmful groups in MXenes, exposed more active sites, and facilitated a more efficient ion exchange between electrolytes and electrodes [93]. Therefore, the composite material exhibits a maximum specific capacitance of 1748.5 mF cm^{-2} and retained 94% capacitance after 15,000 cycles. In addition, supercapacitors based on MXene/CMDF have a specific capacitance of up to 577.5 mF cm^{-2} and a maximum energy density of 80.2 μWh cm^{-2}.

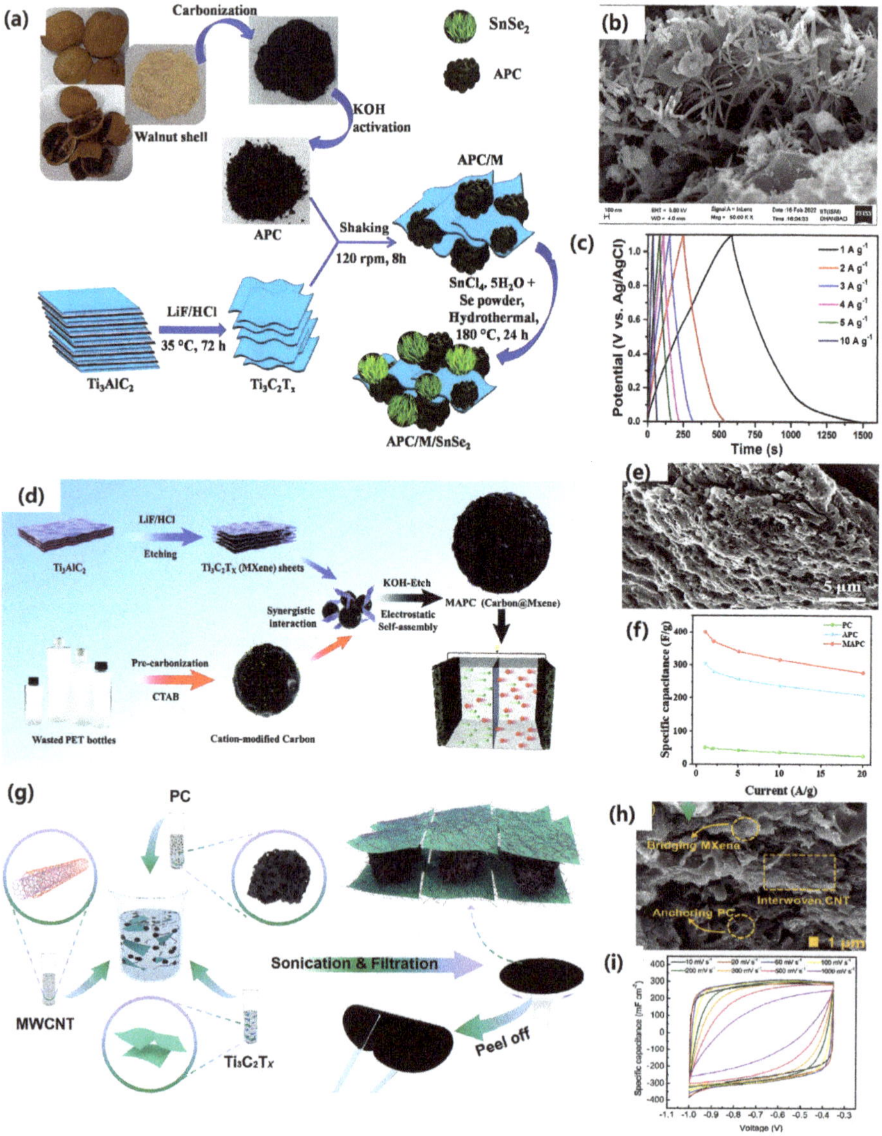

Figure 9. Three-dimensional carbon/MXene composites: (**a**) Schematic diagram, (**b**) SEM, and (**c**) GCD curves of APC/$Ti_3C_2T_x$/$SnSe_2$ [84], copyright 2023, Elsevier. (**d**) Schematic diagram and (**e**) SEM of MAPC, (**f**) capacitance value [86], copyright 2024, Wiley-VCH GmbH. (**g**) Schematic diagram, (**h**) SEM, and (**i**) CV curves of TCP film [89], copyright 2021, Elsevier.

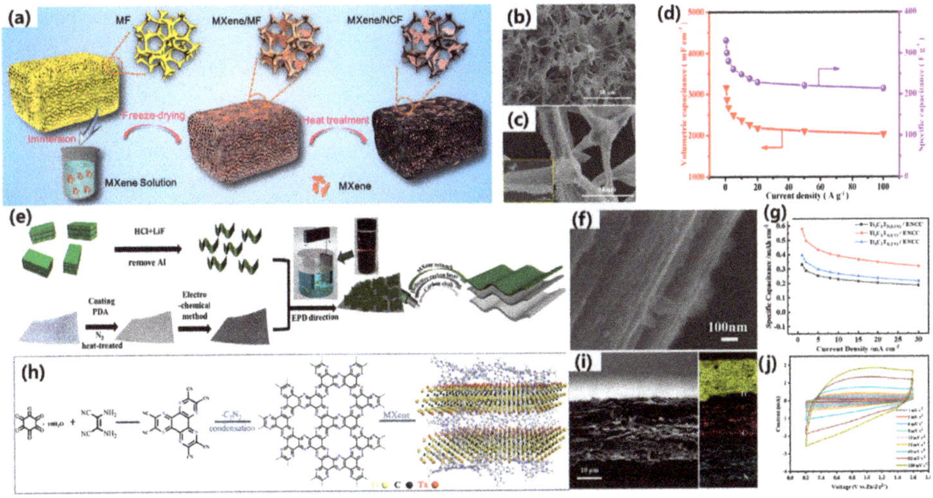

Figure 10. Three-dimensional carbon/MXene composites: (**a**) Schematic diagram, (**b**,**c**) SEM, and (**d**) capacitance value of MXene/NCF [91], copyright 2020, American Chemical Society. (**e**) Schematic diagram and (**f**) SEM of $Ti_3C_2T_x$/ENCC, (**g**) capacitance value [92], copyright 2022, Elsevier. Other carbon/MXenes composites: (**h**) Schematic diagram, (**i**) SEM and mapping, (**j**) CV curves of C_2N@MXene [94], copyright 2024, Elsevier.

3.5. Other Carbon/MXenes Composites

Although 0D, 1D, 2D, and 3D carbon materials combined with MXene exhibit an excellent capacitance performance, researchers have also explored the application of other carbon materials (such as C_2N, graphdiyne nanotube, CNT sponge, and so on)/MXenes composites in supercapacitors, which have demonstrated an outstanding performance. Zhao et al. designed a composite material composed of highly ordered porous nitrogen-rich carbon material C_2N and MXene nanosheets (C_2N@MXene), as shown in Figure 10h–j [94]. The presence of MXene nanosheets enhances conductivity and suppresses the stacking phenomenon of the C_2N layer structure during repeated charge and discharge cycles, while C_2N effectively solves the self-stacking problem of MXene nanosheets. Therefore, the composite material exhibits a high specific capacitance of 240 mAh g^{-1}, a 94% capacitance retention rate after 10,000 cycles, and an energy density of 168 Wh kg^{-1}. In addition, a zinc ion micro-supercapacitor based on a C_2N@MXene electrode still exhibits a high specific capacity of 264 mF cm^{-2} and excellent flexibility characteristics. Wang et al. introduced graphene diene nanotubes (GDY NTs) with planar pores into MXene layers to construct a 3D interconnected, hydrogen-permeable MXene/Graphdiyne nanotube (MG) composite membrane [95]. The capacitance of this composite film can reach 337.4 F g^{-1} and a rate performance of 73%, which is much better than that of a pure $Ti_3C_2T_x$ film (230.8 F g^{-1}, 55%). Moreover, the flexible asymmetric solid-state supercapacitor based on this material exhibits a high energy density of 19.7 Wh kg^{-1} and an 88.2% capacitance retention rate after 10,000 cycles. Yang et al. reported a composite material of $Ti_3C_2T_x$ nanosheets dispersed into an independent porous CNT sponge ($Ti_3C_2T_x$@CNT) [96]. This structure provides a high-speed ion transport pathway, and MXene nanosheets can achieve pseudocapacitive contributions. Therefore, the composite material has a large capacitance value of 468 F g^{-1} and a capacitance retention of 79.8% at 100 mV s^{-1}. Huang et al. prepared a multilayer MXene superlattice with an intercalated monolayer mesoporous carbon framework (MMCF) [97]. MMCF not only increases the interlayer spacing and creates porous channels for ion transport, but also acts as a conductive pillar to guide the transfer of electrons along the z-direction. Therefore, the composite material exhibited a volumetric capacitance of 317 F cm^{-3} and the assembled micro-supercapacitor achieved a surface energy density of 0.10 mWh cm^{-2}.

In summary, the introduction of carbon materials with different structures (0D, 1D, 2D, and 3D) into MXene has different effects on improving their electrochemical performance, as shown in Table 1 [98,99]. Zero-dimensional carbon materials have space saving encapsulation, high active sites, and a large specific surface area, effectively reducing the capacity loss caused by changes in the containment volume and the suppression of the shuttle effect [98]. One-dimensional carbon materials have the characteristic of a linear structure, which can provide a direct electron transfer path and shorten the distance of the electron transfer [19,100,101]. In addition, an excellent conductive network can be formed within the electrode, thereby prolonging electronic conduction and improving the electrochemical performance at high current densities. Two-dimensional carbon materials have a high specific surface area and excellent in-plane conductivity, increasing the in-plane electron transfer and providing sufficient active sites for electrochemical reactions. Moreover, larger lateral dimensions can be easily applied to further enhance the capacitance and rate capability of the corresponding supercapacitors [25]. Three-dimensional carbon materials with large exposed surfaces and adjustable porous structures can accelerate the electron transfer by shortening the conductive path [6]. The high surface area and porous interconnected structure can provide a large contact area with the electrolyte and a continuous ion transport path throughout the space, thereby improving the rate performance, and making it an ideal electrode material for high-performance supercapacitor applications.

Table 1. xD carbon/MXene composites for supercapacitors.

Configuration	Samples	Specific Capacity	Capacity Retention	Ref.
0D carbon/MXene	MXene/carbon dots	688.9 F g^{-1} at 2 A g^{-1}	90% after 10,000 cycles	[59]
	MXene/carbon dots	1244.6 F cm^{-3} at 1 A g^{-1}	93.5% after 30,000 cycles	[60]
	Carbon spheres/MXene	362 F g^{-1} at 0.5 A g^{-1}	93.87% after 10,000 cycles	[61]
	Hollow spherical $NiCo_2S_4$/MXene/carbon	1786 F g^{-1} at 1 A g^{-1}	100% after 10,000 cycles	[63]
1D carbon/MXene	MXene/N-CNT	167.2 F g^{-1} at 0.5 A g^{-1}	73.2% after 10,000 cycles	[65]
	MWCNTs-MXene@CC	114.58 mF cm^{-2} at 5 mV s^{-1}	118% after 16,000 cycles	[66]
	MXene/CF	7.14 F cm^{-3} at 0.5 A g^{-1}	99.8% after 5000 cycles	[69]
	MXene/CF	157 F g^{-1} at 5 mV s^{-1}	94% after 5000 cycles	[70]
	MXene/CNT fiber	295 F g^{-1} at 5 mV s^{-1}	85% after 5000 cycles	[72]
2D carbon/MXene	MXene/rGO	274 mF cm^{-2} at 1 mA cm^{-2}	96% after 100 cycles	[75]
	$Ti_3C_2T_x$/g-C_3N_4	414 F g^{-1} at 1 A g^{-1}	94.93% after 2500 cycles	[78]
	MXene/carbon nanosheet	1825.6 mF cm^{-2} at 5 mA cm^{-2}	99.82% after 10,000 cycles	[79]
	GDY-$Ti_3C_2T_x$	2296 F cm^{-3} at 1 A cm^{-3}	89.2% after 10,000 cycles	[83]
3D carbon/MXene	APC/Ti3C2Tx	815 F g^{-1} at 1 A g^{-1}	85% after 10,000 cycles	[84]
	MXenes/hierarchical porous carbon	404.1 F g^{-1} at 1 A g^{-1}	99% after 15,000 cycles	[86]
	N, S-MXene/3DPC	412.7 F g^{-1} at 1 A g^{-1}	97.5% after 500 cycles	[88]
	MXene/NCF	332 F g^{-1} at 0.5 A g^{-1}	99.2% after 10,000 cycles	[91]
	MXene/3D needled denim felts	1748.5 mF cm^{-2} at 0.5 mA cm^{-2}	94% after 15,000 cycles	[93]

4. Summary and Outlook

Supercapacitors, as a green, safe, and low-cost energy storage device, have received widespread attention in efficient and sustainable energy storage systems due to their unique properties, including fast charging and discharging speeds, high power density, and excellent cycle life compared to many other energy storage solutions. MXenes exhibits diversity in their elemental composition, rich surface chemistry, and atomic-level multiple configurations. Unlike other two-dimensional materials, MXenes exhibit comprehensive strength when combined with a multidimensional carbon matrix. Due to the unavoidable phenomenon of self-stacking and the aggregation of MXenes in practical applications, hybridization can be considered one of the most effective strategies. When MXenes are integrated with carbon materials, hybrid materials have potential synergistic effects, which will push the performance of MXenes-based materials to a new platform. This review summarizes and compares various synthesis routes of carbon materials/MXenes composites and the latest research progress on the hybridization of two-dimensional MXenes with different dimensions of carbon (from 0D CQDs to 3D carbon scaffolds) to achieve an effective improvement in the electrochemical performance, namely xD carbon + 2D Mxenes = ∞. Then, based on the principle of "synthesis structure properties", the structure property relationship of xD carbon/MXenes composite materials was systematically analyzed. However, although carbon/MXenes composites show a good performances and have made great achievements in supercapacitors' applications, there are still many challenges and opportunities for the further development of their design and the corresponding devices, some of which are highlighted below.

Firstly, the electrochemical performance of MXenes is closely related to their surface chemistry (functional groups, atomic defects, and heteroatom doping) and microstructure (interlayer spacing and pore size). MXenes have stable synthetic properties, increased layer spacing, controllable surface functional groups, and regularized atomic defects, which are of great significance for optimizing their electrochemical performance. Further efforts to develop green, safe, economical, and scalable synthesis methods for MXenes and their composite materials, mitigate MXene oxidation, and achieve the preparation of MXenes with controllable and uniform surface ends remain a challenging task, which is crucial for advancing the industrial application of MXene-based nanostructures.

Secondly, MXenes with different chemical compositions and different properties need more exploration to build composite materials with carbon materials. More than 20 different types of MXenes have been produced in experiments. In order to improve the possibility of producing high-quality carbon/MXene electrode materials, it is necessary to conduct more in-depth and substantial research on composites constructed of different MXenes and carbon materials. Meanwhile, the performance and characteristics of each MXenes are compared to maximize the role of MXenes in the composites.

Thirdly, comprehensive research is needed to understand the fundamental mechanisms of synergistic effects between MXenes and carbon components, which will facilitate the rational design of novel composite materials with an enhanced performance and functionality. The stability design of the structure, especially the ability to withstand the volume expansion of carbon nanoparticles of different sizes during intercalation and delamination processes, is very important and worthy of attention. Therefore, in order to improve the stability of MXenes when combined with carbon, it is necessary to thoroughly study the interaction between solvents and compounds, further research and comparisons of MXene/carbon composites, and strengthen the theoretical research.

Fourth, carbon/MXene-based supercapacitors are evaluated according to actual needs. Factors such as the load density, volumetric capacity, structural stability at extremely high/low temperatures, and the safety of active materials should be considered.

Fifth, the tradeoff between mechanical properties and functions must be carefully considered. The development of high-performance, multifunctional, and digital MXenes-based supercapacitors for different application scenarios is essential without compromising their mechanical properties such as their flexibility, strength, and toughness. The development of on-chip energy storage technology for miniaturization and portable electronic devices is rapid. Micro-supercapacitors based on carbon materials/MXene have broad application prospects in future flexible and portable electronic devices due to their small size, high power density, and integration density.

Sixth, the structural evolution and reaction mechanism of MXenes-based materials need to be better understood to further design and synthesize novel MXenes and their composites with desirable properties. The clear structural evolution and charge storage mechanism will guide researchers to further improve the performance of supercapacitor devices. The advanced technology can better characterize these devices and provide unprecedented insights into mechanism elucidation.

Seventh, tailor-made synthesis techniques need to be explored to achieve precise control over the type and number of surface functional groups of Mxenes, such as a low temperature, deoxidation, and the removal of excess salt ions, etc. The surface functional groups of MXenes have a significant impact on their electrochemical performance. However, the current synthetic approach has significant limitations. In order to synthesize MXenes with expected surface functional groups in a standardized procedure, synthesis techniques are crucial. The effective control of surface functional groups can not only improve the stability of MXenes during application, but also promote the large-scale preparation and commercial application of MXenes.

Eighth, machine learning can be used to supplement and assist experimental exploration. Machine learning is based on data-driven methods that collect data from reported research work and then train it to quickly display the best choice when weighing multiple variables.

Funding: This work received funding from the Shaanxi Province Qin Chuangyuan Cited High-Level Innovation and Entrepreneurship Talent Program (Approval No. QCYRCXM-2023-130) and from the Science and Technology Project of the Northwest Institute for Non-ferrous Metals Research (Approval No. 0901YK2316).

Conflicts of Interest: The authors declare no conflicts of interest.

References

1. Kadam, S.A.; Kadam, K.P.; Pradhan, N.R. Advancements in 2D MXene-based supercapacitor electrodes: Synthesis, mechanisms, electronic structure engineering, flexible wearable energy storage for real-world applications, and future prospects. *J. Mater. Chem. A* **2024**, *12*, 17992–18046. [CrossRef]
2. Nidhi; Tyagi, N.; Bhardwaj, V.; Moka, S.; Singh, M.K.; Khanuja, M.; Sharma, G. An overview on synthesis of MXene and MXene based nanocomposites for supercapacitors. *Mater. Today Commun.* **2024**, *41*, 110223. [CrossRef]
3. Wang, R.; Young Jang, W.; Zhang, W.; Venkata Reddy, C.; Kakarla, R.R.; Li, C.; Gupta, V.K.; Shim, J.; Aminabhavi, T.M. Emerging two-dimensional (2D) MXene-based nanostructured materials: Synthesis strategies, properties, and applications as efficient pseudo-supercapacitors. *Chem. Eng. J.* **2023**, *472*, 144913. [CrossRef]
4. Zhou, Y.; Yin, L.; Xiang, S.; Yu, S.; Johnson, H.M.; Wang, S.; Yin, J.; Zhao, J.; Luo, Y.; Chu, P.K. Unleashing the potential of MXene-based flexible materials for high-performance energy storage devices. *Adv. Sci.* **2023**, *11*, 2304874. [CrossRef]
5. Zahra, S.A.; Anasori, B.; Iqbal, M.Z.; Ravaux, F.; Al Tarawneh, M.; Rizwan, S. Enhanced electrochemical performance of vanadium carbide MXene composites for supercapacitors. *APL Mater.* **2022**, *10*, 060901. [CrossRef]
6. Zang, X.; Wang, J.; Qin, Y.; Wang, T.; He, C.; Shao, Q.; Zhu, H.; Cao, N. Enhancing capacitance performance of $Ti_3C_2T_x$ MXene as electrode materials of supercapacitor: From controlled preparation to composite structure construction. *Nano-Micro Lett.* **2020**, *12*, 77. [CrossRef]
7. Yu, Y.; Fan, Q.; Li, Z.; Fu, P. MXene-based electrode materials for supercapacitors: Synthesis, properties, and optimization strategies. *Mater. Today Sustain.* **2023**, *24*, 100551. [CrossRef]
8. Zhu, Y.; Ma, J.; Das, P.; Wang, S.; Wu, Z.S. High-voltage MXene-based supercapacitors: Present status and future perspectives. *Small Methods* **2023**, *7*, 2201609. [CrossRef]
9. Thomas, S.A.; Patra, A.; Al-Shehri, B.M.; Selvaraj, M.; Aravind, A.; Rout, C.S. MXene based hybrid materials for supercapacitors: Recent developments and future perspectives. *J. Energy Storage* **2022**, *55*, 105765. [CrossRef]
10. Prasankumar, T.; Manoharan, K.; Farhana, N.K.; Bashir, S.; Ramesh, K.; Ramesh, S.; Ramachandaramurthy, V.K. Advancements and approaches in developing MXene-based hybrid composites for improved supercapacitor electrodes. *Mater. Today Sustain.* **2024**, *28*, 100963. [CrossRef]
11. An, Y.; Tian, Y.; Shen, H.; Man, Q.; Xiong, S.; Feng, J. Two-dimensional MXenes for flexible energy storage devices. *Energy Environ. Sci.* **2023**, *16*, 4191–4250. [CrossRef]
12. Zan, G.; Li, S.; Chen, P.; Dong, K.; Wu, Q.; Wu, T. Mesoporous cubic nanocages assembled by coupled monolayers with 100% theoretical capacity and robust cycling. *ACS Cent. Sci.* **2024**, *10*, 1283–1294. [CrossRef] [PubMed]
13. Huang, H.; Yang, W. MXene-based micro-supercapacitors: Ink rheology, microelectrode design and integrated system. *ACS Nano* **2024**, *18*, 4651–4682. [CrossRef] [PubMed]
14. Aravind, A.M.; Tomy, M.; Kuttapan, A.; Kakkassery Aippunny, A.M.; Suryabai, X.T. Progress of 2D MXene as an electrode architecture for advanced supercapacitors: A comprehensive review. *ACS Omega* **2023**, *8*, 44375–44394. [CrossRef]
15. Chen, Y.; Yang, H.; Han, Z.; Bo, Z.; Yan, J.; Cen, K.; Ostrikov, K.K. MXene-based electrodes for supercapacitor energy storage. *Energy Fuels* **2022**, *36*, 2390–2406. [CrossRef]
16. Chen, A.; Wang, C.; Abu Ali, O.A.; Mahmoud, S.F.; Shi, Y.; Ji, Y.; Algadi, H.; El-Bahy, S.M.; Huang, M.; Guo, Z.; et al. MXene@nitrogen-doped carbon films for supercapacitor and piezoresistive sensing applications. *Compos. Part A Appl. Sci. Manuf.* **2022**, *163*, 107174. [CrossRef]

17. Siddu, N.K.P.; Jeong, S.M.; Rout, C.S. MXene-carbon based hybrid materials for supercapacitor applications. *Energy Adv.* **2024**, *3*, 341–365. [CrossRef]
18. Zhang, Y.; Feng, Z.; Wang, X.; Hu, H.; Wu, M. MXene/carbon composites for electrochemical energy storage and conversion. *Mater. Today Sustain.* **2023**, *22*, 100350. [CrossRef]
19. Cao, J.-M.; Zatovsky, I.V.; Gu, Z.-Y.; Yang, J.-L.; Zhao, X.-X.; Guo, J.-Z.; Xu, H.; Wu, X.-L. Two-dimensional MXene with multidimensional carbonaceous matrix: A platform for general-purpose functional materials. *Prog. Mater. Sci.* **2023**, *135*, 100350. [CrossRef]
20. Deng, X.; Li, J.; Ma, L.; Sha, J.; Zhao, N. Three-dimensional porous carbon materials and their composites as electrodes for electrochemical energy storage systems. *Mater. Chem. Front.* **2019**, *3*, 2221–2245. [CrossRef]
21. Ling, Z.; Ren, C.E.; Zhao, M.-Q.; Yang, J.; Giammarco, J.M.; Qiu, J.; Barsoum, M.W.; Gogotsi, Y. Flexible and conductive MXene films and nanocomposites with high capacitance. *Proc. Natl. Acad. Sci. USA* **2014**, *111*, 16676–16681. [CrossRef] [PubMed]
22. Sun, N.; Zhu, Q.; Anasori, B.; Zhang, P.; Liu, H.; Gogotsi, Y.; Xu, B. MXene-bonded flexible hard carbon film as anode for stable Na/K-ion storage. *Adv. Funct. Mater.* **2019**, *29*, 1906282. [CrossRef]
23. Cai, Y.; Chen, X.; Xu, Y.; Zhang, Y.; Liu, H.; Zhang, H.; Tang, J. $Ti_3C_2T_x$ MXene/carbon composites for advanced supercapacitors: Synthesis, progress, and perspectives. *Carbon Energy* **2024**, *6*, e501. [CrossRef]
24. Li, L.; Cheng, Q. MXene based nanocomposite films. *Exploration* **2022**, *2*, 20220049. [CrossRef]
25. Muthukutty, B.; Kumar, P.S.; Vivekanandan, A.K.; Sivakumar, M.; Lee, S.; Lee, D. Progress and perspective in harnessing MXene-carbon-based composites (0-3D): Synthesis, performance, and applications. *Chemosphere* **2024**, *355*, e501. [CrossRef]
26. Abid, M.Z.; Rafiq, K.; Aslam, A.; Jin, R.; Hussain, E. Scope, evaluation and current perspectives of MXene synthesis strategies for state-of-the-art applications. *J. Mater. Chem. A* **2024**, *12*, 7351–7395. [CrossRef]
27. Kong, D.; Lv, W.; Liu, R.; He, Y.B.; Wu, D.; Li, F.; Fu, R.; Yang, Q.; Kang, F. Superstructured carbon materials: Design and energy applications. *Energy Mater. Devices* **2023**, *1*, 9370017. [CrossRef]
28. Hasan, M.M.; Hossain, M.M.; Chowdhury, H.K. Two-dimensional MXene-based flexible nanostructures for functional nanodevices: A review. *J. Mater. Chem. A* **2021**, *9*, 3231–3269. [CrossRef]
29. Gao, M.; Wang, F.; Yang, S.; Gaetano Ricciardulli, A.; Yu, F.; Li, J.; Sun, J.; Wang, R.; Huang, Y.; Zhang, P.; et al. Engineered 2D MXene-based materials for advanced supercapacitors and micro-supercapacitors. *Mater. Today* **2024**, *72*, 318–358. [CrossRef]
30. Wang, C.; Chen, S.; Song, L. Tuning 2D MXenes by surface controlling and interlayer engineering: Methods, properties, and synchrotron radiation characterizations. *Adv. Funct. Mater.* **2020**, *30*, 2000869. [CrossRef]
31. Bi, W.; Gao, G.; Li, C.; Wu, G.; Cao, G. Synthesis, properties, and applications of MXenes and their composites for electrical energy storage. *Prog. Mater. Sci.* **2024**, *142*, 101227. [CrossRef]
32. Kshetri, T.; Tran, D.T.; Le, H.T.; Nguyen, D.C.; Hoa, H.V.; Kim, N.H.; Lee, J.H. Recent advances in MXene-based nanocomposites for electrochemical energy storage applications. *Prog. Mater. Sci.* **2021**, *117*, 100733. [CrossRef]
33. Chetana, S.; Mohd Abdah, M.A.A.; Thakur, V.N.; Govinde Gowda, M.S.; Choudhary, P.; Sriramoju, J.B.; Rangappa, D.; Malik, S.; Rustagi, S.; Khalid, M. Progress and prospects of MXene-based hybrid composites for next-generation energy technology. *J. Electrochem. Soc.* **2023**, *170*, 120530. [CrossRef]
34. Yang, G.; Liu, D.; Lei, W. MXene-based nanocomposites for nanofluidic energy conversion: A review. *Adv. Nanocompos.* **2024**, *1*, 94–109. [CrossRef]
35. Deng, Q.; Liu, F.Z.; Wu, X.W.; Li, C.Z.; Zhou, W.B.; Long, B. An aqueous BiI_3-Zn battery with dual mechanisms of Zn^{2+} (de)intercalation and I^-/I_2 redox. *J. Energy Chem.* **2024**, *89*, 670–678. [CrossRef]
36. Sagadevan, S.; Fatimah, I.; Lett, J.A.; Kakavandi, B.; Soga, T.; Oh, W.-C.; Randriamahazaka, H. Exploring the potential of MXene-based aerogels and hybrid nanocomposites for supercapacitor applications. *J. Energy Storage* **2024**, *99*, 113269. [CrossRef]
37. Jiang, S.; Lu, L.; Song, Y. Recent advances of flexible MXene and its composites for supercapacitors. *Chem.-Eur. J.* **2024**, *30*, e202304036. [CrossRef] [PubMed]
38. Liu, Y.; Yu, J.; Guo, D.; Li, Z.; Su, Y. $Ti_3C_2T_x$ MXene/graphene nanocomposites: Synthesis and application in electrochemical energy storage. *J. Alloys Compd.* **2020**, *815*, 152403. [CrossRef]
39. Qi, M.; Li, F.; Zhang, Z.; Lai, Q.; Liu, Y.; Gu, J.; Wang, L. Three-dimensional interconnected ultrathin manganese dioxide nanosheets grown on carbon cloth combined with $Ti_3C_2T_x$ MXene for high-capacity zinc-ion batteries. *J. Colloid Interface Sci.* **2022**, *615*, 151–162. [CrossRef]
40. Shi, X.; Liang, W.; Liu, G.; Chen, B.; Shao, L.; Wu, Y.; Sun, Z.; García, F. Electrode materials for Li/Na storage from mechanochemically synthesised MOFs/MXene Composites: A Solvent-free approach. *Chem. Eng. J.* **2023**, *462*, 142271. [CrossRef]
41. Fu, X.-Y.; Shu, R.-Y.; Ma, C.-J.; Zhang, Y.-Y.; Jiang, H.-B. Laser-induced fabrication of electrodes on graphene oxide-MXene composites for planar supercapacitors. *ACS Appl. Nano Mater.* **2023**, *6*, 4567–4572. [CrossRef]
42. Depijan, M.; Hantanasirisakul, K.; Pakawatpanurut, P. Interfacial engineering of $Ti_3C_2T_x$ MXene electrode using g-C_3N_4 nanosheets for high-performance supercapacitor in neutral electrolyte. *ACS Omega* **2024**, *9*, 22256–22264. [CrossRef] [PubMed]
43. Hong, Z.; Tian, H.; Fang, Z.; Wu, H.; Zhao, F.; Li, Q.; Fan, S.; Wang, J.; Liu, P. Carbon nanotube/MXene composite with a dense regular connective tissue structure and its application in lithium-ion batteries. *ACS Appl. Energy Mater.* **2024**, *7*, 8004–8013. [CrossRef]
44. Yu, L.; Hu, L.; Anasori, B.; Liu, Y.-T.; Zhu, Q.; Zhang, P.; Gogotsi, Y.; Xu, B. MXene-bonded activated carbon as a flexible electrode for high-performance supercapacitors. *ACS Energy Lett.* **2018**, *3*, 1597–1603. [CrossRef]

45. Wang, G.; Li, C.; Estevez, D.; Xu, P.; Peng, M.; Wei, H.; Qin, F. Boosting interfacial polarization through heterointerface engineering in MXene/graphene intercalated-based microspheres for electromagnetic wave absorption. *Nano-Micro Lett.* **2023**, *15*, 152. [CrossRef]
46. Lei, J.; Yu, F.; Xie, H.; Ma, J. $Ti_3C_2T_x$ MXene/carbon nanofiber multifunctional electrode for electrode ionization with antifouling activity. *Chem. Sci.* **2023**, *14*, 3610–3621. [CrossRef]
47. Wang, Z.; Huang, Z.; Wang, H.; Li, W.; Wang, B.; Xu, J.; Xu, T.; Zang, J.; Kong, D.; Li, X.; et al. 3D-printed sodiophilic V_2CT_x/rGO-CNT MXene microgrid aerogel for stable Na metal anode with high areal capacity. *ACS Nano* **2022**, *16*, 9105–9116. [CrossRef]
48. Dai, Y.; Wu, X.; Li, L.; Zhang, Y.; Deng, Z.; Yu, Z.-Z.; Zhang, H.-B. 3D printing of resilient, lightweight and conductive MXene/reduced graphene oxide architectures for broadband electromagnetic interference shielding. *J. Mater. Chem. A* **2022**, *10*, 11375–11385. [CrossRef]
49. Zhu, W.; Zhuang, Y.; Weng, J.; Huang, Q.; Lai, G.; Li, L.; Chen, M.; Xia, K.; Lu, Z.; Wu, M.; et al. Evolution of naturally dried MXene-based composite aerogels with flash joule annealing for large-scale production of highly sensitive customized sensors. *Adv. Mater.* **2024**, *36*, 2407138. [CrossRef]
50. Sangili, A.; Unnikrishnan, B.; Nain, A.; Hsu, Y.-J.; Wu, R.-S.; Huang, C.-C.; Chang, H.-T. Stable carbon encapsulated titanium carbide MXene aqueous ink for fabricating high-performance supercapacitors. *Energy Storage Mater.* **2022**, *53*, 51–61. [CrossRef]
51. Allah, A.E.; Wang, J.; Kaneti, Y.V.; Li, T.; Farghali, A.A.; Khedr, M.H.; Nanjundan, A.K.; Ding, B.; Dou, H.; Zhang, X.; et al. Auto-programmed heteroarchitecturing: Self-assembling ordered mesoporous carbon between two-dimensional $Ti_3C_2T_x$ MXene layers. *Nano Energy* **2019**, *65*, 103991. [CrossRef]
52. Zhang, Y.; Lan, D.; Hou, T.; Jia, M.; Jia, Z.; Gu, J.; Wu, G. Multifunctional electromagnetic wave absorbing carbon fiber/$Ti_3C_2T_x$ MXene fabric with ultra-wide absorption band. *Carbon* **2024**, *230*, 119594. [CrossRef]
53. Ma, C.; Ma, M.G.; Si, C.; Ji, X.X.; Wan, P. Flexible MXene-based composites for wearable devices. *Adv. Funct. Mater.* **2021**, *31*, 2009524. [CrossRef]
54. Sun, M.; Ye, W.; Zhang, J.; Zheng, K. Structure, properties, and preparation of MXene and the application of its composites in supercapacitors. *Inorganics* **2024**, *12*, 112. [CrossRef]
55. Vijayakumar, M.; Elsa, G.; Nirogi, A.; Navaneethan, R.; Sankar, A.B.; Karthik, M. MXenes and their composites for hybrid capacitors and supercapacitors: A critical review. *Emergent Mater.* **2021**, *4*, 655–672. [CrossRef]
56. Forouzandeh, P.; Pillai, S.C. MXenes-based nanocomposites for supercapacitor applications. *Curr. Opin. Chem. Eng.* **2021**, *33*, 100710. [CrossRef]
57. Das, P.; Ganguly, S.; Rosenkranz, A.; Wang, B.; Yu, J.; Srinivasan, S.; Rajabzadeh, A.R. MXene/0D nanocomposite architectures: Design, properties and emerging applications. *Mater. Today Nano* **2023**, *24*, 100428. [CrossRef]
58. Elemike, E.E.; Adeyemi, J.; Onwudiwe, D.C.; Wei, L.; Oyedeji, A.O. The future of energy materials: A case of MXenes-carbon dots nanocomposites. *J. Energy Storage* **2022**, *50*, 104711. [CrossRef]
59. Wang, Y.; Chen, N.; Zhou, B.; Zhou, X.; Pu, B.; Bai, J.; Tang, Q.; Liu, Y.; Yang, W. NH_3-induced in situ etching strategy derived 3D-interconnected porous MXene/carbon dots films for high performance flexible supercapacitors. *Nano-Micro Lett.* **2023**, *15*, 231. [CrossRef]
60. Zhang, P.; Li, J.; Yang, D.; Soomro, R.A.; Xu, B. Flexible carbon dots-intercalated MXene film electrode with outstanding volumetric performance for supercapacitors. *Adv. Funct. Mater.* **2022**, *33*, 2209918. [CrossRef]
61. Wei, L.; Deng, W.; Li, S.; Wu, Z.; Cai, J.; Luo, J. Sandwich-like chitosan porous carbon Spheres/MXene composite with high specific capacitance and rate performance for supercapacitors. *J. Bioresour. Bioprod.* **2022**, *7*, 63–72. [CrossRef]
62. Yang, F.; Lv, K.; Zhao, X.; Kong, D.; Kong, N.; Luo, Z.; Tao, J.; Zhou, J.; Razal, J.M.; Zhang, J. Hierarchical heterostructures of MXene and mesoporous hollow carbon sphere for improved ion accessibility and rate performance. *Chem. Eng. J.* **2024**, *494*, 153426. [CrossRef]
63. Li, B.; Zhang, L.; Zhao, Z.; Zou, Y.; Chen, B.; Fu, X.; Wang, F.; Long, S.; Guo, W.; Liang, J.; et al. Unraveling hierarchical hollow $NiCo_2S_4$/MXene/N-doped carbon microspheres via dual templates for high-performance hybrid supercapacitors. *Chem. Eng. J.* **2024**, *487*, 150730. [CrossRef]
64. Mohajer, F.; Ziarani, G.M.; Badiei, A.; Iravani, S.; Varma, R.S. MXene-carbon nanotube composites: Properties and applications. *Nanomaterials* **2023**, *13*, 345. [CrossRef]
65. Wang, Q.; Yuan, H.; Zhang, M.; Yang, N.; Cong, S.; Zhao, H.; Wang, X.; Xiong, S.; Li, K.; Zhou, A. A highly conductive and supercapacitive MXene/N-CNT electrode material derived from a MXene-Co-melamine precursor. *ACS Appl. Electron. Mater.* **2023**, *5*, 2506–2517. [CrossRef]
66. Li, H.; Chen, R.; Ali, M.; Lee, H.; Ko, M.J. In situ grown MWCNTs/MXenes nanocomposites on carbon cloth for high-performance flexible supercapacitors. *Adv. Funct. Mater.* **2020**, *30*, 2002739. [CrossRef]
67. Hakim, M.W.; Ali, I.; Fatima, S.; Li, H.; Jafri, S.H.M.; Rizwan, S. Enhanced electrochemical performance of MWCNT-assisted molybdenum-titanium carbide MXene as a potential electrode material for energy storage application. *ACS Omega* **2024**, *9*, 8763–8772. [CrossRef]
68. You, M.; Xin, B. MXene nanosheets and carbon nanofiber hybrid membranes for electrochemical energy storage materials. *Fibers Polym.* **2024**, *25*, 3323–3330. [CrossRef]

69. Sun, L.; Fu, Q.; Pan, C. Hierarchical porous "skin/skeleton"-like MXene/biomass derived carbon fibers heterostructure for self-supporting, flexible all solid-state supercapacitors. *J. Hazard. Mater.* **2021**, *410*, 124565. [CrossRef]
70. Dharmasiri, B.; Usman, K.A.S.; Qin, S.A.; Razal, J.M.; Tran, N.T.; Coia, P.; Harte, T.; Henderson, L.C. $Ti_3C_2T_x$ MXene coated carbon fibre electrodes for high performance structural supercapacitors. *Chem. Eng. J.* **2023**, *476*, 146739. [CrossRef]
71. Song, W.; Wang, K.; Lian, X.; Zheng, F.; Niu, H. Non-preoxidation synthesis of MXene integrated flexible carbon film for supercapacitors. *Chem. Eng. J.* **2024**, *493*, 152804. [CrossRef]
72. Zhao, X.; Zhang, J.; Lv, K.; Kong, N.; Shao, Y.; Tao, J. Carbon nanotubes boosts the toughness and conductivity of wet-spun MXene fibers for fiber-shaped super capacitors. *Carbon* **2022**, *200*, 38–46. [CrossRef]
73. Zarepour, A.; Ahmadi, S.; Rabiee, N.; Zarrabi, A.; Iravani, S. Self-healing MXene- and graphene-based composites: Properties and applications. *Nano-Micro Lett.* **2023**, *15*, 100. [CrossRef]
74. Dang, A.; Han, Y.; Sun, Y.; Liu, Y.; Zhao, Z.; Liu, X.; Zada, A.; Han, Y.; Li, T.; Li, J. Mechanically stable reduced graphene oxide/MXene fibers with exceptional volumetric capacitance and energy density mediated by carbon nanotubes for high-performance symmetrical supercapacitors. *ACS Appl. Energy Mater.* **2024**, *7*, 5548–5558. [CrossRef]
75. Qin, Z.; Wang, Z.; Li, D.; Lv, Y.; Zhao, B.; Pan, K. Nanofiber-reinforced MXene/rGO composite aerogel for a high-performance piezoresistive sensor and an all-solid-state supercapacitor electrode material. *ACS Appl. Mater. Interfaces* **2024**, *16*, 32554–32565. [CrossRef] [PubMed]
76. Li, Y.; Xu, L.; Dai, J.; Zhu, H.-X.; Wang, S.-J. Multidimensional nanostructural engineering of MXene-based composite films for high-performance supercapacitors. *Energy Fuels* **2024**, *38*, 5493–5505. [CrossRef]
77. Xu, C.; Tong, L.; Zhang, W.; Zhao, X.; Yang, L.; Yin, S. One-step synthesis of g-C_3N_4/$TiVCT_x$ MXene electrodes for lithium-ion batteries and supercapacitors. *Chem. Eng. J.* **2024**, *497*, 154449. [CrossRef]
78. Zhang, S.; Huang, Y.; Wang, J.; Han, X.; Chen, C.; Sun, X. $Ti_3C_2T_x$/g-C_3N_4 heterostructure films with outstanding capacitance for flexible Solid-state supercapacitors. *Appl. Surf. Sci.* **2022**, *599*, 154015. [CrossRef]
79. Zhang, S.; Huang, Y.; Ruan, Y.; Wang, J.; Han, X.; Sun, X. Electrostatic self-assembly of citrus based carbon nanosheets and MXene: Flexible film electrodes and patterned interdigital electrodes for all-solid supercapacitors. *J. Energy Storage* **2023**, *58*, 106392. [CrossRef]
80. Chen, A.; Wei, H.; Peng, Z.; Wang, Y.; Akinlabi, S.; Guo, Z.; Gao, F.; Duan, S.; He, X.; Jia, C.; et al. MXene/nitrogen-doped carbon nanosheet scaffold electrode toward high-performance solid-state zinc ion supercapacitor. *Small* **2024**, *20*, 2404011. [CrossRef]
81. Mo, T.; Wang, Z.; Zeng, L.; Chen, M.; Kornyshev, A.A.; Zhang, M.; Zhao, Y.; Feng, G. Energy storage mechanism in supercapacitors with porous graphdiynes: Effects of pore topology and electrode metallicity. *Adv. Mater.* **2023**, *35*, 2301118. [CrossRef] [PubMed]
82. Zheng, X.; Chen, S.; Li, J.; Wu, H.; Zhang, C.; Zhang, D.; Chen, X.; Gao, Y.; He, F.; Hui, L.; et al. Two-dimensional carbon graphdiyne: Advances in fundamental and application research. *ACS Nano* **2023**, *17*, 14309–14346. [CrossRef] [PubMed]
83. Wu, D.; Zhang, Y.; Man, Z.; Zhang, H.; Zhu, X.; Ding, J.; Xu, J.; Bao, N.; Lu, W. In situ fabrication of graphdiyne nanoisland anchored $Ti_3C_2T_x$ film to accelerate intercalation pseudocapacitance kinetics. *Adv. Energy Mater.* **2024**, *14*, 2304404. [CrossRef]
84. De, S.; Maity, C.K.; Kim, M.J.; Nayak, G.C. Tin(IV) selenide anchored-biowaste derived porous carbon-$Ti_3C_2T_x$ (MXene) nanohybrid: An ionic electrolyte enhanced high performing flexible supercapacitor electrode. *Electrochim. Acta* **2023**, *463*, 142811. [CrossRef]
85. Yong, S.; Yao, C.; Hillier, N.; Kim, H.; Holicky, M.; Liu, S.; Doherty, R.; Torrisi, F.; Beeby, S. Ti_3C_2 MXene as additive for low-cost textile supercapacitors with enhanced electrical performance. *Adv. Mater. Technol.* **2023**, *9*, 2301266. [CrossRef]
86. Hou, X.; Ren, P.; Tian, W.; Xue, R.; Tong, W.; Chen, Z.; Ren, F.; Jin, Y. Electrostatic self-assembly heterostructured MXenes/wasted PET-derived carbon for superior capacitive energy storage. *Adv. Mater. Technol.* **2024**, *9*, 2301766. [CrossRef]
87. Seenath, J.S.; Biswal, B.P. Construction of MXene-coupled nitrogen-doped porous carbon hybrid from a conjugated microporous polymer for high-performance supercapacitors. *Adv. Energy Sustain. Res.* **2021**, *2*, 2000052. [CrossRef]
88. Huang, Y.; Xu, Y.; Wang, J.; Bao, S.; Zhang, Y.; Yin, Y.; Lu, J. Self-assembly of N,S-MXene/3DPC heterostructure with multiple charge transfer channels for high-performance supercapacitors and sodium-ion batteries. *J. Power Sources* **2024**, *601*, 234312. [CrossRef]
89. Yang, K.; Luo, M.; Zhang, D.; Liu, C.; Li, Z.; Wang, L.; Chen, W.; Zhou, X. $Ti_3C_2T_x$/carbon nanotube/porous carbon film for flexible supercapacitor. *Chem. Eng. J.* **2022**, *427*, 132002. [CrossRef]
90. Yao, W.; Zheng, D.; Li, Z.; Wang, Y.; Tan, H.; Zhang, Y. MXene@ carbonized wood monolithic electrode with hierarchical porous framework for high-performance supercapacitors. *Appl. Surf. Sci.* **2023**, *638*, 158130. [CrossRef]
91. Sun, L.; Song, G.; Sun, Y.; Fu, Q.; Pan, C. MXene/N-doped carbon foam with three-dimensional hollow neuron-like architecture for freestanding, highly compressible all solid-state supercapacitors. *ACS Appl. Mater. Interfaces* **2020**, *12*, 44777–44788. [CrossRef] [PubMed]
92. Li, Z.; Liu, X.; Wang, X.; Wang, H.; Ren, J.; Wang, R. Electrophoretic deposition of $Ti_3C_2T_x$ MXene nanosheet N-carbon cloth as binder-free supercapacitor electrode material. *J. Alloys Compd.* **2022**, *927*, 166934. [CrossRef]
93. Fan, W.; Wang, Q.; Rong, K.; Shi, Y.; Peng, W.; Li, H.; Guo, Z.; Xu, B.B.; Hou, H.; Algadi, H.; et al. MXene enhanced 3D needled waste denim felt for high-performance flexible supercapacitors. *Nano-Micro Lett.* **2023**, *16*, 36. [CrossRef] [PubMed]
94. Zhao, D.; Xu, D.; Wang, T.; Yang, Z. Nitrogen-rich nanoporous carbon with MXene composite for high-performance Zn-ion hybrid capacitors. *Mater. Today Energy* **2024**, *45*, 101671. [CrossRef]

95. Wang, Y.; Chen, N.; Liu, Y.; Zhou, X.; Pu, B.; Qing, Y.; Zhang, M.; Jiang, X.; Huang, J.; Tang, Q.; et al. MXene/graphdiyne nanotube composite films for free-standing and flexible solid-state supercapacitor. *Chem. Eng. J.* **2022**, *450*, 138398. [CrossRef]
96. Yang, R.; Hu, Q.; Yang, S.; Zeng, Z.; Zhang, H.; Cao, A.; Gui, X. Anchoring oxidized MXene nanosheets on porous carbon nanotube sponge for enhancing ion transport and pseudocapacitive performance. *ACS Appl. Mater. Interfaces* **2022**, *14*, 41997–42006. [CrossRef]
97. Huang, X.; Lyu, X.; Wu, G.; Yang, J.; Zhu, R.; Tang, Y.; Li, T.; Wang, Y.; Yang, D.; Dong, A. Multilayer superlattices of monolayer mesoporous carbon framework-intercalated MXene for efficient capacitive energy storage. *Adv. Energy Mater.* **2023**, *14*, 2303417. [CrossRef]
98. Zhao, Y.; Zhang, Y.; Wang, Y.; Cao, D.; Sun, X.; Zhu, H. Versatile zero- to three-dimensional carbon for electrochemical energy storage. *Carbon Energy* **2021**, *3*, 895–915. [CrossRef]
99. Cai, Z.; Ma, Y.F.; Wang, M.; Qian, A.N.; Tong, Z.M.; Xiao, L.T.; Jia, S.T.; Chen, X.Y. Engineering of electrolyte ion channels in MXene/holey graphene electrodes for superior supercapacitive performances. *Rare Met.* **2022**, *41*, 2084–2093. [CrossRef]
100. Long, B.; Liang, X.; Pei, Y.; Wu, X.; Wang, X.; Law, M.K. Free-standing bioi@mwcnts photoelectrodes for photo-rechargeable zinc-ion batteries. *J. Mater. Sci. Technol.* **2024**, *198*, 137–142. [CrossRef]
101. Long, B.; Ma, X.Y.; Chen, L.J.; Song, T.; Pei, Y.; Wang, X.Y.; Wu, X.W. Se Vacancy Activated Bi_2Se_3 Nanodots Encapsulated in Porous Carbon Nanofibers for Aqueous Zinc and Ammonium Ion Batteries. In *Advanced Functional Materials;* Wiley Online Library: Hoboken, NJ, USA, 2024; p. 2411430. [CrossRef]

batteries

MDPI

Article

Composites Based on Poly(ortho-toluidine) and WS_2 Sheets for Applications in the Supercapacitor Field

Teodora Burlanescu [1,2], Ion Smaranda [1], Andreea Androne [1], Cristina Stefania Florica [1], Madalina Cercel [1], Mirela Paraschiv [1], Adelina Udrescu [1], Adam Lőrinczi [1], Petru Palade [1], Andrei Galatanu [1], Catalin Negrila [1], Elena Matei [1], Monica Dinescu [1], Radu Cercel [1] and Mihaela Baibarac [1,*]

[1] National Institute of Materials Physics, Atomistilor Street 405A, 077125 Bucharest, Romania; teodora.burlanescu@infim.ro (T.B.); ion.smaranda@infim.ro (I.S.); andreea.radu@infim.ro (A.A.); stefania.florica@infim.ro (C.S.F.); madalina.chivu@infim.ro (M.C.); mirela.cristea@infim.ro (M.P.); adelina.udrescu@infim.ro (A.U.); lorinczi@infim.ro (A.L.); palade@infim.ro (P.P.); gala@infim.ro (A.G.); catalin.negrila@infim.ro (C.N.); elena.matei@infim.ro (E.M.); monica.daescu@infim.ro (M.D.); radu.cercel@infim.ro (R.C.)

[2] Faculty of Physics, University of Bucharest, Atomistilor Street 405, 077125 Bucharest, Romania

* Correspondence: barac@infim.ro

Abstract: In this work, three methods for the synthesis of composites based on poly(ortho-toluidine) (POT) and WS_2 are reported: (a) the solid-state interaction (SSI) of POT with WS_2 nanoparticles (NPs); (b) the in situ chemical polymerization (ICP) of ortho-toluidine (OT); and (c) the electrochemical polymerization (ECP) of OT. The preparation of WS_2 sheets was performed by the ball milling of the WS_2 NPs followed by ultrasonication in the solvent N,N′-dimethyl formamide. During the synthesis of the POT/WS_2 composites by SSI and ICP, an additional exfoliation of the WS_2 NPs was reported. In this work, we demonstrated the following: (a) the ICP method leads to POT/WS_2 composites, which contain repeating units of POT in the leucoemeraldine salt (LS) state, while (b) the ECP method leads to POT/WS_2 composites, which contain repeating units of POT in the emeraldine salt (ES) state. Capacitances equal to 123.5, 465.76, and 751.6 mF cm^{-2} in the cases of POT-ES/WS_2 composites, synthesized by SSI, ICP, and ECP, respectively, were reported.

Keywords: composites; conducting polymers; transition metal dichalcogenides; sheets; Raman scattering; FTIR spectroscopy; cyclic voltammetry; supercapacitors; energy storage

1. Introduction

Composites based on conducting polymers and transition metal dichalcogenides (TMDs) have been of particular interest as a result of applications in the fields of photocatalysis [1], gas sensors [2], and, last but not least, supercapacitors [3–5]. The methods often used for the synthesis of polyaniline (PANI)/WS_2 composites are as follows: (i) the method of the chemical interaction of nanolayers of WS_2 with PANI in a solid state [3]; (ii) the ECP of a monomer in an anhydrous solution of H_2SO_4, acetonitrile, and WS_2 microparticles [4]; and (iii) the exfoliation of WS_2 microparticles in butyl lithium, followed by the ICP of aniline [6]. The use of TMDs in the bulk state in supercapacitors (SCs) has often been avoided due to their small surface area (SA). The large SA of individual TMD sheets and higher oxidation numbers (e.g., in MoS_2 from +2 to +4) allow TMDs to store electrostatic charge as well as ionic interlacing in the interlayer space, leading to a specific capacitance and an energy density with higher values [7,8]. The main methods for TMD exfoliation are as follows: (a) mechanically; (b) in the liquid phase, using organic solvents or surfactants; (c) ion intercalation; and (d) grinding [9]. Concerning the particular case of WS_2 nano- and microparticles, among the main strategies adopted for the exfoliation of these TMDs are (i) liquid-phase exfoliation assisted by the ultrasonication of the WS_2 crystals in surfactant aqueous solutions [10] or organic solvents [11] as well as superacid (chlorosulfonic acid) [12]; (ii) chemical exfoliation performed in three stages involving magnetic stirring, horn-tip sonication, and shear mixing [13]; (iii) electrochemical exfoliation at a DC potential of 10 V [14]; (iv) solvothermal reaction [15], etc. The methods often used to illustrate TMD exfoliation are X-ray diffraction (XRD), scanning electron microscopy (SEM), Raman scattering, and X-ray photoelectron spectroscopy (XPS), e.g., in [10–15]. A disadvantage of liquid-phase exfoliation assisted by the ultrasonication of WS_2 crystals is the long time taken to obtain the nanolayers as a consequence of the successive ultrasonication and centrifugation stages. In this work, special attention will be paid to obtaining WS_2 sheets by grinding WS_2 NPs followed by ultrasonication for a short period in N,N′-dimethyl formamide (DMF).

The low conductivity in the most stable phases of TMDs prevents their use as potential electrode materials in supercapacitor cells. Conductive polymers such as PANI [16] are often used in the field of supercapacitors as a result of their high specific capacitance, electrical conductivity >0.001 S/cm, manufacturing methods that do not require expensive equipment and high flexibility, ability to be obtained in different oxidation states such as leucoemeraldine, emeraldine, and pernigraniline, which correspond to reduced, semi-oxidized and oxidized states, and obtaining of polymers in undoped and doped states, respectively [17]. The following methods have been used to synthesize PANI/WS_2 composites: (i) the ICP of aniline using WS_2 sheets resulting from the exfoliation of WS_2 powder in water/isopropanol mixtures [18]; (ii) the interaction of the two constituents, e.g., in [3,19]; and (iii) the ECP of WS_2 dispersed in a mixture of aniline and acid solution, e.g., in [4]. According to Ref. [20], a disadvantage of PANI is its rigid structure, which can influence the capacitance of supercapacitors. An improvement in the processability of PANI was reported by replacing aniline with ortho-substituted aniline, with functional groups of the types -CH_3, -OCH_3, etc. [21]. One of the important properties of POT is its conductivity, which varies between 10^{-4} and 2×10^{-2} S/cm [22,23]. In this study, our effort will be given to the synthesis of composites based on WS_2 and POT and the evaluation of their optical, structural, and electrochemical properties. The following strategies will be used for the synthesis of POT/WS_2 composites: (i) the SSI of POT with WS_2 NPs; (iii) the ICP of ortho-toluidine (OT) with the addition of WS_2 NPs; and (iii) the ECP of OT in a semi-aqueous acid solution containing WS_2 sheets in DMF/H_2O (the volumetric ratio was 1:1). Using correlated studies of XRD and SEM, new evidence concerning the exfoliation of WS_2 NPs in the absence and presence of POT will be shown. The chemical interactions at the interface of the two constituents of the POT/WS_2 composites will be highlighted by Raman scattering, FTIR spectroscopy, and X-ray photoelectron spectroscopy (XPS). Using cyclic voltammetry (CV), new information concerning the capacitive and diffusion-controlled intercalation processes and their contributions to the current densities of cyclic voltammograms of the symmetrical supercapacitor cells that have electrodes containing the POT/WS_2 composites as active materials will be shown. An assessment of the dependence of the capacitance of the symmetrical supercapacitors as a function of the potential scan rate and cyclic voltammogram number will also be reported.

2. Materials and Methods

The supplier for the compounds OT, H_2SO_4, DMF, $K_2Cr_2O_7$, NH_4OH, CH_3CN, poly(vinylidene fluoride) (PVDF), dibutyl phthalate (DBP), acetone, diethyl ether, Nafion 117 membrane, and WS_2 NPs was Sigma-Aldrich (St. Louis, MO, USA).

To prepare the WS_2 sheets, a powder containing WS_2 NPs was used, which were subjected to mechanical exfoliation using a ball mill model PM 100—RETSCH and SamplePrep 8000 M shaker mill from SPEX, USA. Using the first ball mill, 1 g of WS_2 NPs was subjected to a grinding time of one hour at 150, 300, and 600 rpm. With the second mill, the grinding of 1 g of WS_2 NPs was carried out for an hour of effective milling time at 900 cycles/min. Milling periods of 5 min were interspersed with 5 min pauses to avoid excessive heating of samples. The changes induced in the crystalline structure of the WS_2 NPs by the exfoliation process were analyzed by XRD. An additional exfoliation after grinding WS_2 NPs for one hour at 900 rpm was performed by ultrasonicating a suspension of WS_2 in DMF for 15 min, leading to a yellowish solution with a concentration of 0.5 mg/mL.

The chemical synthesis of poly(o-toluidine)–emeraldine salt (POT-ES) and poly(o-toluidine)–emeraldine base (POT-EB) involved the use of two solutions of OT 0.18 M and $K_2Cr_2O_7$ 0.05 M, each in 50 mL H_2SO_4 2M. Adding the solution of $K_2Cr_2O_7$ in H_2SO_4 to the solution of OT in H_2SO_4 induces a change in the reaction mixture color from orange to dark green. After two hours, the chemical polymerization reaction led to a green precipitate, which corresponds to POT-ES. The filtration of POT-ES and its interaction with 1 M NH_4OH solution led to obtaining POT-EB, when a change in the color of the precipitate to dark blue occurred. Further, the interaction with 500 mL of distilled water and 100 mL of CH_3CN allowed the removal of oligomers from the reaction product. The polymerization reaction products, i.e., POT-ES and POT-EB, were dried at room temperature for 24 h to constant mass.

The ICP of the POT/WS_2 composites, having TMD concentrations of 20 wt.% and 10 wt.%, was performed as described above; the only difference was the addition of 0.1 g and 0.01 g of WS_2 NPs to the OT solution in H_2SO_4.

Another method for the preparation of the POT-EB/WS_2 composites was the SSI of POT-EB with WS_2 NPs when the TMD concentration in the composite mass was equal to 33 wt.% and 50 wt.%.

Electrochemical synthesis of POT-ES involves the use of a semi-aqueous solution of OT 0.1 M in H_2SO_4 0.5 M with a volumetric ratio of DMF/H_2O equal to 1:1.

The electrosynthesis of the POT-ES/WS_2 composite involved the preparation of (i) a WS_2 solution in DMF (50 mL, 0.5 mg/mL) by ultrasonication of the TMD in DMF for 15 min and (ii) an aqueous solution of 0.1 M OT in 0.5 M H_2SO_4 (50 mL). Subsequently, the solutions were interacted and homogenized by ultrasonication for 5 min. The reaction mixture was placed in a single-compartment cell, which was accessorized with three electrodes as follows: a Pt working electrode with an area of 1 cm^2, an auxiliary electrode made of Pt, and a Ag/AgCl reference electrode. Cyclic voltammograms were recorded in the potential range (−200; +900 mV) vs. Ag/AgCl at a potential scan rate of 50 mV/s, with a Voltalab 80 potentiostat/galvanostat, purchased from Radiometer Analytical.

XRD diagrams of the POT/WS_2 and POT-ES/WS_2 composites were recorded with Bruker's D8 Advance X-ray diffractometer (Bruker, Hamburg, Germany).

The SEM images of WS_2 NPs before and after exfoliation as well as their composites with POT-EB were recorded with the Zeiss Gemini 500 scanning electron microscope (Zeiss, Oberkochen, Germany).

The Raman spectra of WS_2, POT-EB, POT-ES, and the POT-ES/WS_2 and POT-EB/WS_2 composites were recorded using an FTRaman spectrophotometer, MultiRam model, from Bruker (λ_{exc} = 1064 nm, (Ettlingen, Germany)).

To record IR spectra of POT-EB, POT-ES, and the POT-ES/WS_2 and POT-EB/WS_2 composites, we used an FTIR spectrophotometer, model Vertex 80, from Bruker (Billerica, MA, USA), in the transmission geometry mode for compounds prepared by SSI and ICP, while for composites deposited as films on the Pt electrode, the totally attenuated reflection geometry was used.

To record the UV-VIS spectrum of WS_2 sheets, we used a Lambda 950 UV-VIS-NIR spectrophotometer from Perkin-Elmer (PerkinElmer, Inc., Waltham, MA, USA).

The XPS spectra of WS_2, POT-EB, POT-ES, and the POT-ES/WS_2 and POT-EB/WS_2 composites were recorded on a SPECS spectrometer (SPECS Gmbh, Berlin, Germany) with a Phoibos 150 electron energy analyzer and a monochromatic X-ray source of the type Al Kα 1486.74 eV.

The testing of composites as electrode active materials (EAMs) in SC cells was performed by CV. To prepare electrodes, we used a mixture of 80 wt.% EAM, 5 wt.% PVDF, 15 wt.% super-P conductive black carbon, 0.1 mL DBP, and 1 mL acetone, which were mixed for 12 h under magnetic stirring to obtain a homogeneous paste. A film was obtained after the evaporation of the acetone from the paste poured onto the glass slides. To remove DBP from the film to be used as an electrode in SCs, an interaction with diethyl ether was performed. The electrodes for SC cells were in a circular form with a diameter of 8 mm obtained by cutting above the film. The electrolyte used in the SC study was a Nafion 117 membrane treated with a 1 M H_2SO_4 solution [21].

3. Results and Discussion

3.1. Exfoliation of the WS_2 Nanoparticles as Well as Their Optical and Structural Properties

Figure 1(a_1,a_2) show SEM images of WS_2 flakes. According to Figure 1(a_2), WS_2 flakes have a diameter between 260 and 460 nm and a width between 30 and 78 nm. Figure 1 shows that the thinnest WS_2 sheets were obtained when the grinding of WS_2 flakes took place at 900 rpm for one hour.

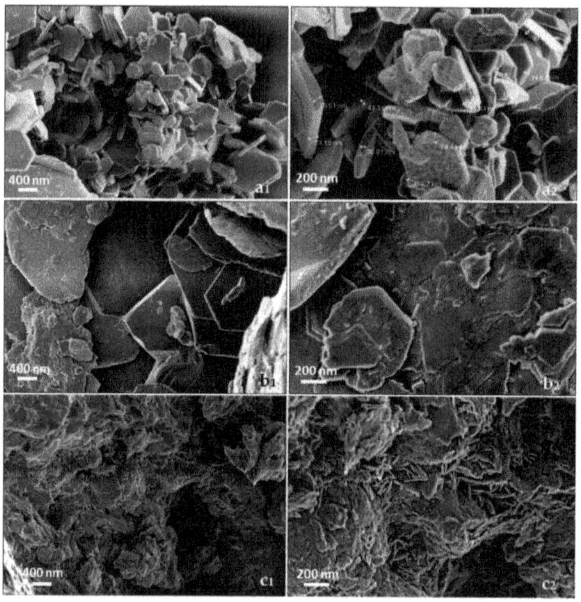

Figure 1. SEM images of WS_2 flakes ($\mathbf{a_1}$,$\mathbf{a_2}$); the WS_2 nanosheets resulting from grinding flakes for one hour at 600 rpm ($\mathbf{b_1}$,$\mathbf{b_2}$) and 900 rpm ($\mathbf{c_1}$,$\mathbf{c_2}$).

Figure 2 shows an SEM image of the WS_2 sheets deposited onto the Si plate after the evaporation of DMF from the solution prepared using WS_2 sheets, resulting from grinding flakes for one hour at 900 rpm, and then its dispersion by ultrasonication in DMF.

Figure 2. SEM image of WS_2 sheets.

Figure 2 shows WS_2 sheets of various sizes, some of which are rolled.

Information confirming the formation of WS_2 sheets is presented by XRD in Figure 3. Figure 3a shows the XRD diagram of WS_2 nanoparticles, where diffraction maxima are observed at 2θ angles of cca. 14.3°, 28.8°, 32.8°, 33.5°, 39.6°, 43.9°, 49.7°, 58.4°, 59.8°, and 60.3°, belonging to the crystalline planes (002), (004), (100), (101), (103), (006), (105), (110), (008), and (112) [PDF 00-008-0237]. An important detail highlighted in Figure 3a is that the peak intensity of the crystalline plane (002) is 3.58×10^5 counts. According to Figure 3c, it is observed in the XRD diagram that mechanical exfoliation induces a significant decrease in the intensity of the maximum corresponding to the crystal plane (002) to cca. 6.53×10^3 counts, accompanied by a diminution in the intensity of the maxima corresponding to the crystalline planes (004), (006), and (008).

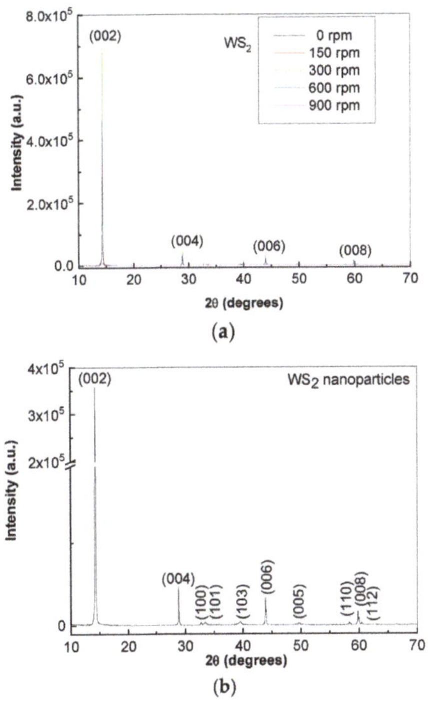

Figure 3. *Cont.*

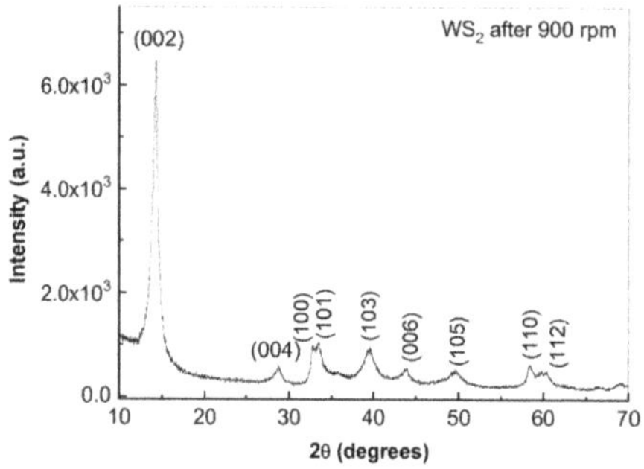

(c)

Figure 3. (**a**) XRD diagrams of WS_2 nanoparticles before and after grinding at 150, 300, 600, and 900 rpm. (**b**,**c**) X-ray diagrams of WS_2 nanoparticles before and after 900 rpm.

These variations are typical for the exfoliation of WS_2 particles, and to support this claim, an example may be the case of exfoliation of WS_2 particles mixed with lithium halides [24] or organic solvents such as ethanol, acetone, methanol, isopropyl alcohol, dimethyl formamide, and N-methyl pyrrolidone [24].

The Raman spectra of WS_2 NPs before and after their exfoliation are shown in Figure 4. According to Figure 4a, two intense Raman lines of WS_2 NPs with maxima of 357 and 422 cm^{-1} are observed. These are attributed to the $E_{2g}{}^1$ and A_{1g} vibrational modes, respectively, that must be understood as related to in- and out-of-plane vibrations of W and S atoms [25].

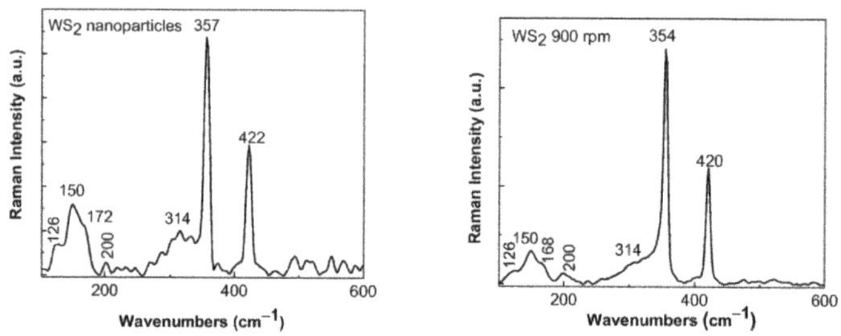

(a) (b)

Figure 4. The Raman spectra of WS_2 NPs before (**a**) and after grinding exfoliation at 900 rpm (**b**).

Raman lines at 126, 168–172, and 314 cm^{-1} were assigned to the vibrational modes A_g, B_u, and A_g of 1T'-WS_2 [26]. The Raman line at 200 cm^{-1} belongs to the vibrational mode E"(M) - ZA(M) [27]. The Raman line that peaks at 150 cm^{-1} is situated no longer in terms of wavelength than that reported at 148 cm^{-1}, which is characteristic of monolayer WS_2 [28]. In Figure 4b, we can observe two Raman lines of WS_2 sheets associated with vibrational modes $E_{2g}{}^1$ and A_{1g}, which show a shift at cca. 354 and 420 cm^{-1}, which has been attributed to the exfoliation of WS_2 microparticles [24]. In the context of the exfoliation process of WS_2 NPs, an important parameter is the distance between the two Raman lines associated with vibrational modes $E_{2g}{}^1$ and A_{1g} and the ratio between their intensities (I_{Eg}/I_{Ag}). Careful analysis of Figure 4a,b reveals that (a) the distance between the Raman lines associated with vibrational modes $E_{2g}{}^1$ and A_{1g} is equal to 65 and 66 cm^{-1}, respectively, and (b) the I_{Eg}/I_{Ag} ratio is equal to 1.8 and 1.1, respectively.

The UV-VIS spectrum of WS_2 sheets mechanically exfoliated at 900 rpm, which were further dispersed into DMF by ultrasonication for 15 min (Figure 5), shows a band at

632 nm, accompanied by another band presenting one shoulder at 530 nm and another one at 454 nm, which were attributed to the exciton band, the transition of the indirect exciton, and the optical transition between the valence and conduction bands specific to the semiconductor character of the direct forbidden band [29].

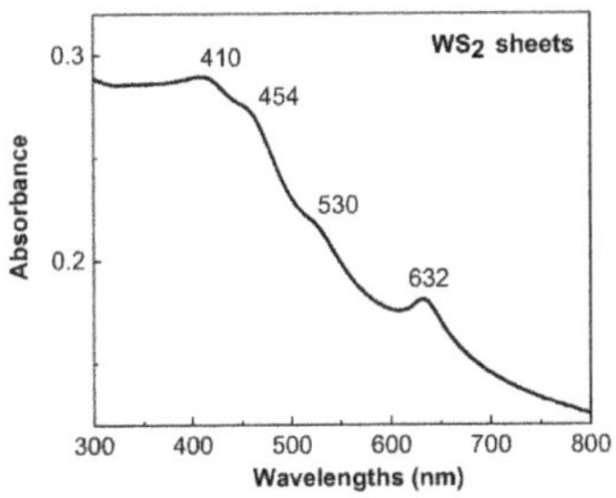

Figure 5. UV-VIS spectrum of WS_2 sheets resulting from mechanical exfoliation at 900 rpm, followed by ultrasonication in DMF.

3.2. POT/TMD Composites Obtained by SSI and ICP

According to Figure 6a, the SSI of the constituents of the POT-EB/WS_2 composite involves the exfoliation of WS_2 nanoparticles, which was evidenced in the XRD by decreasing the peak intensity of the crystal planes (002) from 6.35×10^5 counts/sec (black curve in Figure 6a) to 3.58×10^5 counts/sec (blue curve in Figure 6a) to 1.11×10^5 counts/sec (red curve in Figure 6a).

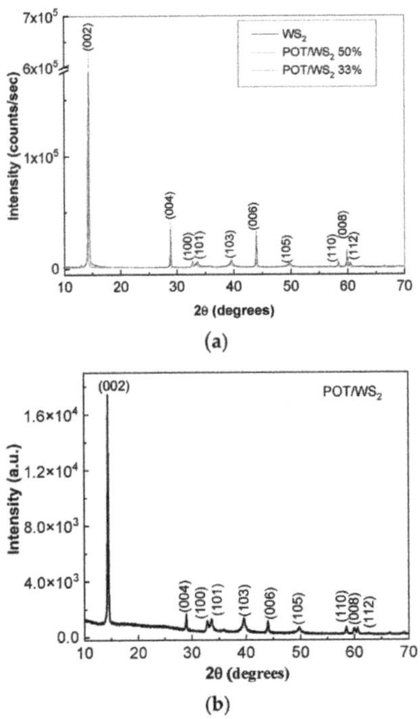

Figure 6. (**a**) XRD diagrams of WS_2 NPs (black curve) and the POT-EB/WS_2 composites obtained by SSI, when the WS_2 concentration in the composite mass is 50 wt.% (blue curve) and 33 wt.% (red curve). (**b**) XRD diagram of POT/WS_2 composite obtained by ICP of OT, when the WS_2 concentration in the composite mass is 20 wt.%.

This exfoliating process of WS_2 nanoparticles is also observed in the ICP of OT assisted by WS_2 NPs. In this context, the diffraction peak intensity of the crystalline plane (002) of WS_2 varies from 6.35×10^5 counts/sec (black curve in Figure 6a) to 1.74×10^4 counts/sec (Figure 6b) for the sample with a WS_2 concentration equal to 20 wt.% in POT-EB/WS_2 composite mass.

The analysis of Figure 7a,b indicates that Raman spectra of the POT-EB and POT/WS_2 composite show Raman lines that belong to (a) WS_2, these being situated at 359 and 421 cm^{-1}, and (b) POT, these being localized at 1119, 1176, 1259, 1329–1363, 1499, and 1620 cm^{-1} and attributed to the vibrational modes of deformation of the benzene ring (B) + C-H bond in the CH_3 group, the C-H bond in B, the stretching C-N bond + stretching C-C + C-H bonds in B, the stretching C-C bond in B, the stretching C-C bond in the quinoid ring (Q) + C-H bond in B, and the stretching C-N bond and stretching C-C + C-H bond in B [30,31]. The Raman lines at 1119, 1259, 1363, and 1620 cm^{-1} are closely situated to those reported in the POT-EB (Figure 7a) that peaked at 1117, 1254, 1367, and 1618 cm^{-1}. In contrast with the Raman line of POT-EB at 1163 cm^{-1} (Figure 7a), the Raman line of the POT/WS_2 composite is situated at 1176 cm^{-1} (Figure 7b) as a consequence of the generation of POT–emeraldine salt (ES).

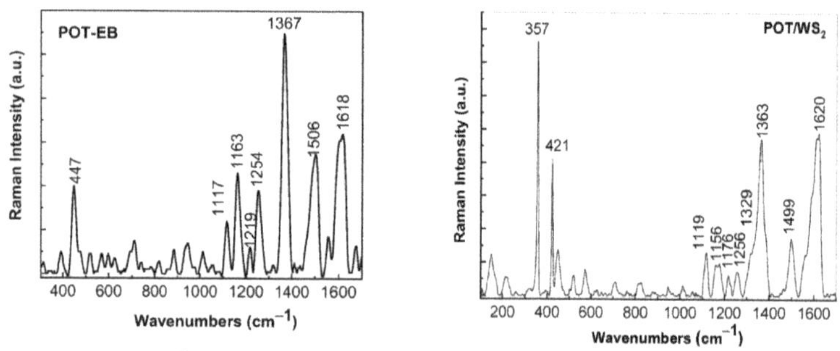

Figure 7. The Raman spectra of POT-EB (**a**) and the POT/WS_2 composite (**b**), with a TMD concentration of 20 wt.%, synthesized by ICP of OT.

The difference between vibrational modes $E_{2g}{}^1$ and A_{1g} of the POT/WS_2 composite is less than that obtained by the mechanical exfoliation of WS_2 NPs followed by ultrasonication in DMF. This fact indicates that the protocol for the preparation of WS_2 sheets, shown in the previous section, is most suitable for preparing composites based on POT and WS_2 sheets.

Information on potential interactions between constituents of the POT-EB/WS_2 composite, synthesized by ICP of OT in the presence of WS_2 NPs, is shown in Figure 8. Figure 8a shows the IR spectrum of POT-EB, which is characterized by IR bands located at 444, 816, 879, 1004, 1111, 1151, 1242, 1306, 1497, and 1601 cm^{-1}, attributed to vibrational modes of B deformation, out-of-plane deformation of the substituted B in the even position, deformation of triple-substituted B, deformation of B, deformation of the B + C-H bond in B, the C-H bond in-plane deformation of Q, stretching of the C-N bond in the B-N=Q structure, stretching of the C-N bond in Q, stretching of C-C + C-H + N-H bonds, and the structure of N=Q=N [32]. According to Figure 8b, the following vibrational changes are induced in POT by WS_2: (a) the IR bands are down-shifted from 1111, 1242, 1497, and 1601 cm^{-1} (Figure 8a) to 1105, 1213, 1490, and 1587 cm^{-1} (Figure 8b), accompanied by an up-shift in the IR band from 1306 cm^{-1} (Figure 8a) to 1313 cm^{-1} (Figure 8b); (b) a change in the absorbances ratios of the IR bands situated at (b_1) 1242–1213 cm^{-1} and 1497 cm^{-1} ($I_{1242\text{-}1213}/I_{1497}$) as well as ($b_1$) 1497 cm^{-1} and 1587–1601 cm^{-1} ($I_{1497}/I_{1587\text{-}1601}$) from 1.16 and 1.76 to 3.73 and 1, respectively. The IR bands that peaked at 3381 and 3730 cm^{-1} (Figure 8a) were assigned to the N-H symmetric stretching [33] and free OH group [34] vibrational modes. In contrast to Figure 8a, the FTIR spectrum of the POT/WS_2 composite highlights the two IR bands situated at 3371 and 3729 cm^{-1}, the last one being accompanied by a new IR band at 3628 cm^{-1} assigned to the stretching OH vibrational mode [35]. An explanation for the OH vibrational mode must consider the generation of new hydrogen bonds between POT/WS_2 and H_2O molecules, according to Scheme 1.

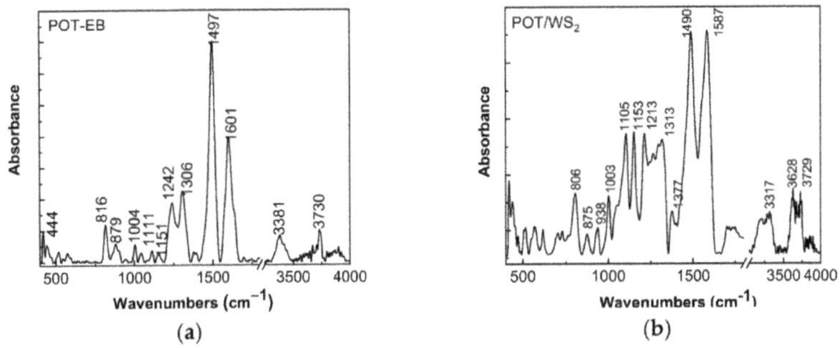

Figure 8. FTIR spectra of POT-EB (**a**) and POT/WS_2 composite (**b**), obtained by ICP of OT, with a WS_2 concentration in composite mass equal to 20 wt.%.

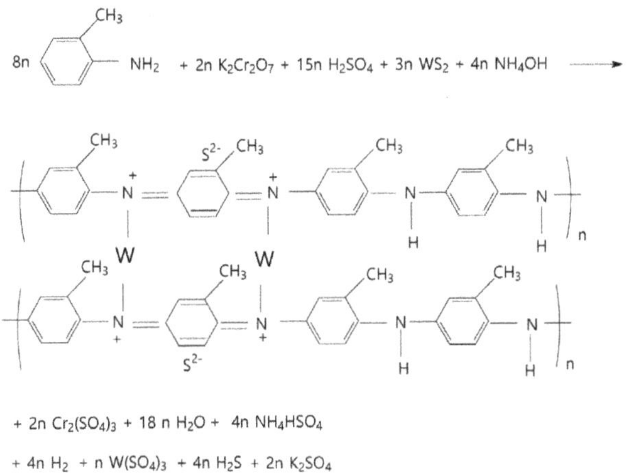

Scheme 1. Reaction of ICP of OT in the presence of WS_2 NPs.

In the case of the chemically synthesized POT/WS_2 composite, the following is observed by deconvolution: (a) In the XPS C1s spectrum, there are two bands of high intensity, with the maximum at 284.5 and 286.3 eV, which are accompanied by another five bands of lower intensity located at 287.3, 285.3, 284.8, 288.9, and 291 eV belonging to POT-EB (Figure 9a). The band from 291 eV corresponds to the π-π^* transition of the aromatic ring of POT [36]. The bands that peak at 284.5, 284.8, 285.3 and 286.3 eV are assigned to bonds C=C (C hybridized sp^2), C-C/C-H, C-N/C=N, and C-O-H [37]. The bands that peak at 287.6 and 288.9 eV are assigned to the C=O and -COOR bonds existing as contaminants in the XPS installation. (b) In the XPS N1s spectrum, there are three bands located at 399.2, 399.9, and 401.4 eV (Figure 9b) that are assigned to bonds C-N=C [37], C-NH-C [37] and C-N^+ [38] of POT. (c) In the XPS W4f spectrum, three bands at 32.8, 34.9, and 35.9 eV (Figure 9c) are highlighted. The intense bands from 32.8 and 34.9 eV correspond to the doublet W $4f^{7/2}$ and W $4f^{5/2}$, which are separated by cca. 2.1 eV, the ratio of the areas of the two bands being 4:3 [31]. The low-intensity band from 36.7 eV is associated with the $4f^{7/2}$ band of the tungsten sulfate adsorbed from the manufacturing process. (d) In the XPS S2p spectrum, four bands with maxima at 162.4, 163.6, 168.4, and 169.6 eV (Figure 9d) are remarked. The first two bands correspond to the $2p^{3/2}$ and $2p^{1/2}$ states, which are separated by cca. 1.18 eV, the ratio of the areas of the two bands being 2:1. The doublet from 162 to 163 eV represents 78% S from WS_2, while the doublet from 169 to 170 eV corresponds to 22% S from $W(SO_4)_3$. This fact indicates the presence of $W(SO_4)_3$ as a reaction byproduct in the WS_2 sample [31].

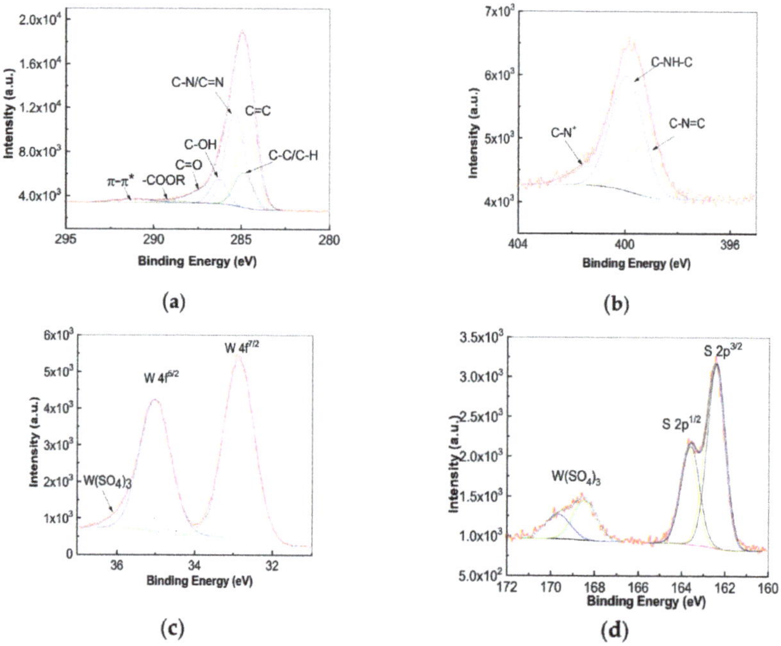

Figure 9. XPS C1s (**a**), W4f (**c**), S2p (**d**), and N1s (**b**) spectra of the POT/WS_2 composite with a WS_2 concentration in composite mass equal to 20 wt.%, synthesized by ICP of OT.

Considering the variations presented above, the total reaction of ICP of OT is shown in Scheme 1.

In our opinion, Scheme 1 explains the existence of the C-N=C, C-NH-C, and $C-N^+$ bonds in the POT-EB/WS_2 composites resulting from the ICP of OT in the presence of WS_2 NPs.

3.3. Electrochemical Synthesis of POT/WS_2 Composite and Its Optical Properties

Figure 10 shows the first 50 cyclic voltammograms recorded in the potential range (−200; +900) mV vs. Ag/AgCl, using a semi-aqueous solution of 0.1 M OT in 0.5 M H_2SO_4 (the volumetric ratio of DMF/H_2O was 1:1) and 0.5 mg/mL WS_2 sheets exfoliated according to the protocol reported in this work.

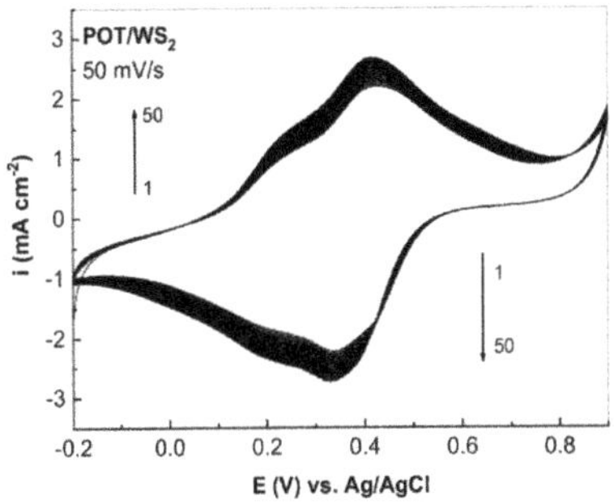

Figure 10. The first 50 cyclic voltammograms of ECP of OT in a solution of H_2SO_4 and WS_2 sheets, recorded at a potential scan rate of 50 mV/s.

The main changes in the cyclic voltammogram profile whilst increasing the cycle number recorded on the working electrode surface (Figure 10) can be quantified in the case of the following: (a) the first cyclic voltammogram, when the potential scanning takes place from (a_1) −200 to +900 mV vs. Ag/AgCl at the oxidation maximum with a potential of +418 mV, which is accompanied by another one with a potential of +246 mV, with the current densities of the two oxidation maxima at +418 and +246 mV equal to 2.12 mA/cm^2 and 1.06 mA/cm^2, respectively, and (a_2) +900 mV to −200 mV vs. Ag/AgCl, by a reduction maximum characterized by a potential equal to +334 mV, which is accompanied by another one with a potential of +200 mV; the two reduction maxima from +334 and +200 mV are characterized by current densities equal to 2.25 and 1.83 mA/cm^2, respectively; (b) the 50th cyclic voltammogram, when the potential scanning occurs from (b_1) −200 to +900 mV vs. Ag/AgCl, by an oxidation maximum with potential of +414 mV that is accompanied by another maximum with potential of +248 mV (the current densities of the two oxidation maxima at +414 and +248 mV are equal to 2.67 mA/cm^2 and 1.48 mA/cm^2), and (b_2) +900 mV to −200 mV vs. Ag/AgCl, where a reduction maximum characterized by a potential equal to +328 mV is observed, which is accompanied by another one with a potential of +198 mV; the two reduction maxima from +328 and +198 mV are characterized by current densities equal to 2.73 and 2.33 mA/cm^2, respectively. Regardless of the number of cyclic voltammograms recorded at the electrolyte/electrode interface, it is observed that two oxidation–reduction processes take place, which are evidenced by oxidation and reduction maxima with anodic and cathodic potentials located at (i) +418 mV and +334 mV as well as +246 mV and +200 mV in the first cyclic voltammogram, and (ii) +414 mV and +328 mV as well as +248 mV and +198 mV, respectively, in the 50th cyclic voltammogram. The above variations are accompanied by a progressive increase in the density of anodic and cathodic currents as the cyclic voltammogram number increases. The two oxidation–reduction processes are characterized by a potential difference of $\Delta E = E_a - E_c$, (where E_a and E_c are anodic and cathodic peak potential), equal to (i) 84 mV and 46 mV for the first cyclic voltammogram and (ii) 88 mV and 50 mV in the case of the 50th cyclic voltammogram. The ratio between the peak current densities of the anode to cathode (i_a/i_c) is equal to (i) 0.94 and 0.58 for the first cyclic voltammogram and (ii) 0.97 and 0.63 in the case of the 50th cyclic voltammogram. Analyzing the above results, it is noticed that there are oxidation–reduction processes for which ΔE differs by 56.5 mV/n, where n is the number of electrons involved in the oxidation–reduction process, and/or the ratio of anodic to cathodic current densities (i_a/i_c) differs from one, facts which allows us to conclude that the ECP of OT onto a Pt electrode in the presence of WS_2 sheets is an irreversible process. After 50 cyclic voltammograms, a dark green film is generated on the Pt electrode surface as a consequence of the generation of the POT-ES/WS_2 composite. To confirm the formation of this composite, Figure 11 shows the Raman spectrum of the POT-ES/WS_2 composite synthesized electrochemically on the Pt electrode surface, after recording the 50th cyclic voltammogram.

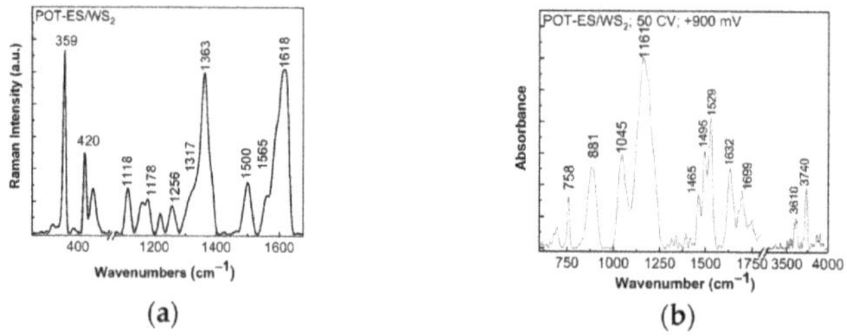

Figure 11. Raman spectrum (a) and FTIR spectrum (b) of the POT-ES/WS_2 composite synthesized by ECP on the Pt electrode surface after recording 50 cyclic voltammograms.

According to Figure 11a, the Raman spectrum of the POT-ES/WS_2 composite is characterized by Raman lines located at (a) 359 and 420 cm^{-1}, coming from WS_2 sheets, being related to vibrational modes $E_{2g}{}^1$ and A_{1g} [24], and (b) 1118, 1178, 1256, 1317, 1363, 1500, and 1618 cm^{-1}, belonging to POT-ES [31,32]. The FTIR spectrum of the electrochemically synthesized POT-ES/WS_2 composite after recording 50 cyclic voltammograms in the range (−200; +900) mV (Figure 11b) is characterized by IR bands situated at 758, 881, 1045, 1161, 1465 −1495, 1529, and 1632 cm^{-1}. According to Ref. [39], the FTIR spectrum of POT-ES is characterized by IR bands localized at 754, 889, 1045, 1145–1185, 1481–1489, and 1610 cm^{-1} that are attributed to the following vibrational modes: B deformation, Q deformation, out-of-plane deformation of triple-substituted B, (B)-NH^+=(Q), N-B-N, and N=Q. The differences concerning the wavenumbers of IR bands associated with the vibrational modes of B–NH^+=Q and N=Q can be explained by steric hindrance effects due to the covalent coordination bonds established between W atoms in WS_2 and amine groups of POT-ES. According to Figure 11b, the IR bands assigned to the stretching OH and free OH group vibrational modes peaked at 3610 and 3740 cm^{-1}. A careful analysis of Figures 8b and 11b highlights that the ratio between the absorbance of the OH stretching vibrational mode and free OH groups is equal to 1 and 0.5, respectively. The low absorbance of the IR band assigned to OH bond vibration indicates a small proportion of hydrogen bonds and a low degree of connectivity of molecules through hydrogen bonds for the samples resulting from ECP. More information about the bonds existing in the electrochemically synthesized POT-ES/WS_2 composite is presented in the following by XPS. Figure 12 illustrates the XPS spectra of the electrochemically synthesized POT/WS_2 composite, for which the deconvolution of (a) the XPS C1s spectrum highlights a band of high intensity at 284.3 eV and another three bands of lower intensity situated at 285.1, 285.7, and 286.8 eV belonging to POT-ES (Figure 12a). The bands from 284.3, 285.1, 285.7, and 286.8 eV have been assigned to the bonds C=C (C hybridized sp^2), C-C/C-H, C-N/C=N, and C-O-H [34]. (b) The XPS N1s spectrum highlights three bands located at 399.2, 399.9, and 401.4 eV (Figure 12b), which have been assigned to bonds C-N=C [37], C-NH-C [37], and C-N^+ [38]. (c) The XPS W4f spectrum highlights two bands at 32.6 and 34.7 eV (Figure 12c), corresponding to the doublet W $4f^{7/2}$ and W $4f^{5/2}$, the two bands being separated with cca. 2.1 eV, the ratio of areas being 4:3. (d) The XPS S2p spectrum leads to four bands with maxima at 162.1, 163.3, 168, and 169.2 eV (Figure 12d).

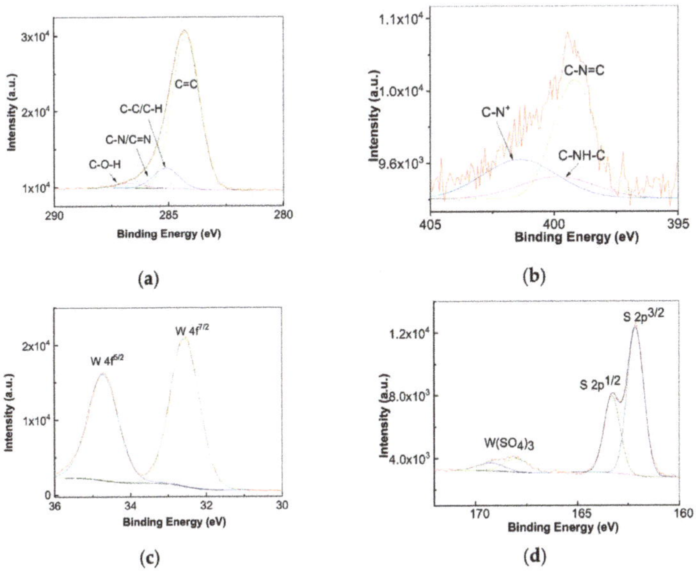

Figure 12. XPS spectra C1s (**a**), N1s (**b**), W4f (**c**), and S2p (**d**) of the POT-ES/WS_2 composite synthesized by the electrochemical method.

The first two bands correspond to the $2p^{3/2}$ and $2p^{1/2}$ states, which are separated by cca. 1.18 eV, the ratio of the areas of the two bands being 2:1. The doublet from 162 to 163 eV represents S from WS_2, while the doublet from 168 to 169 eV corresponds to S derived from $W(SO_4)_3$. This fact indicates that the polymerization reaction leads to a byproduct of the type $W(SO_4)_3$. All these changes clearly prove the generation of POT-ES/WS_2 composites by the ECP method. The chemical mechanism that can explain the electrochemical generation of the POT/WS_2 composite on the Pt electrode surface takes into account (i) the oxidation reaction of OT when a cation radical of OT is produced, which is unstable and consequently reacts with itself, forming the dimer of OT; (ii) further, macromolecular chain growth involves a propagation reaction when the dimer is successively transformed into a trimer, oligomer, and polymer, with POT thus obtained being in a leucoemeraldine base (LB) state, which reacts with H_2SO_4 and leads to POT in leucoemeraldine salt (LS). According to Scheme 2, the oxidation reaction of POT-LS leads to the generation of POT- ES, which chemically interacts with WS_2, resulting in the POT-ES/WS_2 composite.

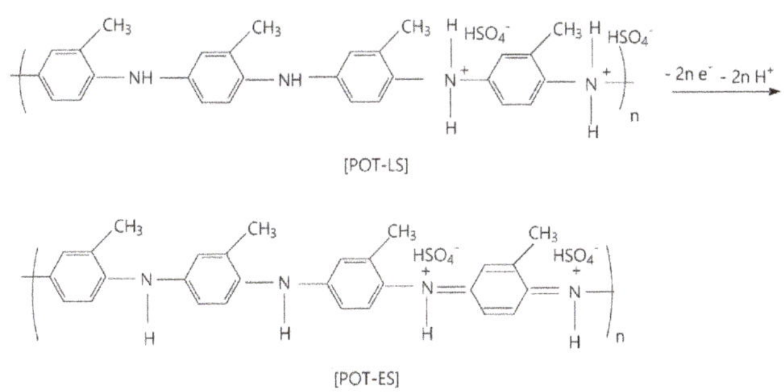

(1)

Scheme 2. *Cont.*

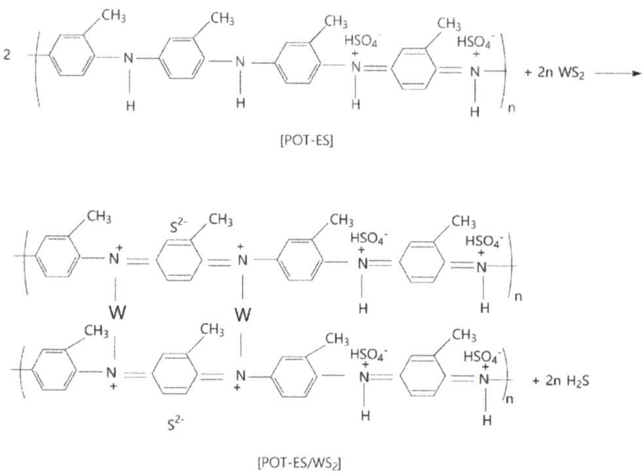

(2)

Scheme 2. The mechanism of EC synthesis of the POT/WS_2 composite: (**1**) oxidation reaction of POT-LS with the generation of POT-ES; (**2**) chemical interaction of POT-ES with WS_2.

3.4. *Electrochemical Properties of the POT/WS_2 Composites as Promising EAM in Supercapacitor Cells*

Figure 13 shows cyclic voltammograms of the POT-EB/WS_2 and POT-ES/WS_2 composites, prepared by SSI and ICP, respectively, when a Nafion membrane that interacted with H_2SO_4 1M was used as an electrolyte. A careful analysis of Figure 13a,b highlights higher values of the voltametric output currents of the POT-EB/WS_2 and POT-ES/WS_2 composites than of POT-EB and POT-ES. These results allow us to anticipate that the high capacitance value of the POT-EB/WS_2 and POT-ES/WS_2 composites would be obtained.

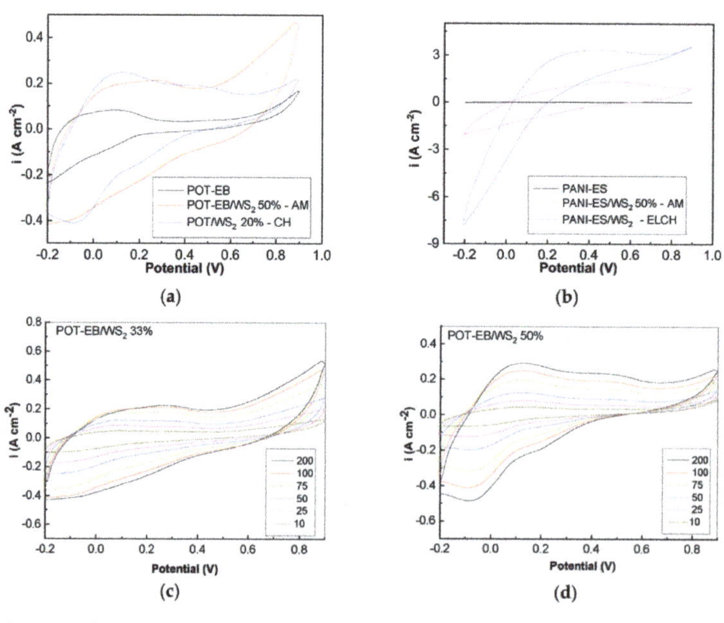

Figure 13. *Cont.*

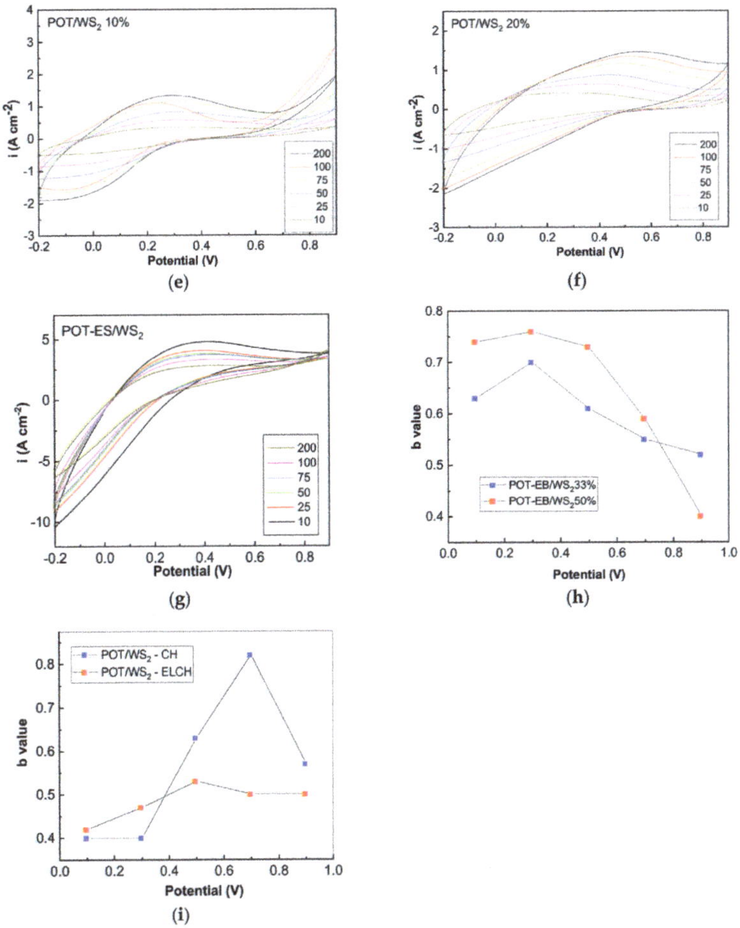

Figure 13. The 5th cyclic voltammogram of (**a**) POT-EB (black curve), the POT-EB/WS_2 composite synthesized by chemical polymerization (with a WS_2 concentration in composite mass equal to 50 wt.%, red curve), and the POT-EB/WS_2 composite resulting from SSI of the two compounds when the WS_2 concentration in the composite mass is equal to 20 wt.% (blue curve) and the scan rate is 100 mV s^{-1}; (**b**) POT-ES (black curve), the electrochemically synthesized POT-ES/WS_2 composite (blue curve), and the POT-ES/WS_2 composite obtained by SSI when the WS_2 concentration in the composite mass is 20 wt.% (red curve), when the scan rate is 100 mV s^{-1}; the POT-EB/WS_2 composites with a WS_2 concentration equal to 33 wt.% (**c**) and 50 wt.% (**d**); the POT/WS_2 composites, synthesized by ICP, when the WS_2 concentration in the composite mass is equal to 10 wt.% (**e**) and 20 wt.% (**f**); and the POT-ES/WS_2 composite synthesized by the ECP method (**g**). Black, red, green, blue, magenta, and dark yellow curves in (**c–g**) show the cyclic voltammograms recorded at 200, 100, 75, 50, 25, and 10 mV s^{-1}. (**h**) The variation in the b value with voltage in the case of the POT-EB/WS_2 composites synthesized by SSI of POT-EB with WS_2 nanoparticles, when the WS_2 concentration is equal to 33 wt.% (blue square) and 50 wt.% (red square); (**i**) the dependence of the b value on voltage for the POT/WS_2 composites with the WS_2 concentration of 20 wt.%, synthesized by chemical polymerization of OT assisted by WS_2 (blue square), and the POT-ES/WS_2 composites synthesized by the ECP method (red square).

Using the equation $i(V) = a \times v^b$, where I, v, a, and b correspond to current density (A cm^{-2}), potential scan rate (mV s^{-1}), and the two constants, we calculate the contribution of the capacitive process due to the double-layer charge at the electrolyte/electrode interface and diffusion processes [40]. According to Ref. [40], depending on the value of the b constant, (a) a redox reaction at the electrode surface and the charge/discharge processes of the electric double-layer capacitors (EDLCs) occur when b = 1; (b) a pseudocapacitive process takes place when $0.8 < b < 1$; and (c) a behavior of the type battery is invoked when $0.5 < b < 0.8$. With all these in mind, from plot log(i) to log(v), we have determined that all samples show battery behavior. The contribution of the surface capacitive and diffusion-controlled intercalation (DCI) processes in cyclic voltammograms can be assessed using the protocol published in Ref. [31]. Thus, using the equation $i(V) = k_1v + k_2v^{1/2}$, where i(V), k_1v, and $k_2v^{1/2}$ are current density for a fixed potential, the weight of the capacitive effect, and the intercalation process, respectively, and applying a linear fit in the case of the plot $i(V)/v^{1/2}$ vs. $v^{1/2}$, the values of k_1 and k_2 were determined. The contribution of the two processes is shown in Figures 14 and 15.

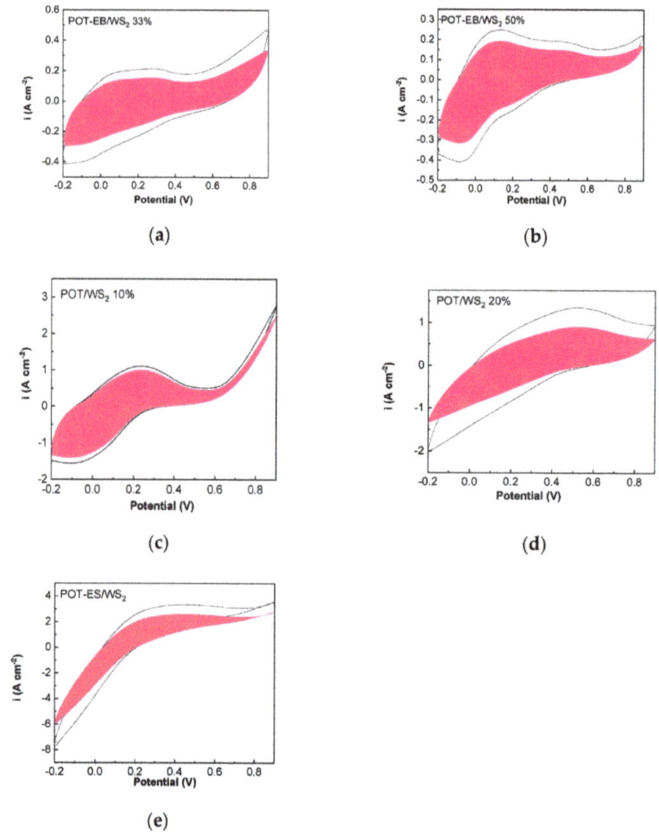

Figure 14. Deconvoluted intercalation (pink area) and capacitance (white area) processes of the POT-EB/WS_2 composites with WS_2 concentrations of 33 wt.% (**a**) and 50 wt.% (**b**); the POT/WS_2 composite with a WS_2 concentration in the composite mass equal to 10 wt.% (**c**) and 20 wt.% (**d**), prepared by the ICP method; and the POT-ES/WS_2 composite obtained by the ECP method (**e**), when the cyclic voltammograms were recorded at a scan rate of 100 mV s^{-1}.

According to Figures 14 and 15, all samples show a significant contribution given the diffusion-controlled intercalation process, which is accompanied by a capacitive process. Scheme 3 shows the reactions at the electrolyte/electrode interface when the electrodes contain the composites prepared by ICP and ECP.

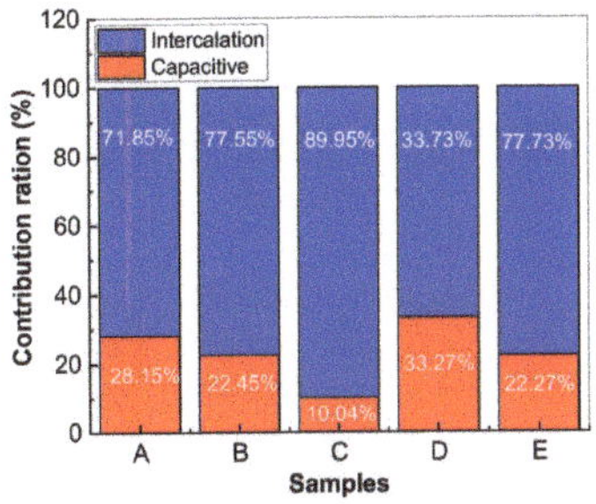

Figure 15. Contribution ratio between capacitive and diffusion processes of POT-EB/WS_2 composites with WS_2 concentrations of 33 wt.% (sample A) and 50 wt.% (sample B), the POT/WS_2 composites synthesized by the ICP method (WS_2 concentration in composite mass is 10 wt.% (sample C) and 20 wt.% (sample D)), and the POT-ES/WS_2 composite synthesized by the ECP method (sample E).

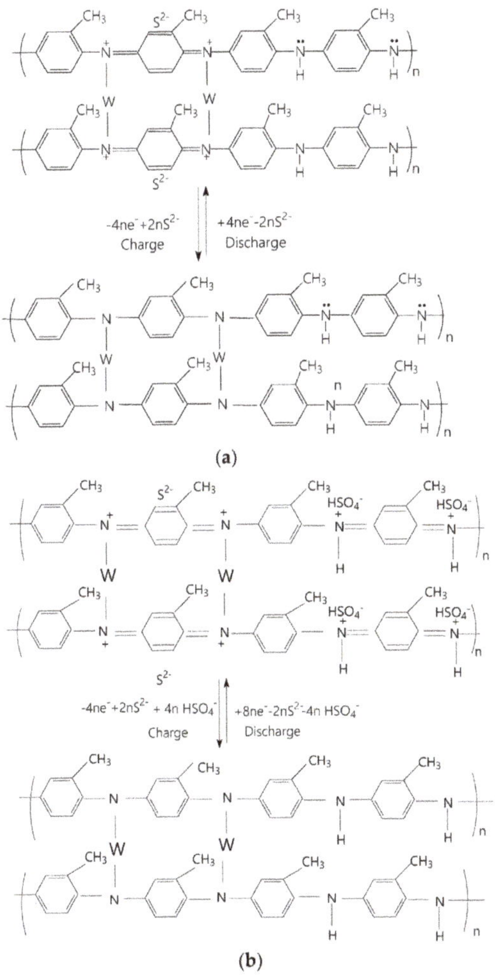

Scheme 3. The reactions at the electrolyte/electrode interface of the electrodes containing the POT/WS_2 composites synthesized by ICP (**a**) and the POT-ES/WS_2 composite synthesized by ECP (**b**).

Figure 16 shows the variation in capacitance vs. scan rate of the electrodes prepared in this study.

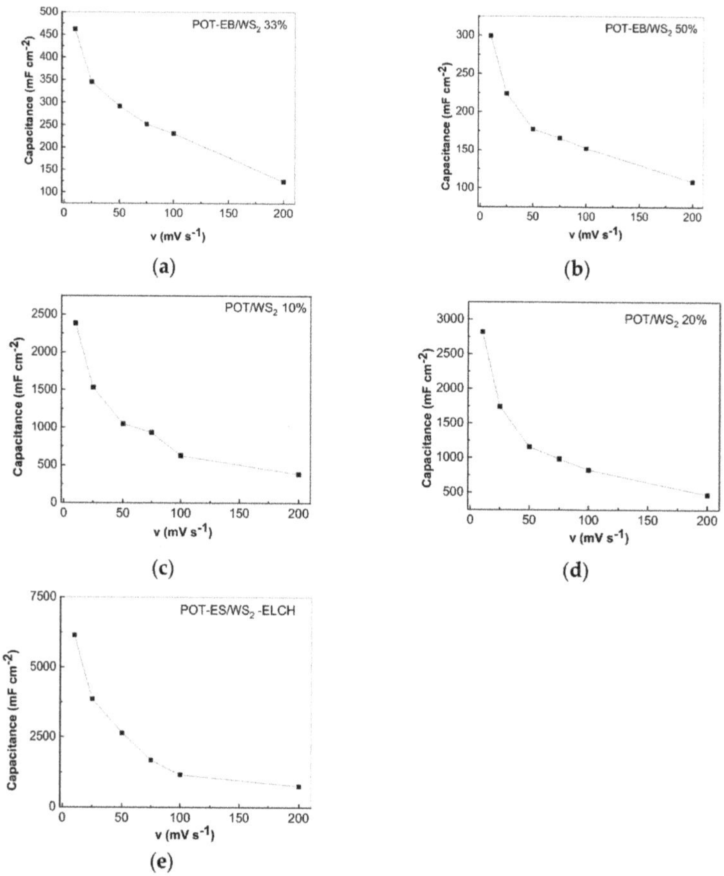

Figure 16. The dependence of capacitance vs. potential scan rate of electrodes based on the POT-EB/WS_2 composites resulting from SSI with WS_2 concentrations equal to 33 wt.% (**a**) and 50 wt.% (**b**); the POT-EB/WS_2 composites synthesized by ICP with WS_2 concentrations equal to 10 wt.% (**c**) and 20 wt.% (**d**); and the POT-ES/WS_2 composite synthesized by the ECP method (**e**).

At a scan rate of 10 mV s^{-1}, the capacitance of the SCs with composites prepared by ICP and ECP was equal to 2.82 and 12.5 F cm^{-2}. A decrease in capacitance when increasing the potential scan rate from 10 to 200 mV/s takes place for (i) the POT-EB/WS_2 composites resulting from the SSI method, which have WS_2 concentrations equal to 33 wt.% and 50 wt.%, from 462.18 and 299.68 mF cm^{-2} to 123.52 and 108.16 mF cm^{-2}; (ii) the POT/WS_2 composites, synthesized by the ICP method, when the TMD concentration in the composite mass is 10 wt.% and 20 wt.%, from 2386.2 and 2820.71 mF cm^{-2} to 382.69 and 465.76 mF cm^{-2}; and (iii) the POT-ES/WS_2 composite synthesized by the ECP method from 6136.75 to 751.6 mF cm^{-2}. Figure 17 shows the variation in the capacitance of the SCs with the cyclic voltammogram numbers, in the case of electrodes based on (i) the POT-EB/WS_2 composite resulting from the interaction of the two constituents when the WS_2 concentration is 50 wt.%; (ii) the POT/WS_2 composite, synthesized by the chemical polymerization of OT, when the WS_2 concentration is 20 wt.%; and (iii) the electrochemical synthesized POT-ES/WS_2 composite.

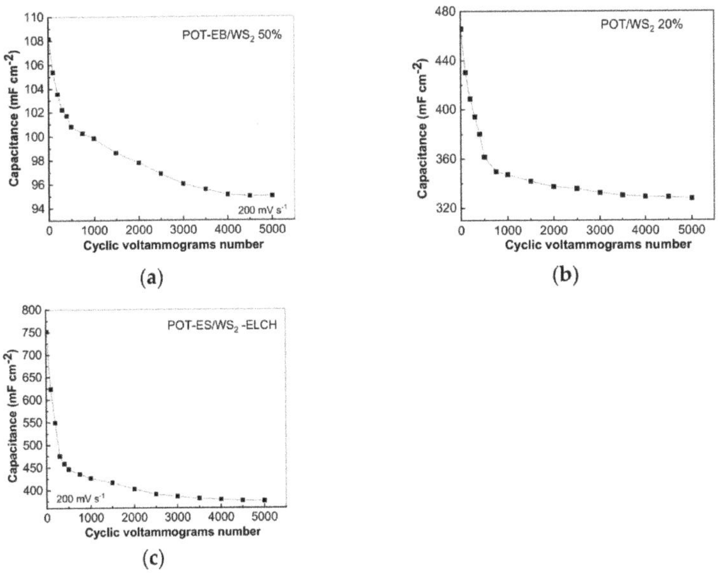

Figure 17. The variation in capacitance vs. number of cyclic voltammograms recorded at 200 mV s^{-1} of electrodes based on (**a**) the POT-EB/WS_2 composite prepared by the SSI method, when the WS_2 concentration in composite mass was 50 wt.%; (**b**) the POT/WS_2 composite, synthesized by ICP (the WS_2 concentration in the composite mass was 20 wt.%); and (**c**) the POT-ES/WS_2 composite, synthesized by the ECP method.

According to Figure 17, an important decrease in the capacitance takes place in the first 750 cyclic voltammograms for the POT/WS_2 composites prepared by the three synthesis methods. This behavior should be explained by considering various expansions and contractions of the volume of the electrode as a consequence of the expulsion and insertion of ions during discharge and charge processes, as shown in Schemes 3 and 4.

Scheme 3 highlights, based on Equations (1) and (2), that during the discharge half-cycle of the first cyclic voltammogram, the HSO_4^- and S^{2-} ions leave the POT-ES/WS_2 composites, with the macromolecular compounds having an LB structure. In the charge half-cycle of the first cyclic voltammogram, both the HSO_4^- and S^{2-} ions will inserted into the composite matrix. Scheme 4 shows the chemical reaction that can occur after the discharge half-cycle of the first cyclic voltammogram considering the H^+ and HSO_4^- ions of the Nafion membrane treated with H_2SO_4, which interact with the amine groups according to Equation (1) when POT-LB/WS_2 is transformed into POT-LS/WS_2. This fact indicates that, in the following cyclic voltammograms, the charge/discharge processes take place according to Equation (2). Increasing the number of cyclic voltammograms will induce more and more amine groups to react with H_2SO_4 according to Equation (1), which, in our opinion, may explain the decrease in SC capacitance in the first 750 cyclic voltammograms and the degradation of EAM. After recording 5000 cyclic voltammograms, the values of the capacitance of the supercapacitor cells with electrodes based on the POT/WS_2 composites obtained by SSI, ICP, and ECP are equal to 95.03 (Figure 17a), 327.71 (Figure 17b), and 375.33 mF/cm^2 (Figure 17c).

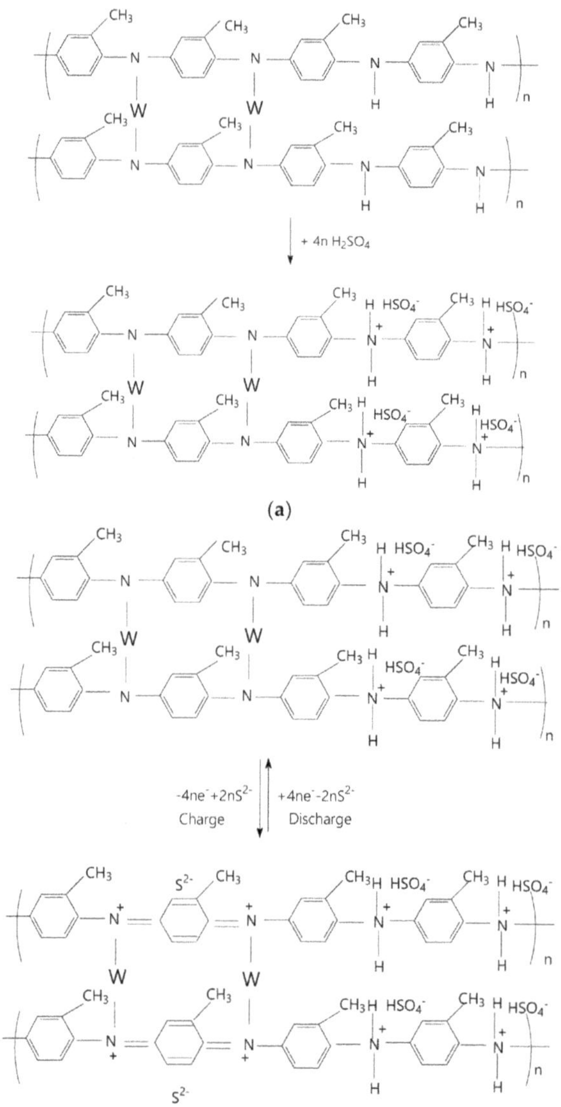

(**a**) (**b**)

Scheme 4. Reactions during discharge and charge processes in SC cell containing POT-ES/WS_2 composite as EAM: (**a**) the reaction of the POT-LB/WS_2 composite with H_2SO_4 (Equation (1)) and (**b**) charge/discharge processes of the POT-LS/WS_2 composite (Equation (2)).

The differences between the capacitance values of supercapacitors containing the PANI/WS_2 and POT/WS_2 composites, prepared in different ways and using various synthesis methods or suppliers for WS_2 particles, are summarized in Table 1 [41–46]. According to Table 1, the capacitance value depends on the shape of the WS_2 particles, the synthesis method of the composites used as active materials in the SC cells, the type of electrolyte, and the applied current densities. Table 1 further presents the performance of SCs with PANI/MnO_2 and polybenzidine/MnO_2 composites as EAMs. The comparison between SCs with composites of the type PANI/WS_2 [42,44] and γ-MnO_2/PANI [45] as the EAM indicates that the former shows higher values of capacitance. In the case of SCs based on POT/WS_2 composites, which were prepared by ICP and ECP, higher capacitance values of 2.82 and 12.5 F cm^{-2} at a potential scan rate of 10 mV s^{-1} are reported in comparison with the SC containing polybenzidine/MnO_2 as the EAM [46]. In this stage of our investigations, to optimize the SCs developed in this work, we plan to test SC cells by preparing new POT/WS_2 composites, in which WS_2 sheets will be replaced by WS_2 rods. Considering the work reported by Zhang et al [16], which demonstrated that by doping PANI with $HClO_4$ and p-toluenesulfonic acid (p-TSA), respectively, higher energy density values can be obtained only in the former case, another parameter to be considered for improving the performance of SCs containing POT/WS_2 is co-doping with a suitable dopant agent such as $HClO_4$. An electrolyte often used in the SC field containing sulfides [47,48] or MnO_2 [44] is KOH, which will be considered by us in future studies.

Table 1. The performance of SCs based on PANI/WS_2, POT/WS_2, PANI/MnO_2, and polybenzidine/MnO_2 composites.

EAMs; Inorganic Particle Synthesis Method or Supplier	Synthesis Method of Composite	Electrolyte	Current Density or Potential Scan Rate	Capacitance	Ref.
PANI/WS_2/C; Merck	ICP	1M Na_2SO_4	10 mVs^{-1}	464 F g^{-1}	[41]
PANI nanofibers/WS_2; Sigma Aldrich	Electrodeposition by CV	1M Na_2SO_4	1Ag^{-1}; 5000 V s^{-1}	72.27 F g^{-1}; 331 mF cm^{-2}	[41]
PANI/WS_2; hydrothermal	ICP	1M H_2SO_4	1Ag^{-1}	560 F g^{-1}	[42]
PANI/WS_2; mechano-chemically assisted NaCl	Mechano-chemical reaction	1M H_2SO_4	1Ag^{-1}	580 F g^{-1}	[43]
POT/WS_2; ball milling	SSI	1M H_2SO_4	50 mV s^{-1}	177 mF cm^{-2}	This work
POT/WS_2; ball milling	ICP	1M H_2SO_4	50 mV s^{-1}	1.15 F cm^{-2}	This work
POT/WS_2; ball milling	ECP by CV	1M H_2SO_4	50 mV s^{-1}	2.64 F cm^{-2}	This work
MnO_2 Nanorods@PANI; hydrothermal	ICP	1M KOH	50 mVs^{-1}	259 F g^{-1}	[44]
γ-MnO_2/PANI; from precursor of $KMnO_4$	ICP	0.5M H_2SO_4	0.5 A g^{-1}	493 F g^{-1}	[45]
Polybenzidine/MnO_2	Chemical interaction	1M H_2SO_4	0.5 $mAcm^{-2}$	950 mF cm^{-2}	[46]
POT/WS_2; ball milling	ICP	1M H_2SO_4	10 mVs^{-1}	2.82 F cm^{-2}	This work
POT/WS_2; ball milling	ECP by CV	1M H_2SO_4	10 mVs^{-1}	12.5 F cm^{-2}	This work

4. Conclusions

The synthesis methods of the POT/WS_2 composites, their optical and structural properties, and their applications in the symmetrical SC field were reported. Using SEM, XRD, FTIR spectroscopy, Raman scattering, XPS, UV-VIS spectroscopy, and CV, the following conclusions are highlighted:

(i) Mechanical exfoliation by grinding WS_2 NPs using a ball mill followed by exfoliation in DMF allows the transformation of nanoparticles of WS_2 in sheets and induces a diminution in the intensity of the crystalline plane (002), a change in the distance between the Raman lines related to vibrational modes $E_{2g}{}^1$ and A_{1g}, and the presence of three bands situated at 632, 530, and 454 nm in the UV-VIS spectrum.

(ii) SSI of POT-EB with WS_2 induces an exfoliation of the WS_2 NPs.

(iii) ICP of OT in the presence of WS_2 sheets induces the formation of covalent coordination bonds between the imine groups of EB and the transition metal W, when a POT/WS_2 composite is created, simultaneously with the generation of a byproduct such as tungsten sulfate.

(iv) ECP of OT in the presence of WS_2 sheets involves oxidation and reduction reactions at the electrolyte/electrode interface that have an irreversible character as a result of the generation of the POT-ES/WS_2 composite when the formation of covalent coordination bonds between imine groups of POT and W atoms of TMD takes place.

(v) The electrochemical processes at the electrode/electrolyte interface in symmetrical SCs show a significant contribution given the performance of a diffusion-controlled intercalation process and an additional capacitive-type process; the best values of capacitance of the SCs with EAMs of the composites prepared by ICP and ECP were reported at 10 mV s^{-1}, these being equal to 2.82 and 12.5 F cm^{-2}; the increase in the scan rate at 200 mV s^{-1} led to values of capacitance equal to 123.5, 465.76, and 751.6 mF cm^{-2}, respectively, for the POT/WS_2 composites prepared by SSI, ICP, and ECP. After recording 5000 cyclic voltammograms at 200 mV s^{-1}, a decrease in the capacitance of the three SCs to 95.03, 327.71, and 375.33 mF cm^{-2} was reported.

Author Contributions: Conceptualization, M.B.; methodology, M.B. and A.U.; validation, M.B.; investigation, T.B., I.S., A.A., C.S.F., M.C., M.P., A.U., A.L., P.P., A.G., C.N., E.M., M.D., R.C. and M.B.; writing—original draft preparation, M.B.; writing—review and editing, M.B., A.U., A.L., C.N. and P.P.; visualization, T.B., I.S., A.A., C.S.F., M.C., M.P., A.U., A.L., P.P., A.G., C.N., E.M., M.D. and R.C.; supervision, M.B. All authors have read and agreed to the published version of the manuscript.

Funding: This work was funded by the Core Program of the National Institute of Materials Physics, granted by the Romanian Ministry of Research, Innovation and Digitization through the Project PC3-PN23080303.

Data Availability Statement: The original contributions presented in this study are included in the article. Further inquiries can be directed to the corresponding author.

Conflicts of Interest: The authors declare no conflicts of interest.

Abbreviations

The following abbreviations are used in this manuscript:

POT	Poly(o-toluidine)
OT	O-toluidine
NPs	Nanoparticles
XRD	X-ray diffraction
XPS	X-ray photoelectron spectroscopy
UV-VIS	Ultraviolet–visible
FTIR	Fourier transform infrared
LS	Leucoemeraldine salt
ES	Emeraldine salt
EB	Emeraldine base
TMDs	Transition metal dichalcogenides
PANI	Polyaniline
SEM	Scanning electron microscopy
DMF	N,N′-dimethyl formamide
PVDF	Poly(vinylidene fluoride)
DBP	Dibutyl phthalate
CV	Cyclic voltammetry
SSI	Solid-state interaction
ICP	In situ chemical polymerization
ECP	Electrochemical polymerization
SCs	Supercapacitors
SA	Surface area
EAMs	Electrode active materials
B	Benzene ring
Q	Quinoid ring

DCI	Diffusion-controlled intercalation
EC	Electrochemical
λ_{exc}	Excitation wavelength

References

1. Shahabuddin, S.; Mehmood, S.; Ahmad, I.; Sridewi, N. Synthesis and Characterization of 2D-WS2 Incorporated Polyaniline Nanocomposites as Photo Catalyst for Methylene Blue Degradation. *Nanomaterials* **2022**, *12*, 2090. [CrossRef] [PubMed]
2. Parangusan, H.; Bhadra, J.; Al-Qudah, R.A.; Elhadrami, E.C.; Al-Thani, N.J. Comparative Study on Gas-Sensing Properties of 2D (MoS_2,WS_2)/PANI Nanocosmposites-Based Sensor. *Nanomaterials* **2022**, *12*, 4423. [CrossRef] [PubMed]
3. Iqbal, M.; Saykar, N.G.; Alegaonkar, P.S.; Mahapatra, S.K. Synergistically modified WS_2@PANI binary nanocomposite-based all-solid-state symmetric supercapacitor with high energy density. *New J. Chem.* **2022**, *46*, 7043–7054. [CrossRef]
4. De Adhikari, A.; Shauloff, N.; Turkulets, Y.; Shalish, I.; Jelinek, R. Tungsten-Disulfide/Polyaniline High Frequency Supercapacitors. *Adv. Electron. Mater.* **2021**, *7*, 2100025. [CrossRef]
5. Zeng, R.; Li, Z.C.; Li, L.B.; Li, Y.Z.; Huang, J.; Xiao, Y.B.; Yuan, K.; Chen, Y.W. Covalent Connection of Polyaniline with MoS2 Nanosheets toward Ultrahigh Rate Capability Supercapacitors. *ACS Sustain. Chem. Eng.* **2019**, *7*, 11540–11549. [CrossRef]
6. Sajedi-Moghaddam, A.; Mayorga-Martinez, C.C.; Saievar-Iranizad, E.; Sofer, Z.; Pumera, M. Exfoliated transition metal dichalcogenide (MX_2; M = Mo, W; X = S, Se, Te) nanosheets and their composites with polyaniline nanofibers for electrochemical capacitors. *Appl. Mater. Today* **2019**, *16*, 280–289. [CrossRef]
7. Choi, W.; Choudhary, N.; Han, G.H.; Park, J.; Akinwande, D.; Lee, Y.H. Recent development of two-dimensional transition metal dichalcogenides and their applications. *Mater. Today* **2017**, *20*, 116–130. [CrossRef]
8. Nam, M.S.; Patil, U.; Park, B.; Sim, H.B.; Jun, S.C. A binder free synthesis of 1D PANI and 2D MoS_2 nanostructured hybrid composite electrodes by the electrophoretic deposition (EPD) method for supercapacitor application. *RSC Adv.* **2016**, *6*, 101592–101601. [CrossRef]
9. Joseph, N.; Shafi, P.M.; Bose, A.C. Recent Advances in 2D-MoS_2 and its Composite Nanostructures for Supercapacitor Electrode Application. *Energ. Fuel.* **2020**, *34*, 6558–6597. [CrossRef]
10. Zhao, X.; Ma, X.; Sun, J.; Li, D.H.; Yang, X.R. Enhanced Catalytic Activities of Surfactant-Assisted Exfoliated WS_2 Nanodots for Hydrogen Evolution. *ACS Nano* **2016**, *10*, 2159–2166. [CrossRef] [PubMed]
11. Coleman, J.N.; Lotya, M.; O'Neill, A.; Bergin, S.D.; King, P.J.; Khan, U.; Young, K.; Gaucher, A.; De, S.; Smith, J.R.; et al. Two-Dimensional Nanosheets Produced by Liquid Exfoliation of Layered Materials. *Science* **2011**, *331*, 568–571. [CrossRef] [PubMed]
12. Bhandavat, R.; David, L.; Singh, G. Synthesis of Surface-Functionalized WS_2 Nanosheets and Performance as Li-Ion Battery Anodes. *J. Phys. Chem. Lett.* **2012**, *3*, 1523–1530. [CrossRef]
13. Desai, J.A.; Adhikari, N.; Kaul, A.B. Chemical exfoliation efficacy of semiconducting WS_2 and its use in an additively manufactured heterostructure graphene–WS_2–graphene photodiode. *RSC Adv.* **2019**, *9*, 25805–25816. [CrossRef] [PubMed]
14. Leong, S.X.; Mayorga-Martinez, C.C.; Chia, X.; Luxa, J.; Sofer, Z.; Pumera, M. 2h → 1T Phase Change in Direct Synthesis of WS_2 Nanosheets via Solution-Based Electrochemical Exfoliation and Their Catalytic Properties. *ACS Appl. Mater. Interfaces* **2017**, *9*, 26350–26356. [CrossRef]
15. Huang, F.; Jian, J.K.; Wu, R. Few-layer thick WS_2 nanosheets produced by intercalation/exfoliation route. *J. Mater. Sci.* **2016**, *51*, 10160–10165. [CrossRef]
16. Zhang, M.; Nautiyal, A.; Du, H.; Wei, Z.; Zhang, X.; Wang, R. Electropolymerization of polyaniline as high-performance binder free electrodes for flexible supercapacitor. *Electrochim. Acta* **2021**, *376*, 138037. [CrossRef]
17. Dhandapani, E.; Thangarasu, S.; Ramesh, S.; Ramesh, K.; Vasudevan, R.; Duraisamy, N. Recent development and prospective of carbonaceous material, conducting polymer and their composite electrode materials for supercapacitor-A review. *J. Energy Storage* **2022**, *52*, 104937. [CrossRef]
18. Crisci, M.; Boll, F.; Merola, L.; Pflug, J.J.; Liu, Z.M.; Gallego, J.; Lamberti, F.; Gatti, T. Nanostructured 2D WS_2@PANI nanohybrids for electrochemical energy storage. *Front. Chem.* **2022**, *10*, 1000910. [CrossRef] [PubMed]
19. Abdelnasir, S.; Mungroo, M.R.; Shahabuddin, S.; Siddiqui, R.; Khan, N.A.; Ahmad, I.; Anwar, A. Polyaniline (PANI)-conjugated tungsten disulphide (WS_2) nanoparticles as potential therapeutics against brain-eating amoebae. *Appl. Microbiol. Biot.* **2022**, *106*, 3279–3291. [CrossRef]
20. Basnayaka, P.A.; Ram, M.K.; Stefanakos, L.; Kumar, A. High performance graphene-poly(o-anisidine) nanocomposite for supercapacitor applications. *Mater. Chem. Phys.* **2013**, *141*, 263–271. [CrossRef]
21. Kuwertz, R.; Kirstein, C.; Turek, T.; Kunz, U. Influence of acid pretreatment on ionic conductivity of Nafion® membranes. *J. Membrane. Sci.* **2016**, *500*, 225–235. [CrossRef]

22. Wan, M.; Yang, J. Studies on the structure and electrical properties of poly (ortho-toluidine). *Synth. Met.* **1995**, *73*, 201–204. [CrossRef]
23. Wan, M.; Li, J. Tubular poly(ortho-toluidine) synthesized by a template-free method. *Polym. Adv. Technol.* **2003**, *14*, 320–325. [CrossRef]
24. Ghorai, A.; Midya, A.; Maiti, R.; Ray, S.K. Exfoliation of WS_2 in the semiconducting phase using a group of lithium halides: A new method of Li intercalation. *Dalton Trans.* **2016**, *45*, 14979–14987. [CrossRef] [PubMed]
25. Mitioglu, A.A.; Plochocka, P.; Deligeorgis, G.; Anghel, S.; Kulyuk, L.; Maude, D.K. Second-order resonant Raman scattering in single-layer tungsten disulfide WS_2. *Phys. Rev. B* **2014**, *89*, 245442. [CrossRef]
26. Lai, Z.C.; He, Q.Y.; Ha Tran, T.H.; Repaka, D.V.M.; Zhou, D.D.; Sun, Y.; Xi, S.B.; Li, Y.X.; Chaturvedi, A.; Tan, C.L.; et al. Metastable $1T'$-phase group VIB transition metal dichalcogenide crystals. *Nat. Mater.* **2021**, *20*, 1113–1120. [CrossRef] [PubMed]
27. Agyapong, A.D.; Cooley, K.A.; Mohney, S.E. Reactivity of contact metals on monolayer WS_2. *J. Appl. Phys.* **2020**, *128*, 055306. [CrossRef]
28. Shi, W.; Lin, M.L.; Tan, Q.H.; Qiao, X.F.; Zhang, J.; Tan, P.H. Raman and photoluminescence spectra of two-dimensional nanocrystallites of monolayer WS_2 and WSe_2. *2D Mater.* **2016**, *3*, 025016. [CrossRef]
29. Pagona, G.; Bittencourt, C.; Arenal, R.; Tagmatarchis, N. Exfoliated semiconducting pure 2H-MoS_2 and 2H-WS_2 assisted by chlorosulfonic acid. *Chem. Commun.* **2015**, *51*, 12950–12953. [CrossRef] [PubMed]
30. Baibarac, M.; Cochet, M.; Łapkowski, M.; Mihut, L.; Lefrant, S.; Baltog, I. SERS spectra of polyaniline thin films deposited on rough Ag, Au and Cu. Polymer film thickness and roughness parameter dependence of SERS spectra. *Synthetic. Met.* **1998**, *96*, 63–70. [CrossRef]
31. Bowen, R.D.; Edwards, H.G.M.; Varnali, T. Influence of a methyl substituent on the Raman spectrum of but-3-enyl methyl ether. *Spectrochim. Acta A* **2012**, *93*, 26–32. [CrossRef] [PubMed]
32. Quillard, S.; Louarn, G.; Lefrant, S.; MacDiarmid, A.G. Vibrational analysis of polyaniline: A comparative study of leucoemeraldine, emeraldine, and pernigraniline bases. *Phys. Rev. B* **1994**, *50*, 12496–12508. [CrossRef] [PubMed]
33. Boyer, M.I.; Quillard, S.; Rebourt, E.; Louarn, G.; Buisson, J.P.; Monkman, A.; Lefrant, S. Vibrational analysis of polyaniline: A model compound approach. *J. Phys. Chem. B* **1998**, *102*, 7382–7392. [CrossRef]
34. Dai, F.W.; Zhuang, Q.Y.; Huang, G.; Deng, H.Z.; Zhang, X. Infrared Spectrum Characteristics and Quantification of OH Groups in Coal. *ACS Omega* **2023**, *8*, 17064–17076. [CrossRef] [PubMed]
35. Li, G.; Zhang, Y.Y.; Li, Q.M.; Wang, C.; Yu, Y.; Zhang, B.B.; Hu, H.S.; Zhang, W.Q.; Dai, D.X.; Wu, G.R.; et al. Infrared spectroscopic study of hydrogen bonding topologies in the smallest ice cube. *Nat. Commun.* **2020**, *11*, 5449. [CrossRef] [PubMed]
36. Golczak, S.; Kanciurzewska, A.; Fahlman, M.; Langer, K.; Langer, J.J. Comparative XPS surface study of polyaniline thin films. *Solid State Ion.* **2008**, *179*, 2234–2239. [CrossRef]
37. Mahat, M.M.; Mawad, D.; Nelson, G.W.; Fearn, S.; Palgrave, R.G.; Payne, D.J.; Stevens, M.M. Elucidating the deprotonation of polyaniline films by X-ray photoelectron spectroscopy. *J. Mater. Chem. C* **2015**, *3*, 7180–7186. [CrossRef]
38. Wu, M.S.; Wen, T.C.; Gopalan, A. Electrochemical copolymerization of diphenylamine and anthranilic acid with various feed ratios. *J. Electrochem. Soc.* **2001**, *148*, D65–D73. [CrossRef]
39. Trchová, M.; Stejskal, J.; Prokeš, J. Infrared spectroscopic study of solid-state protonation and oxidation of polyaniline. *Synthetic. Met.* **1999**, *101*, 840–841. [CrossRef]
40. Cho, W.H.; Cheng, I.C.; Chen, J.Z. Performance Comparison of Reduced Graphene Oxide (rGO)-polyaniline (PANI) Supercapacitors with LiCl, Li_2SO_4, and H_2SO_4 Electrolytes. *J. Electrochem. Soc.* **2023**, *170*, 010532. [CrossRef]
41. Yuvika Sood, A.; Lokhande, P.E.; Kadam, V.; Jagtap, C.; Mudila, H.; Rednam, U.; Kumar, D.; Ansar, S.; Kumar, Y.A.; Aepuru, R.; et al. Nanocomposites combining 2D WS2 and 1D polyaniline for enhanced high-performance supercapacitors. *J. Alloys Compd.* **2024**, *1005*, 1760. [CrossRef]
42. Visakh, C.; Mohan, V.; Rakhi, R.B. WS_2/Conducting polymer nanocomposite-based flexible and binder-free electrodes for high-performance supercapacitors. *Electrochim. Acta* **2024**, *498*, 144657. [CrossRef]
43. Kozarenko, D.O.; Dyadyun, V.S.; Kotenko, I.E.; Posudievsky, O.Y.; Koshechko, V.G.; Pokhodenko, V.D. Mechanochemically prepared nanocomposites based on polyaniline and molybdenum and tungsten disulfides as electrode materials for supercapacitors. *Theor. Exp. Chem.* **2018**, *54*, 85–91. [CrossRef]
44. Chevulamaddi, H.; Kalagadda, V.R. The improved electrochemical performance of MnO_2 nanorods decorated with polyaniline as efficient electrode towards high performance supercapacitors. *J. Mater. Sci. Mater. Electron.* **2024**, *35*, 2273. [CrossRef]
45. Hekmat, F.; Shahrokhian, S.; Taghavinia, N. Ultralight, flexible asymmetric supercapacitors based on manganese dioxide–polyaniline nanocomposite and reduced graphene oxide electrodes directly deposited on foldable cellulose papers. *J. Phys. Chem. C* **2018**, *122*, 27156–27168. [CrossRef]
46. Dadashi, R.; Farhadi, K.; Bahram, M. Polybenzidine-MnO_2 nanocomposite on anodized graphite sheet as a novel system for high-performance supercapacitors. *Diam. Relat. Mater.* **2024**, *143*, 110923. [CrossRef]

47. Zhang, M.; Du, H.; Wei, Z.; Zhang, X.; Wang, R. Facile electrodeposition of Mn-CoP nanosheets on Ni foam as high-rate and ultrastable electrodes for supercapacitors. *ACS Appl. Energy Mater.* **2022**, *5*, 186–195. [CrossRef]
48. Zhag, M.; Du, H.; Wei, Z.; Zhang, X.; Wang, R. Ultrafast Microwave Synthesis of Nickel-Cobalt Sulfide/Graphene Hybrid Electrodes for High-Performance Asymmetrical Supercapacitors. *ACS Appl. Energy Mater.* **2021**, *4*, 8262–8274. [CrossRef]

batteries

MDPI

Article

Preparation of Mesoporous Boron-Doped Porous Carbon Derived from Coffee Grounds via Hybrid Activation for Carbon Capture and Storage

Hyeon Hye Kim [1,2], Kay-Hyeok An [3,*] and Byung-Joo Kim [1,3,*]

1 Material Application Research Institute, Jeonju University, Jeonju 55069, Republic of Korea; hyeon_h@jj.ac.kr
2 School of Polymer Science and Engineering, Chonnam National University, Gwangju 61186, Republic of Korea
3 Department of Materials Science and Chemical Engineering, Jeonju University, Jeonju 55069, Republic of Korea
* Correspondence: khandragon@jj.ac.kr (K.-H.A.); kimbyungjoo@jj.ac.kr (B.-J.K.)

Abstract: The increasing concentration of carbon dioxide (CO_2) in the atmosphere necessitates the development of efficient carbon capture and storage (CCS) technologies. Among these, adsorption-based methods using porous carbon (PC) have attracted considerable attention due to their low energy requirements and cost-effectiveness. Biomass waste-derived porous carbon is particularly attractive as a sustainable alternative, offering environmental benefits and high-value applications with low costs. In this study, coffee grounds (CGs) were selected as a precursor due to their abundance and cost-effectiveness compared with other biomass wastes. To improve the pore characteristics of CG-derived carbon (CCG), boric acid treatment was applied during carbonization followed by steam activation to prepare boron-doped CG-derived porous carbon (B-PCG). The N_2/77K adsorption–desorption isotherms revealed a significant increase in the specific surface area and total pore volume of B-PCG from 1590 m^2/g and 0.71 cm^3/g to 2060 m^2/g and 1.01 cm^3/g, respectively, compared with PCG. Furthermore, high pressure CO_2 adsorption analysis at 298 K up to 50 bar showed an approximately 50% improvement in CO_2 adsorption capacity for B-PCG compared with PCG. These results suggest that boron doping is an effective strategy to optimize the pore structure and adsorption performance of biomass-derived porous carbon materials for CCS application.

Keywords: carbon capture and storage; porous carbon; coffee grounds; low cost; boron

1. Introduction

Globally, anthropogenic activities related to energy production and consumption are among the major contributors to global warming. In particular, the increasing concentration of greenhouse gases (GHGs) in the atmosphere has intensified the frequency and magnitude of climate change phenomena, including abnormal temperatures, rising sea levels, and extreme weather events [1–3]. Addressing global warming has thus emerged as a critical challenge for the international community and the global scientific community. In response, the European Commission has established an energy roadmap to achieve climate neutrality by 2050, emphasizing the importance of renewable energy sources in decarbonizing the energy sector [4,5]. Currently, fossil fuels such as coal, oil, and natural gas account for approximately 82% of the global energy consumption. However, their combustion significantly increases GHG emissions, making them a major cause of environmental issues. Consequently, carbon capture and storage (CCS) technology has gained

https://doi.org/10.3390/batteries11040158

increasing attention as a sustainable approach to drastically reduce the concentration of excessive GHGs, such as CO_2, emitted from large-scale energy production systems [6,7].

Considering the depletion of fossil-based resources and their significant impact on environmental pollution, the development of sustainable and environmentally friendly alternative resources is imperative. In this context, the use of waste biomass offers significant environmental and economic benefits [8,9]. Agricultural biomass waste, including crop stalks, leaves, roots, coffee cherries, fruit peels, seeds, and nutshells, is typically disposed of in landfills or incinerated, which can have detrimental environmental effects [10]. Biomass waste is rich in cellulose, hemicellulose, and lignin, making it applicable to various industrial sectors such as fuel production, polymer manufacturing, and construction materials [11–13]. However, the high-value utilization of biomass wastes remains limited, except for certain materials such as rice husks, sugarcane bagasse, and wheat straw [14]. Moreover, only a small fraction of biomass waste is used as feedstock for industrial applications and power production generation, while the majority is either left untreated or incinerated, resulting in groundwater, air, and soil contamination, increased pest infestation, and adverse effects on human health. Notably, the decomposition of unregulated biomass waste in soils can release nitrogen oxides (NO and N_2O), which have a significantly higher global warming potential than CO_2 [15,16]. Worldwide, over 2 gigatons (Gt) of agricultural waste are incinerated annually, contributing approximately 18% of global CO_2 emissions and releasing substantial amounts of particulate matter and black carbon [17].

Coffee is one of the most popular beverages in modern society, with an annual production exceeding 5 million tons [18]. Over the past three decades, the global demand for coffee has steadily increased, leading to the expansion of coffee production and exports worldwide. As a result, global coffee production has risen by more than 60%. Consequently, the quantity of biomass waste in the form of coffee grounds (CGs) has also increased, with over 10.5 million tons generated annually [19]. Typically, CGs are disposed of in landfills or by incineration, with the latter process emitting approximately 338 kg of CO_2 per ton of CGs burned. Additionally, methane gas generated during landfilling has a greenhouse effect approximately 25 times greater than that of CO_2, underscoring the urgent need for efficient CG utilization strategies [20,21].

Reuse and recycling of agricultural biomass wastes offer potential benefits for carbon neutrality but also pose complex challenges, including land use changes, soil nutrient depletion, and environmental and health concerns [22]. To address these issues, research efforts have focused on converting low-cost biomass wastes into environmentally friendly and high-value materials. In developed countries, initiatives to minimize biomass waste disposal and promote its use in energy and heat production have been increasing. For instance, in Europe, open burning of agricultural waste has been prohibited, and wheat straw has become a primary feedstock for bioenergy production. Many countries are striving to recover and utilize biomass waste as a resource while avoiding incineration and landfilling [23–25]. However, the increased utilization of biomass waste requires specific management strategies to address the challenges associated with the significant amount of ash produced during combustion and pyrolysis processes [26].

This study aims to address the growing problem of biomass waste and maximize the potential of carbon-based materials by recycling CGs into high-value porous carbon materials. CGs are particularly advantageous as a precursor for porous carbon synthesis due to their abundant supply and low raw material cost [27–29]. Additionally, during the coffee extraction process, intrinsic molecules such as caffeine, tannins, and polyphenols are leached out, naturally creating multiple voids. These internal voids can be transformed into a hierarchical porous structure through thermal treatment, and surface modifications can introduce various functional groups, enhancing their potential for energy storage applications [30,31].

Furthermore, CG-derived porous carbon materials exhibit excellent adsorption properties, making them suitable for water purification and gas adsorption applications. Previous studies have reported that coffee by-products can be converted into high-performance adsorbents via carbonization and chemical modification processes, demonstrating their effectiveness in pollutant removal [32]. The development of CG-derived porous carbon materials holds significant promise for various applications, including electrode materials for energy storage devices and adsorbents for environmental remediation [33–35]. In particular, this study focuses on optimizing the structural properties of CG-based porous carbon materials through activation processes and enhancing CO_2 selectivity by heteroatom doping as a surface modification strategy. In general, porous carbon materials possess well-developed pore structures, large surface areas, excellent electrical conductivity, and remarkable chemical stability. These properties make them promising candidates for gas adsorption applications, and their performance can be significantly enhanced by surface treatments. However, conventional porous carbons typically suffer from hydrophobic surfaces and a limited number of specific active sites, which restrict their effectiveness in particulate CO_2 capture applications [36,37].

To address these limitations, considerable research has been devoted to the heteroatom doping of carbon materials with elements such as B, N, O, and P, which can improve their electrochemical properties, CO_2 adsorption capacity, and selectivity [38–40]. Among these, B is particularly effective, as it can substitute carbon atoms within the lattice, acting as an electron acceptor due to its three valence electrons. This substitution modifies the electronic structure of the doped carbon by shifting the Fermi level toward the conduction band. B-containing functional groups incorporated on the CG surface are expected to enhance CO_2 capture performance by providing additional active sites for efficient CO_2 molecules.

Therefore, this study proposes the synthesis of boron-doped porous carbon using CGs as a precursor and its application in CCS. This approach aims to provide a high-value utilization pathway for biomass waste while contributing to greenhouse gas reduction and sustainable environmental protection.

2. Materials and Methods

2.1. Materials and Preparation of Boric Acid-Treated Coffee Grounds

The CGs used in this study were purchased from Mega Coffee (Seoul, Republic of Korea). To remove residual moisture, the CGs were dried in an oven at 100 °C for 12 h. To remove impurities, 200 g of CGs were immersed in 1500 mL of 1 M hydrogen chloride (HCl) solution at room temperature for 1 h. The samples were then thoroughly washed with distilled water until the pH reached 7.0 and then dried in an oven at 100 °C for 24 h. For dry boric acid pretreatment, 15 g of dried CGs were mixed with 5 g of boric acid (Daejung Chemical & Metals Co., Siheung, Republic of Korea) using a mortar and pestle to ensure homogeneity. The prepared samples were designated as 'B-CGs' and 'CGs' according to the presence or absence of boric acid in the mixture.

2.2. Carbonization and Activation of CGs and B-CGs

A uniformly mixed 20 g quantity of CGs and B-CGs was placed in an alumina boat and loaded into a cylindrical alumina tube furnace (diameter: 90 mm, length: 1000 mm). The samples were carbonized at temperatures ranging from 800 to 1000 °C for 1 h under a nitrogen (N_2) atmosphere (99.99%, 500 cc/min) at a heating rate of 5 °C/min. The resulting carbonized CG (CCG) and boric acid-treated CCG (B-CCG) were subjected to vacuum-assisted washing with distilled water to remove residual borates formed during carbonization. Subsequently, the samples were thoroughly dried in an oven for 24 h. After complete drying, 3.0 g of CCG and B-CCG were placed in an alumina boat and loaded into a cylindrical stainless steel tube furnace (diameter: 80 mm, length: 1200 mm). The samples were heated to 900 °C under an N_2 atmosphere (99.9999%, 200 mL/min) at a heating rate of 10 °C/min. After reaching 900 °C, the N_2 atmosphere was switched to a steam environment (0.5 mL/min water supply), and the activation process was maintained for 60 min. The synthesized porous carbon derived from coffee grounds (PCG) was designated as 'boric acid treatment (B)–porous carbon derived from coffee grounds (PCG)–carbonization temperature (C8, C9, and C10)'.

2.3. Characterization of PCG

The crystalline structure of PCG was analyzed by X-ray diffraction (XRD; MiniFlex, Rigaku, Tokyo, Japan). XRD analysis was performed with Cu-Kα radiation (0.1542 nm) at a scanning rate of 2°/min over a 2θ range of 5–60°. Crystallite size and interplanar spacing were calculated using the Scherrer and Bragg equations, respectively [41].

The textural properties of PCG were characterized using an N_2/77K adsorption–desorption isothermal analyzer (BELSORP Max II, MicrotracBEL, Tokyo, Japan). Prior to measurement, the samples were degassed at 300 °C under a residual pressure of less than 10^{-3} bar for 12 h to ensure the complete removal of adsorbed impurities and gases. The specific surface area of PCG was determined by the Brunauer–Emmett–Teller (BET) method based on the N_2/77K adsorption–desorption isotherm curves [42]. The pore size distribution (PSD) was calculated using the Barrett–Joyner–Halenda (BJH) method and the non-local density functional theory (NLDFT) [43,44]. Additionally, the micropore volume of PCG was estimated from the intercept value of the t-plot.

Elemental composition and surface functional group analysis of PCG were performed by X-ray photoelectron spectroscopy (XPS; Thermo Scientific NEXSA, Watertown, NY, USA). The samples were analyzed under vacuum conditions at pressures below 3×10^{-7} Pa using Al-Kα radiation (1486.6 eV).

3. Results

3.1. Textural Properties of PCG and B-PCG by Carbonization Temperature

The pore formation and adsorption behavior of PCG and B-PCG as a function of carbonization temperature were analyzed using the N_2/77K adsorption–desorption method, as shown in Figure 1. In Figure 1a, all PCG samples exhibited a type I (a) isotherm according to the classification of the International Union of Pure and Applied Chemistry (IUPAC) [45]. Type I (a) isotherms are typically observed when nitrogen adsorption primarily occurs within micropores at low relative pressures ($P/P_0 \leq 0.1$). This indicates that PCG-C8 to PCG-C10 undergo monolayer adsorption predominantly driven by micropore development. In contrast, B-PCG exhibited a transition to a type I (b) isotherm with a larger specific surface area compared with PCG. The type I (b) adsorption isotherm is generally characteristic of porous materials with both micropores and mesopores.

The basal planes of carbon materials are formed through an sp^2 hybridized bond, while the crystallite edges are composed of sp^3 hybridized bonds. In this study, it was confirmed that doping elements, specifically boron (B), can be introduced in various functionalized forms at the crystallite edges of carbon materials via a dry thermal treatment process utilizing differences in elemental bonding energy. In particular, the transformation of the B-PCG adsorption isotherm to a type I (b) profile suggests that various boron functional groups were incorporated at the crystallite edges of B-CCG during carbonization. Additionally, defects were generated in regions of the basal planes where boron was not incorporated, leading to a gradual increase in pore diameter throughout the activation process. Furthermore, in the relative pressure ranges of $(P/P_0) \leq 0.1$ and $(P/P_0) \leq 0.3$, B-PCG-C8 exhibited higher N_2 adsorption behavior compared with B-PCG-C9 and B-PCG-C10. This can be attributed to the preferential oxidation of amorphous regions and small crystallites within the structure of B-PCG-C8 at the relatively low carbonization temperature of 800 °C. Generally, during the activation process, oxidation of carbon materials preferentially occurs at amorphous regions and crystallite edges [46]. Consequently, the higher proportion of amorphous structures and small crystallites in B-PCG-C8 facilitates more active oxidation, resulting in the dominant development of micropores and mesopores, as well as an increased specific surface area, as summarized in Table 1.

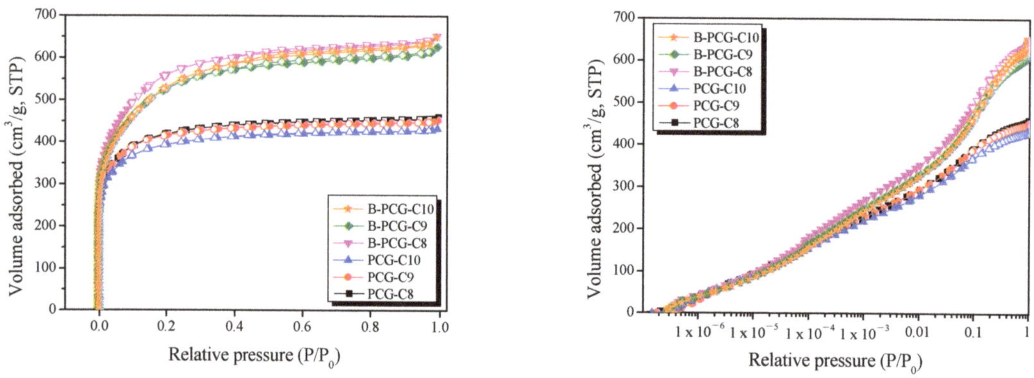

(a) (b)

Figure 1. N_2/77K adsorption–desorption isotherm curves of PCG and B-PCG as a function of various carbonization temperatures; (**a**) normal and (**b**) logarithmic.

Table 1. Texture Properties of PCG and B-PCG as a Function of Various Carbonization Temperature.

Sample	S_{BET} [1] (m^2/g)	V_{Total} [2] (cm^3/g)	V_{Micro} [3] (cm^3/g)	V_{meso} [4] (cm^3/g)	Micropore Ratio [5] (%)	Mesopore Ratio [6] (%)
PCG-C8	1590	0.71	0.65	0.06	91.5	8.5
PCG-C9	1580	0.70	0.64	0.06	91.4	8.6
PCG-C10	1490	0.67	0.61	0.06	91.0	9.0
B-PCG-C8	2060	1.01	0.87	0.14	86.1	13.9
B-PCG-C9	1960	0.97	0.82	0.15	84.5	15.5
B-PCG-C10	1940	1.00	0.83	0.17	83.0	17.0

[1] S_{BET}: Specific surface area; Brunauer–Emmett–Teller (BET) method; $\frac{P}{v(P_0-P)} = \frac{1}{v_m} + \frac{c-1}{v_m c} \cdot \frac{P}{P_0}$. [2] V_{Total}: Total pore volume; the amount of adsorbed $P/P_0 = 0.99$. [3] V_{Micro}: Micropore volume; t-plot method. [4] V_{Meso}: [2] V_{Total} − [3] V_{Micro}. [5] Micropore ratio: $\frac{V_{Micro}}{V_{Total}} \times 100$. [6] Mesopore ratio: $\frac{V_{Meso}}{V_{Total}} \times 100$.

Figure 2a,b indicate the PSD curves of B-PCG obtained by the BJH and NLDFT methods, respectively. In Figure 2a, B-PCG exhibited an increase in mesopores within the 2–5 nm range compared with PCG, regardless of the carbonization temperature. Similarly, in Figure 2b, the PSD curve of B-PCG revealed a greater presence of micropores (1.3–2 nm) along with an increase in mesopores (2–5 nm) compared with PCG. These results are consistent with the N_2/77K adsorption–desorption isotherms in Figure 1, confirming that B-PCG has a more developed pore structure than PCG, not only in the micropore region but also for mesopores smaller than 5 nm.

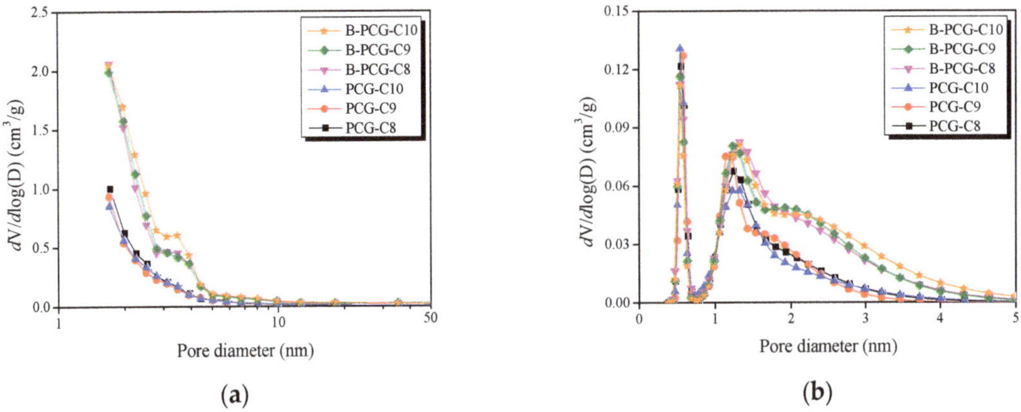

Figure 2. Pore size distribution of PCG and B-PCG as a function of various carbonization temperatures; (**a**) BJH equation and (**b**) NLDFT method.

In general, the mechanism of pore development in porous carbon can be categorized into two processes: (i) pore drilling, where oxidation at crystallite edges increases pore diameter, and (ii) pore deepening, where oxidation of amorphous regions increases pore depth without significant changes in pore diameter [47]. Based on these results, it is inferred that various boron functional groups were incorporated at the crystallite edges of B-PCG. Among them, weakly bonded functional groups such as B–OH and B–H, which are incompletely bonded to carbon, are likely to be decomposed during heat treatment, leading to an increase in pore diameter. In other words, the thermal decomposition of boron functional groups introduced at the crystallite edges of B-PCG promoted pore formation via the pore drilling mechanism. Consequently, boron doping in PCG was found to facilitate the development of both micropores and mesopores, which significantly influenced the formation of a porous structure.

3.2. *Crystal Structure of Coffee Grounds Under Various Processing Conditions*

To investigate the structural changes in CCG and B-CCG after activation under identical conditions at different carbonization temperatures, XRD pattern analysis was performed, as shown in Figures 3 and 4. As shown in Figure 3a, both CCG and B-CCG exhibited similar trends regardless of carbonization temperature, and no significant structural transformation was observed. For a more quantitative assessment of the structural changes, the crystallite diameter (L_a) and the crystallite height (L_c) were analyzed and are shown in Figure 3b. The results indicate that both the L_a and the L_c increased with increasing carbonization temperature. In particular, the L_a showed a particularly dramatic increase in both CCG and B-CCG. This trend is attributed to the growth of crystallites due to graphitization at higher carbonization temperatures, as well as the influence of boric acid incorporated within the coffee grounds. These effects are further supported by the decrease in interlayer spacing (d_{002}) and the increase in interplanar crystalline spacing (d_{10l}), as shown in Figure 3c. Typically, the d_{002} corresponds to the interlayer spacing between graphitic planes. A decrease in d_{002} suggests that the carbon layers are stacked more closely, which is indicative of an increased degree of graphitization and improved structural ordering. Meanwhile, the increase in the (10*l*) peak intensity reflects enhanced in-plane crystallinity and larger crystallite sizes, further supporting the growth of graphitic domains.

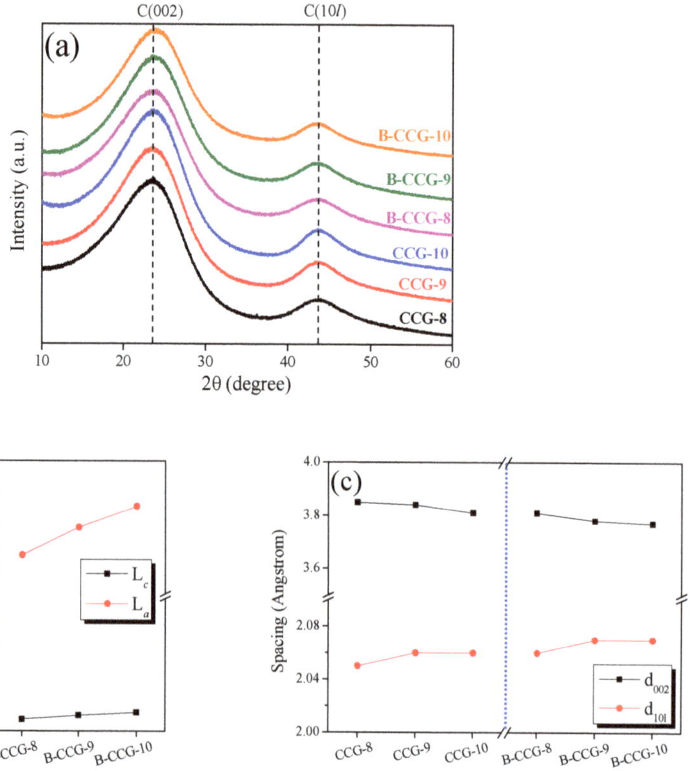

Figure 3. (**a**) X-ray diffraction patterns; (**b**) structural parameters; and (**c**) interplanar distances of PCG prepared with and without boric acid treatment as a function of carbonization temperature.

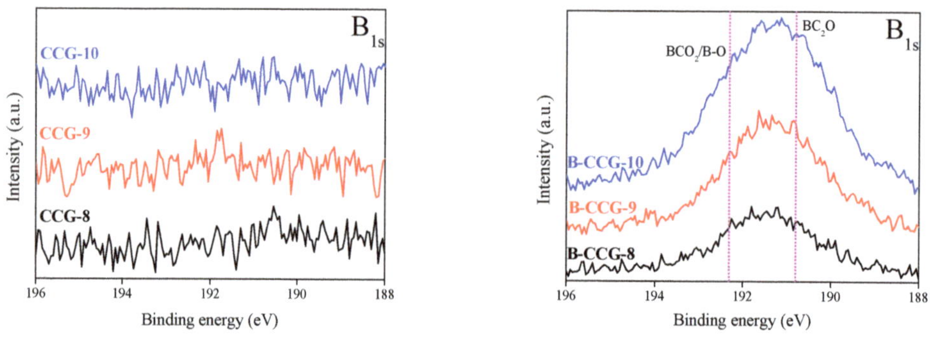

Figure 4. B_{1s} spectra of CCG as a function of various carbonization temperatures with and without boric acid treatment; (**a**) CCG and (**b**) B-CCG.

Additionally, the C(002) peak of carbon materials appears near 26.5°, corresponding to the formation of a hexagonal honeycomb-like planar structure (2D) of carbon atoms via sp^2 hybridization. These planar carbon structures are stacked along the vertical direction, forming a three-dimensional structure where weak π–π interactions (Van der Waals forces) occur between the layers. The C(002) peak indicates the interlayer spacing of the vertically stacked planar structure, while the C(10*l*) peak represents the atomic arrangement of carbon within the planar structure [48]. In other words, the (002) peak of carbon materials corresponds to the interlayer distance within the crystalline structure, while the (10*l*) peak indicates the interplanar crystalline spacing within the hierarchical structure.

Based on this, the increase in the L_a observed in Figure 3b and the decrease in d_{002} in Figure 3c can be attributed to grain growth in CCG and B-CCG as a result of increasing carbonization temperature. Notably, B-CCG exhibited a higher L_a value compared with CCG, which is likely due to the presence of boron introduced into the CCG. During the carbonization process, boron undergoes substitution reactions with carbon in CCG, forming functional groups such as B–C and B–O, thereby promoting enhanced grain growth in B-CCG. These findings are further corroborated by the XPS results of CCG and B-CCG presented in Figure 4.

Figure 4 shows the result of XPS analysis carried out to investigate the functional group bonding and surface characteristics of CCG following boron doping. The XPS survey spectra of the samples subjected to different carbonization temperatures and boric acid treatments are shown in Figure S3a, where C_{1s}, O_{1s}, and a small amount of N_{1s} peaks were identified in all samples. Due to the relatively low detection amount of the B_{1s} peak compared with other elements, high resolution spectral analysis was performed, as shown in Figure 4a,b.

The results indicate that while no B_{1s} peak was observed in CCG, clear detection of boron functional groups, including B–C and B–O bonds, was confirmed in B-CCG. Based on these findings, it can be inferred that boron atoms were successfully incorporated into the CCG structure by dry boric acid treatment. Notably, B-CCG-C10 exhibited the strongest B_{1s} peak, suggesting that boron doping was more effectively facilitated at higher carbonization temperatures. The chemical structure of the boron functional groups introduced into CCG is illustrated in Figure S4.

Figure 5 indicates the analysis of crystalline structure changes in PCG and B-PCG after activation treatment under the same experimental conditions as in Figure 3. As shown in Figure 5b, both the L_a and the L_c of PCG and B-PCG increase with increasing carbonization temperature, exhibiting a similar trend to that observed in Figure 3b. However, differences arise due to the distinct mechanisms governing crystallite growth during the carbonization and activation processes. While carbonization primarily promotes crystallite growth, the activation process induces pore development within the crystalline structure through oxidation reactions. As previously mentioned, the oxidation reaction of carbon materials preferentially occurs in amorphous carbon and at crystallite edges. Therefore, the observed increase in the L_a and the L_c of PCG and B-PCG can be attributed not only to crystallite growth at higher carbonization temperatures but also to the relative increase in crystallite proportion due to the oxidation of amorphous carbon and smaller crystallites during activation. Furthermore, a more pronounced increase in the L_a was observed in B-PCG compared with PCG, with B-PCG-C10 exhibiting the highest L_a value. Since boron can form both sp^2 and sp^3 hybridizations within the carbon structure, B-PCG is expected to undergo relatively disordered structure and exhibit a higher proportion of amorphous regions than PCG. This is supported by the lower L_a value and higher d_{002} observed for B-PCG-C8 compared with PCG-C8, as shown in Table S1 and Figure 4b. Additionally, the enhanced N_2 adsorption behavior of B-PCG in the $(P/P_0) < 0.01$ region of Figure 1 further confirms the relative increase in amorphous regions. During the carbonization process, boron incorporation influences the crystallite growth of CCG, leading to more significant grain development at higher carbonization temperatures (Figure 3b). In other words, B-CCG-C10 forms larger crystallites than other B-PCG samples and contains a relatively less amorphous region within the same volume. Therefore, the sharp increase in the L_a observed for B-PCG-C10 can be attributed to the preferential oxidation of amorphous regions during activation, resulting in a relative increase in the proportion of crystallites. These results are further supported by the marked decrease in d_{002} of B-PCG with increasing carbonization temperature, as shown in Figure 5c.

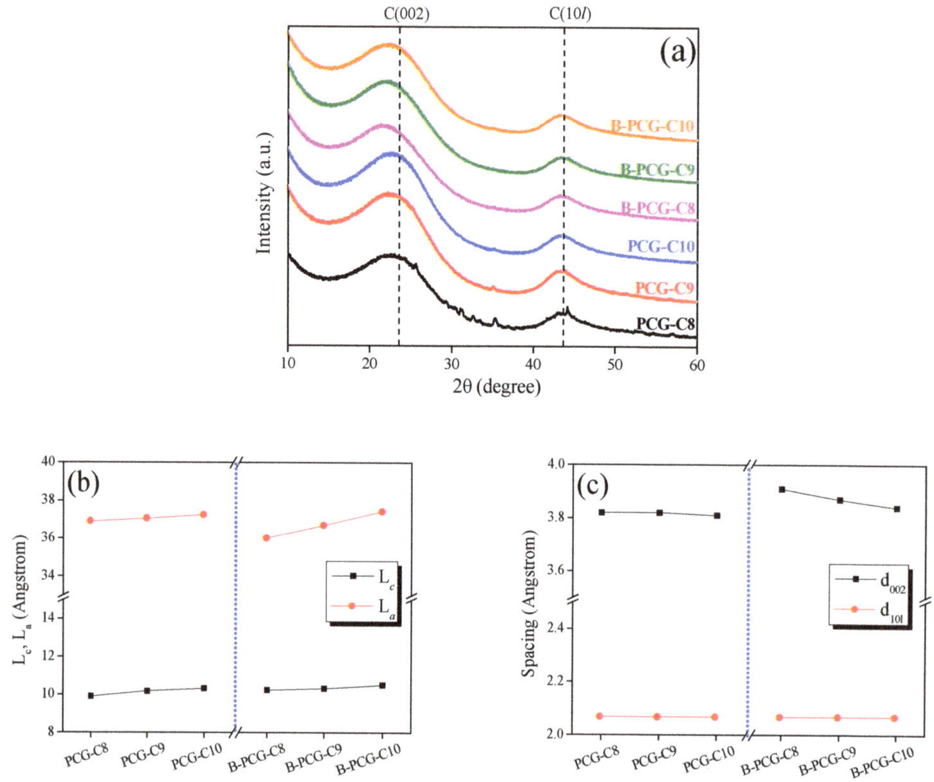

Figure 5. (**a**) X-ray diffraction patterns; (**b**) structural parameters; and (**c**) interplanar distances of PCG prepared with and without boric acid treatment as a function of activation temperature.

In Figure 5c, while the d_{002} value of PCG remains nearly constant with varying carbonization temperatures, B-PCG exhibits a significant decrease in d_{002}. This difference can be attributed not only to crystallite growth and the oxidation of amorphous regions but also to the influence of boron doping on the crystalline structure of PCG during activation. In particular, the oxidation reaction at the crystallite edges appears to be more pronounced in B-PCG compared with PCG. This is likely due to the presence of various boron functional groups (e.g., B–OH, B=O, B–C–O) at the crystallite edges, which facilitate oxidation reactions and induce structural changes during activation [49]. As the amorphous regions preferentially oxidize in this process, the alignment of the B-PCG crystalline structure improves, ultimately leading to a reduction in the d_{002} value.

Figure 6 shows the XPS analysis results of the surface functional group changes in PCG and B-PCG following activation treatment. The analysis confirmed the presence of boron functional groups exclusively in B-PCG. However, when compared with the results in Figure 4, a significant decrease in the intensity of boron functional groups was observed in B-PCG compared with B-CCG. This finding suggests that the boron functional groups bonded to the CCG surface acted as an "activation mediator" during the activation process. Mild boric acid treatment of the CCG surface by a dry method leads to the formation of various functional groups during carbonization (Figure 4b).

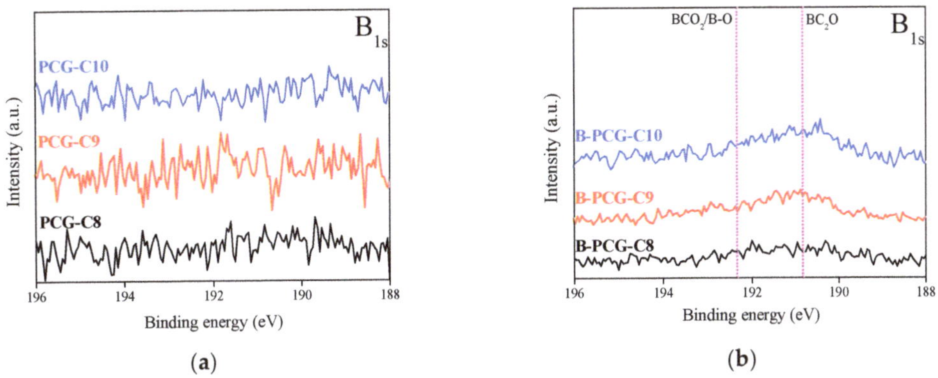

Figure 6. B_{1s} spectra of PCG as a function of various carbonization temperatures with and without boric acid treatment; (**a**) PCG and (**b**) B-PCG.

Since boron has a lower electronegativity than carbon, its incorporation into the carbon structure can modulate the electron density, thereby increasing the bonding energy. In particular, B–C sp^2 bonds enhance the thermal stability of B-CCG and contribute to the structural stability during the heat treatment process of carbonization. During the activation process, B-CCG undergoes partial oxidation of carbon, resulting in the formation of pores and its conversion into B-PCG. At this stage, boron present within the carbon structure can react with steam to form oxides such as B_2O_3. Since the resulting boron oxides have a high hydrophilicity, they readily react with steam during activation, subsequently transforming into intermediate oxides (B_xO_y), which volatilize and are eventually removed [50,51]. This pore development mechanism, facilitated by boron as an activation mediator, is supported by the increase in the specific surface area and mesopore formation in B-PCG with increasing carbonization temperature (Figure 1), as well as the reduction in boron functional groups (Figure 6b). As shown in Table 1, the proportion of mesopores in B-PCG gradually increases from 13.9% in B-PCG-C8 to 17.0% in B-PCG-C10. Thus, with increasing carbonization temperature, boron bonded to B-CCG not only promotes crystallite growth but also acts as an activation mediator to improve the overall pore characteristics during the activation process. Consequently, this mechanism can be described as a "hybrid activation" approach, in which chemically bonded boron facilitates pore development through a physical activation process. This approach demonstrates that it is possible to achieve a large specific surface area and well-developed porosity comparable to conventional chemical activation methods (Table S2).

3.3. *Carbon Dioxide Capture Performance*

Figure 7 presents the CO_2 adsorption isotherms of B-CCG and B-PCG measured at 0.5 bar. The CO_2 adsorption curves of undoped CCG and PCG are shown in Figure S5. In Figure 7a, the CO_2 adsorption capacity of B-CCG gradually decreases with increasing carbonization temperature. In contrast, as shown in Figure 7b (left), B-PCG exhibits approximately four times higher CO_2 adsorption capacity compared with B-CCG. This result suggests that boron doping facilitates the activation process of CCG by influencing the formation of micropores and mesopores in B-PCG (Figure 1 and Table 1). The resulting changes in pore structure increase the available physical adsorption sites for CO_2 molecules, ultimately contributing to an improved CO_2 storage capacity. Furthermore, the CO_2 adsorption capacity of PCG, shown in Figure S5b (left), is significantly lower than that of B-CCG. This phenomenon is attributed to the effect of boron doping, where boron-containing functional groups such as B–CO, B–CO_2, and B–C_2O, identified in Figure 6b, function as a determining factor in enhancing the interactions with CO_2 molecules.

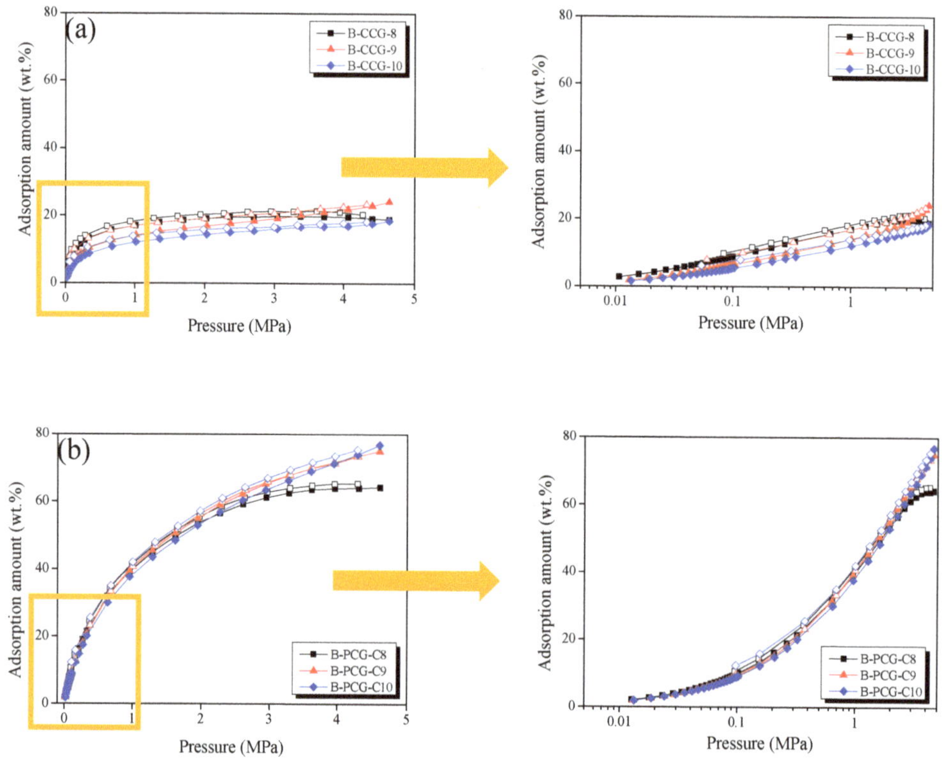

Figure 7. CO_2 storage capacity of boron-doped CGs under different heat treatment conditions as a function of carbonization temperature at 0.5 bar/298 K; (**a**) B-CCG and (**b**) B-PCG.

In particular, the B–CO_2 bond can facilitate the selective adsorption of CO_2 through Lewis acid–base interactions. This surface chemistry property enhances the CO_2 affinity of B-PCG, thereby positively influencing the CO_2 capture performance. This trend can be confirmed by comparing PCG in Figure S5b (left) with B-PCG in Figure 7b (left). Unlike PCG, for which CO_2 adsorption performance decreases with increasing carbonization temperature, B-PCG was observed to tend to increase CO_2 adsorption performance due to the presence of boron-containing functional groups on the surface. Therefore, boron doping plays a crucial role in modulating the surface chemistry and pore structure of carbon materials, ultimately improving their affinity for CO_2 molecules and enhancing CO_2 capture efficiency.

4. Conclusions

In this study, boron-doped porous carbon derived from coffee grounds (B-PCG) was synthesized using biomass waste as a precursor, and the effects of boron doping on the pore structure and surface chemical properties during carbonization and activation processes were systematically evaluated. During physical activation, boron functioned as an activation mediator to facilitate the development of porosity. Based on these findings, this study proposes a hybrid activation mechanism that utilizes the synergistic effects of chemical boron doping and physical activation. Compared with PCG, B-PCG exhibited significantly enhanced specific surface area and pore development (Figure 1), which can be attributed to the thermal decomposition of boron-containing functional groups incorporated at the crystallite edges of PCG during carbonization. During activation, boron-doped carbonized carbon derived from coffee grounds (B-CCG) underwent oxidative reactions that resulted in the formation of hierarchical porosity and its subsequent conversion to B-PCG. At this step in the process, boron species present on the carbon surface reacted with steam to form boron oxides (e.g., B_2O_3), which acted as activation promoters. As a result, the specific surface area and mesopore formation increased, which correlated with the reduction of boron-containing functional groups, as confirmed in the XPS results (Figure 6).

The proposed hybrid activation mechanism was suggested in order to induce a level of porosity and surface area comparable to that achieved by conventional chemical activation. Consequently, B-PCG exhibited a 50% improvement in CO_2 adsorption capacity compared with PCG. This improvement is attributed to the increased interaction between boron functional groups and CO_2 molecules, as well as the development of mesopores that facilitate physical adsorption. Therefore, the hybrid activation-assisted boron doping represents a promising strategy for carbon capture and storage (CCS) applications.

Supplementary Materials: The following supporting information can be downloaded at: https://www.mdpi.com/article/10.3390/batteries11040158/s1, Figure S1: N_2/77K isothermal adsorption–desorption curves of B-CCG as a function of carbonization temperature; (a) normal and (b) logarithmic; Figure S2: Pore size distribution of B-CCG as a function of carbonization temperature; (a) calculated by BJH method, and (b) calculated by NLDFT method; Figure S3: XPS survey spectra of (a) CCGs and (b) PCG as a function of various processes conditions with and without boric acid treatment; Figure S4: Schematic of the boron doping process in CGs and the chemical structure of the introduced boron functional groups; Figure S5: CO_2 storage capacity of CGs as a function of various heat treatments at 0.5 bar/298 K; (a) CCG and (b) PCG; Table S1: Textural Properties of B-CCGs as a Function of Carbonization Temperature. Table S2. Specific surface area of porous carbon derived from waste coffee grounds as a function of the activation process. References [52–67] are cited in the supplementary materials.

Author Contributions: Conceptualization, H.H.K.; methodology, H.H.K.; software, H.H.K.; validation, H.H.K.; formal analysis, H.H.K.; investigation, H.H.K.; resources, K.-H.A. and B.-J.K.; data curation, H.H.K.; writing—original draft preparation, H.H.K.; writing—review and editing, H.H.K. and B.-J.K.; visualization, H.H.K.; supervision, B.-J.K.; project administration, K.-H.A. and B.-J.K.; funding acquisition, K.-H.A. All authors have read and agreed to the published version of the manuscript.

Funding: This research was funded by the "Carbon Industry Foundation project" no. 20016795 (Development of Manufacturing Technology Independence of Advanced Activated Carbons and Application for High-Performance Supercapacitor). In addition, this study was funded by the "Material Parts Technology Development Project" no. 20017563 (Development of Single-Walled Carbon Nanotube-Binder Integrated Conductive Material and its Application to High Energy-Density Secondary Battery Technology). This research was supported by the Jeonju University Research Year (K.H. An).

Data Availability Statement: The original contributions presented in the study are included in the article, further inquiries can be directed to the corresponding author.

Conflicts of Interest: The authors declare no conflicts of interest.

References

1. Dhillon, R.S.; Wuehlisch, G.V. Mitigation of global warming through renewable biomass. *Biomass Bioenerg.* **2013**, *48*, 75–89. [CrossRef]
2. Wienchol, P.; Szlek, A.; Ditaranto, M. Waste-to-energy technology integrated with carbon capture—Challenges and opportunities. *Energy* **2020**, *198*, 117352. [CrossRef]

3. Montagnini, F.; Nair, P.K.R. Carbon sequestration: An underexploited environmental benefit of agroforestry systems. *Agrofor. Syst.* **2004**, *61*, 281–295. [CrossRef]
4. Gill, L.; Bernardo, J. An approach to energy and climate issues aiming at carbon neutrality. *Renew. Energy Focus* **2020**, *33*, 37–42. [CrossRef]
5. Perissi, L.; Jones, A. Investigating european union decarbonization strategies: Evaluating the pathway to carbon neutrality by 2050. *Sustainability* **2022**, *48*, 4728. [CrossRef]
6. Paltsev, S.; Morris, J.; Kheshgi, H.; Herzog, H. Hard-to-abate sectors: The role of industrial carbon capture and storage (CCS) in emission mitigation. *Appl. Energy* **2021**, *300*, 117322. [CrossRef]
7. Shen, Y. Preparation of renewable porous carbons for CO_2 capture—A review. *Fuel Process. Technol.* **2022**, *236*, 107437. [CrossRef]
8. Makepa, D.C.; Chihobo, C.H. Sustainable pathways for biomass production and utilization in carbon capture and storage—A review. *Biomass Conv. Bioref.* **2024**, *16*, 1–22. [CrossRef]
9. Yin, Y.; Liu, Q.; Wang, J.; Zhao, Y. Recent insights in synthesis and energy storage applications of porous carbon derived from biomass waste: A review. *Int. J. Hydrogen Energy* **2022**, *47*, 39338–39363. [CrossRef]
10. Roussos, S.; de los Angeles Aquiáhuatl, M.; del Refugio Trejo-Hernández, M.; Gaime Perraud, I.; Favela, E.; Ramakrishna, M.; Raimbault, M.; Viniegra-González, G. Biotechnological Management Of Coffee Pulp—Isolation, Screening Characterization, Selection of Caffeine Degrading Fungi And Natural Microflora Present In Coffee Pulp And Husk. *Appl. Microbiol. Biotechnol.* **1995**, *42*, 756–762. [CrossRef]
11. Koul, B.; Yakoob, M.; Shah, M.P. Agricultural waste management strategies for environmental sustainability. *Environ. Res.* **2022**, *206*, 112285. [CrossRef]
12. Chakraborty, R.; Vilya, K.; Pradhan, M.; Nayak, A.K. Recent advancement of biomass-derived porous carbon based materials for energy and environmental remediation applications. *J. Mater. Chem. A* **2022**, *10*, 6965–7005. [CrossRef]
13. Jiang, T.; Zhang, Y.; Olayiwola, S.; Lau, C.; Fan, M.; Ng, K.; Tan, G. Biomass-derived porous carbons support in phase change materials for building energy efficiency: A review. *Mater. Today Energy* **2022**, *23*, 100905. [CrossRef]
14. Singh, Y.; Shdhu, H.S. Management of Cereal Crop Residues for Sustainable Rice-Wheat Production System in the Indo-Gangetic Plains of India. *Proc. Indian Natl. Sci. Acad.* **2014**, *80*, 95–114. [CrossRef]
15. Searchinger, T.; Heimlich, R.; Houghton, R.A.; Dong, F.; Elobid, A.; Fabiosa, J.; Tokgoz, S.; Hayes, D.; Yu, T.H. Use of US croplands for biofuels increases greenhouse gases through emissions from land-use change. *Science* **2008**, *319*, 1238–1240. [CrossRef]
16. Kaab, A.; Sharifi, M.; Mobli, H.; Nabavi-Pelesaraei, A.; Chau, K.W. Combined lifecycle assessment and artificial intelligence for prediction of output energy and environmental impacts of sugarcane production. *Sci. Total Environ.* **2019**, *664*, 1005–1019. [CrossRef] [PubMed]
17. Rivera, X.C.S.; Gallego-Schmid, A.; Najdanovic-Visak, V.; Azapagic, A. Life cycle environmental sustainability of valorization routes for spent coffee grounds: From waste to resource. *Resour. Conserv. Recycl.* **2020**, *157*, 104751. [CrossRef]
18. Liu, X.; Zhang, S.; Wen, X.; Chen, X.; Wen, Y.; Shi, X.; Mijowska, E. High yield conversion of biowaste coffee grounds into hierarchical porous carbon for superior capacitive energy storage. *Sci. Rep.* **2020**, *10*, 3518. [CrossRef]
19. Forcina, A.; Petrillo, A.; Travaglioni, M.; Chiara, S.; Felice, F. A comparative life cycle assessment of different spent coffee ground reuse strategies and a sensitivity analysis for verifying the environmental convenience based on the location of sites. *J. Clean. Prod.* **2023**, *385*, 135727. [CrossRef]
20. Bevilacqua, E.; Cruzat, V.; Singh, I.; Rose'Meyer, R.B.; Panchal, S.K.; Brown, L. The potential of spent coffee grounds in functional food development. *Nutrients* **2023**, *15*, 994. [CrossRef]
21. LeBouf, R.F.; Aldridge, M. Carbon monoxide emission rates from roasted whole bean and ground coffee. *J. Air Waste Manag. Assoc.* **2019**, *69*, 89–96. [CrossRef] [PubMed]
22. Cho, E.J.; Trinh, L.T.P.; Song, Y.; Lee, Y.G.; Bae, H.J. Bioconversion of biomass waste into high value chemicals. *Bioresour. Technol.* **2020**, *298*, 122386. [CrossRef] [PubMed]
23. Igliński, B.; Kujawski, W.; Kiełkowska, U. Pyrolysis of Waste Biomass: Technical and Process Achievements, and Future Development—A Review. *Energies* **2023**, *16*, 1829. [CrossRef]
24. Kalak, T. Potential Use of Industrial Biomass Waste as a Sustainable Energy Source in the Future. *Energies* **2023**, *16*, 1783. [CrossRef]
25. Tripathi, N.; Hills, C.D.; Singh, R.S.; Singh, J.S. Offsetting anthropogenic carbon emissions from biomass waste and mineralised carbon dioxide. *Sci. Rep.* **2020**, *10*, 958. [CrossRef]
26. Sri Shalini, S.; Palanivelu, K.; Ramachandran, A.; Raghavan, V. Biochar from biomass waste as a renewable carbon material for climate change mitigation in reducing greenhouse gas emissions—A review. *Biomass Conv. Bioref.* **2021**, *11*, 2247–2267. [CrossRef]
27. Kemp, K.C.; Baek, S.B.; Lee, W.G.; Meyyappan, M.; Kim, K.S. Activated carbon derived from waste coffee grounds for stable methane storage. *Nanotechnology* **2015**, *26*, 385602. [CrossRef]
28. Adan-Mas, A.; Alcaraz, L.; Arévalo-Cid, P.; López-Gómez, F.A.; Montemor, F. Coffee-derived activated carbon from second biowaste for supercapacitor applications. *Waste. Manag.* **2021**, *120*, 280–289. [CrossRef]

29. Chung, D.Y.; Son, Y.J.; Yoo, J.M.; Kang, J.S.; Ahn, C.Y.; Park, S.; Sung, Y.E. Coffee waste-derived hierarchical porous carbon as a highly active and durable electrocatalyst for electrochemical energy applications. *ACS Appl. Mater. Interfaces* **2017**, *9*, 41303–41313. [CrossRef]
30. Stylianou, M.; Agapiou, A.; Omirou, M.; Vyrides, I.; Ioannis, M.I.; Maratheftis, G.; Fasoula, D. Converting environmental risks to benefits by using spent coffee grounds (SCG) as a valuable resource. *Environ. Sci. Pollut. Res.* **2018**, *25*, 35776–35790. [CrossRef]
31. Kumari, R.; Singh, V.; Kant, C.R. Enhanced performance of activated carbon-based supercapacitor derived from waste soybean oil with coffee ground additives. *Mater. Chem. Phys.* **2023**, *305*, 127882. [CrossRef]
32. Kwak, H.S.; Kim, J.K.; Chun, Y. A review: Preparation and characterization of high-performance spent coffee grounds based adsorbent of for water treatment. *J. Korea Acad. Ind. Coop. Soc.* **2024**, *25*, 316–322. [CrossRef]
33. Hwang, J.; Kim, D. Development of Sustainable Packaging Materials Using Coffee Silverskin and Spent Coffee Grounds: A Comprehensive Review. *Korean J. Packag. Sci. Tech.* **2024**, *30*, 1–14. [CrossRef]
34. Ariharan, A.; Kim, S.K. Bioinspired sustainable Sheetlike porous carbon derived from Cassia fistula flower petal as an electrode for high-performance supercapacitors. *Energy Fuels* **2022**, *36*, 9337–9346. [CrossRef]
35. Ariharan, A.; Viswanathan, B. Porous activated carbon material derived from sustainable bio-resource of peanut shell for H2 and CO_2 storage applications. *Indian J. Chem. Technol.* **2018**, *25*, 140–149.
36. Chen, H.; Zhou, M.; Wang, Z.; Zhao, S.Y.; Guan, S.Y. Rich nitrogen-doped ord *Energy Fuels* ered mesoporous phenolic resin-based carbon for supercapacitors. *Electrochimica Acta* **2014**, *148*, 187–194. [CrossRef]
37. Wei, J.; Zhou, D.D.; Sun, Z.K.; Deng, Y.H.; Xia, Y.Y.; Zhao, D.Y. A controllable synthesis of rich nitrogen-doped ordered mesoporous carbon for CO_2 capture and supercapacitors. *Adv. Funct. Mater.* **2013**, *23*, 2322–2328. [CrossRef]
38. Ariharan, A.; Viswanathan, B.; Nandhakumar, V. Hydrogen storage on boron substituted carbon materials. *Int. J. Hydrogen Energy* **2016**, *41*, 3527–3536. [CrossRef]
39. Ariharan, A.; Viswanathan, B.; Nandhakumar, V. Heteroatom doped multi-layered graphene material for hydrogen storage application. *Graphene* **2016**, *5*, 39–50. [CrossRef]
40. Ariharan, A.; Viswanathan, B.; Nandhakumar, V. Nitrogen-incorporated carbon nanotube derived from polystyrene and polypyrrole as hydrogen storage material. *Int. J. Hydrogen Energy* **2018**, *43*, 5077–5088. [CrossRef]
41. Biscoe, J.; Warren, B.E. An X-Ray study on carbon black. *Int. J. Appl. Phys.* **1942**, *13*, 364–371. [CrossRef]
42. Urita, K.; Urita, C.; Fujita, K.; Horio, K.; Yoshida, M.; Moriguchi, I. The ideal porous structure of EDLC carbon electrodes with extremely high capacitance. *Nanaoscale* **2017**, *9*, 15643–15649. [CrossRef]
43. Lewandowski, W.M.; Ryms, M.; Kosakowski, W. Thermal Biomass Conversion: A Review. *Processes* **2020**, *8*, 516. [CrossRef]
44. Wang, Q.; Tian, D.; Hu, J.; Shen, F.; Yang, G.; Zhang, Y.; Deng, S.; Zhang, J.; Zeng, Y.; Hu, Y. Fates of hemicellulose, lignin and cellulose in concentrated phosphoric acid with hydrogen peroxide (PHP) pretreatment. *RSC Adv.* **2018**, *8*, 12714–12723. [CrossRef] [PubMed]
45. Sing, K.S.W. Reporting physisorption data for gas/solid systems with special reference to the determination of surface area and porosity. *Pure Appl. Chem.* **1985**, *57*, 603–619. [CrossRef]
46. Kim, J.H.; Kim, S.H.; Kim, B.J.; Lee, H.M. Effects of oxygen-containing functional groups on the electrochemical performance of activated carbon for EDLCs. *Nanomaterials* **2023**, *13*, 262. [CrossRef] [PubMed]
47. Kim, J.H.; Jung, S.C.; Lee, H.M.; Kim, B.J. Comparison of pore structures of cellulose-based activation carbon fibers and their applications for electrode materials. *Int. J. Mol. Sci.* **2023**, *23*, 3680. [CrossRef]
48. Kim, J.H.; Lee, H.M.; Jung, S.C.; Chung, D.C.; Kim, B.J. Bamboo-based mesoporous activated carbon for high-power-density electric double-layer capacitors. *Nanomaerials* **2021**, *1*, 2750. [CrossRef]
49. Enterría, M.; Pereira, M.F.R.; Martins, J.I.; Figueiredo, J.L. Hydrothermal functionalization of ordered mesoporous carbons: The effect of boron on supercapacitor performance. *Carbon* **2015**, *95*, 72–83. [CrossRef]
50. Kong, S.; Xiang, X.; Jin, B.; Guo, X.; Wang, H.; Zhang, G.; Huang, H.; Cheng, K. B, O and N codoped biomass-derived hierarchical porous carbon for high-performance electrochemical energy storage. *Nanomaterials* **2022**, *12*, 1720. [CrossRef]
51. Li, P.H.; Wu, W.J. Co-doping mechanism of biomass-derived nitrogen-boron porous carbon and its applications in energy storage and environmental purification. *J. Alloys Compd.* **2024**, *1002*, 175098. [CrossRef]
52. Wang, C.H.; Wen, W.C.; Hsu, H.C.; Yao, B.Y. High-capacitance KOH-activated nitrogen-containing porous carbon material from waste coffee grounds in supercapacitor. *Adv. Power Technol.* **2016**, *27*, 1387–1395. [CrossRef]
53. Poochai, C.; Srikhaow, A.; Lohitkarn, J.; Kongthong, T.; Tuantranont, S.; Tuantranont, S.; Primpray, V.; Maeboonruan, N.; Wisitsoraat, A.; Sriprachuabwong, C. Waste coffee grounds derived nanoporous carbon incorporated with carbon nanotubes composites for electrochemical double-layer capacitors in organic electrolyte. *J. Energy Storage* **2021**, *43*, 103169. [CrossRef]
54. Hadebe, L.; Cele, Z.; Gumbi, B. Properties of porous carbon electrode material derived from biomass of coffee waste grounds for capacitive deionization. *Mater. Today Proc.* **2022**, *56*, 2178–2183. [CrossRef]
55. Mengesha, D.N.; Abebe, M.W.; Appiah-Ntiamoah, R.; Kim, H. Ground coffee waste-derived carbon for adsorptive removal of caffeine: Effect of surface chemistry and porous structure. *Sci. Total Environ.* **2022**, *818*, 151669. [CrossRef]

56. Chiang, C.; Chen, J.; Lin, J. Preparation of pore-size tunable activated carbon derived from waste coffee grounds for high adsorption capacities of organic dyes. *J. Environ. Chem. Eng.* **2020**, *8*, 103929. [CrossRef]
57. Qian, M.; Xuan, X.Y.; Pan, L.K.; Gong, S.Q. Porous carbon electrodes from activated wasted coffee grounds for capacitive deionization. *Ionics* **2019**, *25*, 3443–3452. [CrossRef]
58. Liu, Y.; Lee, D.J.; Ahn, H.J.; Nam, S.Y.; Cho, K.K.; Ahn, J.H. Waste coffee grounds-derived carbon: Nanoarchitectured pore-structure regulation for sustainable room-temperature sodium–sulfur batteries. *Renew. Energy* **2013**, *212*, 856–874. [CrossRef]
59. Liu, S.H.; Huang, Y.Y. Valorization of coffee grounds to biochar-derived adsorbents for CO_2 adsorption. *J. Clean. Prod.* **2018**, *175*, 354–360. [CrossRef]
60. Park, M.H.; Yun, Y.S.; Cho, S.Y.; Kim, N.R.; Jin, H.J. Waste coffee grounds-derived nanoporous carbon nanosheets for supercapacitors. *Carbon Lett.* **2016**, *19*, 66–71. [CrossRef]
61. Pandey, K.; Jeong, H.K. Coffee waste-derived porous carbon based flexible supercapacitors. *Chem. Phys. Lett.* **2022**, *809*, 140173. [CrossRef]
62. Rufford, T.E.; Hulicova-jurcakova, D.; Zhu, Z.; Lu, G.Q. Nanoporous carbon electrode from waste coffee beans for high performance supercapacitors. *Electrochem. Commun.* **2008**, *10*, 1594–1597. [CrossRef]
63. Jutakridsada, P.; Prajaksud, C.; Kuboonya-Aruk, L.; Theerakulpisut, S.; Kamwilaisak, K. Adsorption characteristics of activated carbon prepared from spent ground coffee. *Clean. Technol. Environ.* **2016**, *18*, 639–645. [CrossRef]
64. Hung, Y.H.; Liu, T.Y.; Chen, H.Y. Renewable Coffee Waste-Derived Porous Carbons as Anode Materials for High-Performance Sustainable Microbial Fuel Cells. *ACS Sustainable Chem. Eng.* **2019**, *7*, 16991–16999. [CrossRef]
65. Hossain, R.; Nekouei, R.K.; Mansuri, I.; Sahajwalla, V. In-situ O/N-heteroatom enriched activated carbon by sustainable thermal transformation of waste coffee grounds for supercapacitor material. *J. Energy Storage* **2021**, *33*, 102113. [CrossRef]
66. Hsieh, T.H.; Wang, H.L.; Yu, G.T.; Huang, G.M.; Lin, J.H. Meso-pore dominant activated carbon from spent coffee grounds for high-performance electrochemical capacitors in organic electrolyte. *J. Environ. Chem. Eng.* **2021**, *9*, 106418. [CrossRef]
67. Kim, M.J.; Choi, S.W.; Kim, H.; Mun, S.; Lee, K.B. Simple synthesis of spent coffee ground-based microporous carbons using K_2CO_3 as an activation agent and their application to CO_2 capture. *Chem. Eng. J.* **2020**, *397*, 125404. [CrossRef]

batteries

MDPI

Article

Interfacial Layer ("Interlayer") Addition to Improve Active Material Utilisation in Lithium–Sulfur Batteries: Use of a Phenylsulfonated MWCNT Film

Luke D. J. Barter [1], Steven J. Hinder [2], John F. Watts [2], Robert C. T. Slade [1] and Carol Crean [1,*]

[1] School of Chemistry and Chemical Engineering, University of Surrey, Guildford GU2 7XH, UK
[2] School of Mechanical Engineering Sciences, University of Surrey, Guildford GU2 7XH, UK
* Correspondence: c.crean@surrey.ac.uk

Abstract

Films of functionalised multiwalled carbon nanotubes (MWCNTs) were fabricated as interlayers (interfacial layers between the cathode and separator) in a lithium–sulfur battery (LSB). Phenylsulfonate functionalisation of commercial MWCNTs was achieved via diazotisation to attach lithium phenylsulfonate groups and was characterised by IR and XPS spectroscopies. SEM-EDX showed sulfur and oxygen colocations due to the sulfonate groups on the interlayer surface. However, CHNS elemental microstudies showed a low degree of functionalisation. Without an interlayer, the LSB produced stable cycling at a capacity of 600 mA h $g^{-1}{}_{sulfur}$ at 0.05 C for 40 cycles. Using an unfunctionalised interlayer as a control gave a capacity of 1400 mA h $g^{-1}{}_{sulfur}$ for the first cycle but rapidly decayed to the same 600 mA h $g^{-1}{}_{sulfur}$ at the 40th cycle at 0.05 C, suggesting a high degree of polysulfide shuttling. Adding a lithium phenylsulfonated interlayer gave an initial capacity increase to 1100 mA h $g^{-1}{}_{sulfur}$ that lowered to 800 mA h $g^{-1}{}_{sulfur}$ at 0.05 C by the 40th cycle, showing an increase in charge storage (33%) relative to the other cells. This performance increase has been attributed to lessened polysulfide shuttling due to repulsion by the phenylsulfonate groups, increased conductivity at the separator-cathode interface and an increase in surface area.

Keywords: lithium–sulfur; battery; MWCNT; interlayer; film; phenylsulfonate; funtionalisation; diazotisation

1. Introduction

Lithium–sulfur batteries (LSBs) have a theoretically high specific energy, reaching 2567 W h kg^{-1}, which is due to the high specific capacity afforded to sulfur cathodes (1675 mA h $g^{-1}{}_{sulfur}$) at high active material loadings ($\geq 70\%_{mass}$ sulfur) [1,2]. However, the achievable specific capacities and cycle stability are lower than that predicted theoretically from the lithium–sulfur electrochemistry. While currently inferior to the cost-effectiveness and cycle stability of lithium-ion batteries (LIBs), LSB improvement paves the way for a future after LIBs/in tandem with LIBs. The most effective applications of LSBs are, therefore, applications that are gravimetrically limited such as in aircraft and spacecraft, owing to their need to be lightweight [3].

LSBs are assembled in the fully charged state, with sulfur in the S_8 form, which is reduced to Li_2S in the fully discharged state [4]. However, this is not a trivial process due to the 16-electron conversion and the series of phase transformations that occur. Elemental

 https://doi.org/10.3390/batteries11070266

sulfur (S_8) is a non-polar, electrically non-conductive, and insoluble material in common LSB (ether/carbonate-based) electrolytes, and, conversely, lithium sulfide (Li_2S) is highly polar but also electrically non-conductive and insoluble in the same LSB electrolytes. This low conductivity can be surmounted by adding components that assist with charge mediation, such as electrically conductive carbonaceous materials [5].

During discharge from S_8 to Li_2S, the active material undergoes conversion through various lithium polysulfide species (Li_2S_n, where $3 \leq n \leq 8$) as well as lithium disulfide (Li_2S_2). Lithium disulfide is only slightly soluble in organic electrolytes and exhibits properties similar to lithium sulfide [6,7]. Polysulfides, by comparison, have covalent bridging sulfur ($S_B{}^0$) bonds between two sulfur atoms in a chain as well as charged sulfur chain termini ($S_T{}^{1-}$), and this mix of ionic and covalent centres makes these molecular ions polar, electronically conductive and soluble in ether/carbonate-based electrolytes [8,9]. This rapid increase in cathode material solubility can allow migration of the active material from the cathode to the electrolyte and separator across to the anode, that migration infamously known in the battery community as the polysulfide shuttle [10]. If these polysulfides continue past the separator, and react and deposit on the anode, the capacity is lowered, and the cycling stability is threatened due to the internal chemical short circuit [11].

It is, however, well established that LSB performance can be improved by the addition of various films and coatings to the surface of a cell's cathode [12,13]. Carbonaceous interlayers typically have three modi operandi: improving the cathode surfaces' conductivities through charge mediation, modifying the polarity of the cathode surface during cycling, or a moderation of surface polarity and surface conductivity improvements [14]. Such interlayers facilitate performance enhancements via a greater utilisation of active material, greater cycling stability or both [12].

Carbon nanotubes (CNTs) are high-surface-area conductive nanomaterials. Decorating CNTs with functional groups (to make the CNT surfaces polar) prior to fabrication into a film could add much-needed polarity at the cathode–separator interface as well as imparting additional conductivity at the cathode surface [15]. Table 1 provides a summary of different literature works where functionalised carbons were used in the assembly of LSBs, including the format of the addition, the type of functionalisation and synthesis technique, and the specific capacities obtained and the improvements relative to the control cells. However, the electrolyte-to-sulfur (E/S) ratios used are typically very large and, consequentially, are often omitted from published works (as seen in Table 1).

Lithium phenylsulfonate groups can be added to carbon surfaces by diazotisation of sulfanilic acid, with heating to decompose the in situ formed diazonium salt and subsequent work-up with lithium hydroxide to deprotonate the sulfonic acid group [16]. Lithium sulfonate functionalisation of the porous carbon sulfur-host has been shown to hinder the polysulfide-shuttle effect [17]. A phenylsulfonate group is thought to be even more beneficial than a sulfonate group as the aromatic ring enhances the sulfonic acid's pKa from -1.9 (as R-CH_2SO_3H) to -2.8 (as R-$C_6H_4SO_3H$), leading to greater acidity and dissociation, as well as being large groups which theoretically can size exclude polysulfide losses to a better extent [18].

Table 1. A literature review comparison of functionalised carbonaceous materials added in the assembly of lithium–sulfur batteries. The comparison includes the format of the carbon addition, the type of functionalisation added and its method of addition, the discharge improvement and the C-rate of the improvement, if the E/S ratio of the cell is given, and the associated reference of the work.

Format	Functionalisation	Method of Addition	Cell Capacity (mA h $g^{-1}{}_{sulfur}$)	Improvement (mA h $g^{-1}{}_{sulfur}$)	C-Rate (h^{-1})	Refs
Cathode	Methylsulfonation	Strecker sulfite alkylation	~820 (@ 100th cycle)	+ ~270 (@ 100th cycle)	0.1	[17]
Cathode	PEI +	Grafted to CNT surface	~750 (@ 100th cycle)	+ ~350 (@ 100th cycle)	0.5	[19] *
Cathode	N-doped CNTs grown on Co_3O_4	Spray pyrolysis on Co_3O_4	~700 (@ 400th cycle)	+ ~460 (@ 400th cycle)	1.0	[20] *
Cathode	N-doped	Melamine addition during polymerisation	~963 (@ 100th cycle)	+ ~281 (@ 100th cycle)	1.0	[21] *
Cathode/ Interlayer	O or H-doping	Annealing	~750 (@ 250th cycle)	+ ~480 (@ 250th cycle)	0.5	[22] *
Separator coating	Phenylsulfonation	Diazotisation of rGO ‡ with sulfamic acid	~930 (@ 100th cycle)	+ ~286 (@ 100th cycle)	0.5	[23] *
Separator coating	rGO ‡-PEDOT:PSS ◊	Air-controlled Electrospray	~813 (@ 100th cycle)	+ ~516 (@ 100th cycle)	0.5	[24] *
Separator coating	Mixing with PANi [0] †	Filtering a dispersion	~400 (@ 50th cycle)	+ ~50 (@ 50th cycle)	0.2	[25] *

* E/S (μL $mg^{-1}{}_{sulfur}$) omitted and/or missing values to calculate from other parameters. + Polyethylenimine. [0] Polyaniline. † Nanofiber. ‡ Reduced graphene oxide. ◊ Poly(3,4-ethylenedioxythiophene):polystyrene sulfonate.

2. Materials and Methods

2.1. Chemicals and Electrolyte Composition

The sulfur-hosts used in this study were porous carbons that were derived from resole-type phenol-formaldehyde resins; synthesis conditions are fully discussed and specified in our previous paper [26]. The starting chemicals used in this study were phenol (99%, ACS grade, Merck, London, UK); formaldehyde (37% in water [10% methanol], ACS grade, Merck); lithium carbonate (99%, ACS grade, Merck); CTAB [cetyltrimethyl ammonium bromide] (98%, Alfa Aesar, Lancashire, UK); hydrochloric acid (37% in water, Fisher, Leicestershire, UK); sulfur (99.98%, Merck); oxalic acid dihydrate (99%, ACS grade, Merck); ammonium thiosulfate (98%, Merck); MWCNTs [multiwalled carbon nanotubes] (95+% carbon content, 3100 thin MWCNTs, Nanocyl, Sambreville, Belgium); Triton X-100 (laboratory grade, Merck); ethanol (absolute, Fisher); sulfanilic acid (99%, ACS grade, Sigma-Aldrich, Dorset, UK); sodium nitrite, $NaNO_2$ (97+%, ACS grade, Sigma-Aldrich); sulfuric acid (at least $95\%_{mass}$, Fisher); lithium hydroxide monohydrate (98%, Thermoscientific, Leicestershire, UK); Super P® carbon (Timcal, Congleton, UK); NMP [N-methyl 2-pyrolidone] (99%, anhydrous, Merck); PVDF [polyvinylidene difluoride] (~534,000 molecular weight by GPC, Merck); lithium ribbon (99.9, 0.75 mm thick, Merck).

The electrolyte was dissolved in a solvent mixture of 50:50 v/v of DME [1,2-dimethoxyethane] (99.5%, anhydrous, inhibitor-free, Merck) and DOL [1,3-dioxolane] (99.8%, with 75 ppm BHT [butylated hydroxytoluene] inhibitor, anhydrous, Merck); each solvent was dried using molecular sieves (4 Å, 8–12 mesh, Merck). LiTFSI [lithium bis(trifluromethanesulfonyl)imide] (99.95%, Merck) had been dried for 3 days at 120 °C on a Schlenk line before dissolving in the mixed solvents at to a concentration of 1.0 mol dm^{-3}. Lithium nitrate (99.99%, Merck) had also been dried in a vacuum oven overnight at 80 °C. This was added to the electrolyte to a 0.8 mol dm^{-3} concentration. The filtration membrane used was 0.22 μm pore PVDF membrane (Merck).

The cathode current collector was a carbon-coated aluminium foil (MTI Corporation). Coin cell (2032 format) assembly used stainless-steel 304 components with Celgard 2400 polypropylene film separator membrane.

2.2. Functionalisation of Multiwalled Carbon Nanotubes

A quantity of 60 mg of multiwalled carbon nanotubes (MWCNTs) was weighed into a round-bottomed flask in a high-efficiency particulate air (HEPA)-filtered fume hood, and 3.46 g sulfanilic acid and 1.38 g sodium nitrite ($NaNO_2$) were added.

The mixture was magnetically stirred to mix the powders with a stopper in place. The stopper was then removed and 10 cm^3 of concentrated sulfuric acid was next added dropwise. After the highly exothermic acidification step, the black paste cooled for 20 min, and the paste was then stirred at 60 °C for 1 h (Figure 1).

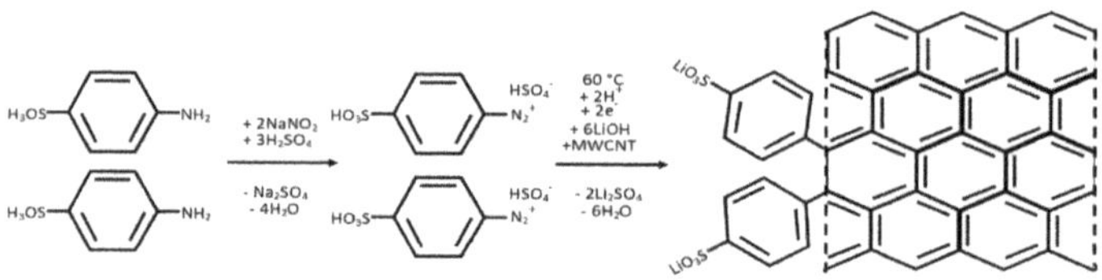

Figure 1. Lithium phenylsulfonation (grafting of the phenylsulfonate moiety) (via diazotisation) of MWCNTs.

When cooled, the paste was vacuum-filtered using a PVDF membrane (0.22 μm) and washed with acetone. Washing was continued until the filtrate passed through the membrane clear and colourless. The modified MWCNTs were washed with 100 cm^3 of 1.0 mol dm^{-3} lithium hydroxide to remove aqueous impurities and neutralise surface sulfonic acid groups, followed by deionised water. The functionalised MWCNTs were collected (while still damp) and ultrasonicated (stainless-steel sonic horn till) for 30 min in 80 cm^3 of deionised water to form a homogeneous dispersion.

2.3. Fabrication of a Functionalised MWCNT Interlayer Film

Once ultrasonicated, the dispersion above was re-filtered through fresh PVDF membrane and washed with 50 cm^3 lithium hydroxide (1 mol dm^{-3}) to further allow conversion of the sulfanilic acid groups to lithium sulfonate groups. After leaving to dry on the pumped Buchner funnel for a further 2 h, the entire PVDF membrane (topped with a thin and even layer of MWCNTs) was dried overnight in an oven at 40 °C. The PVDF membrane was subsequently peeled away, yielding a film of functionalised MWCNTs.

2.4. Fabrication of an Unfunctionalised MWCNT Interlayer Film

As a control, a film of unfunctionalised MWCNTs was made by dispersing 60 mg MWCNTs in $1\%_{mass}$ aqueous Triton X-100 and sonication as above for 30 min. The homogeneous dispersion was vacuum-filtered as before, and the filtrate was washed using deionised water and ethanol to remove the Triton X-100 surfactant. The MWCNT film was dried in an oven at 40 °C overnight, followed by removal of the PVDF membrane to reveal a self-supportive film of unfunctionalised MWCNTs with a thickness of ~100 μm.

Circular cuttings (14 mm diameter, with masses of ~3 mg) of the carbon films (functionalised and unfunctionalised) were used in electrochemical studies of lithium–sulfur coin cells, the films being placed between the cathode (the syntheses of which were given in previous work) and the separator membrane in each case [26].

2.5. Cell Production

The cathodes were a composition of sulfur, porous carbon (Reg-600, see previous work), Super P® and PVDF of 70:10:10:10 ($\%_{mass}$), respectively [26]. The porous carbon and sulfur were in the form of a 1:7 mass composite (by melting or depositing [via thiosulfate acidification] the sulfur into the carbon) of porous carbon (pyrolysed resole) to sulfur. This, in turn, would constitute 80$\%_{mass}$ of the final coating. The super P® conductive additive was as a 4.2$\%_{mass}$ ink that was dispersed using NMP (10$\%_{mass}$ of the dried film) [26]. The coatings were applied to a carbon-coated aluminium foil and dried to a thickness of 400 μm followed by vacuum-drying at 50 °C overnight. The resulting cathodes were cut to 14 mm diameter discs, at areal loadings of sulfur of 2.1–3.7 mg_{sulfur} cm^{-2}, to be used in CR2032 coin cells.

Cell assembly: All cell assembly occurred in an MBRAUN UNIlab pro dry box that was in an argon atmosphere (H_2O and O_2, both <0.1 ppm). CR2032 coin cells (with 304 stainless-steel casing, spring and spacer) used lithium electrodes with a 14 mm diameter, and a 16 mm diameter Cellgard 2400 membrane was placed atop the lithium. The electrolyte was syringed on the separator at a project-specific (the Faraday Institution's LiSTAR project) E/S ratio of 10 μL $mg^{-1}{}_{sulfur}$, based on the sulfur content on the cathode coating irrespective of interlayer addition (i.e., 10 μL $mg^{-1}{}_{sulfur}$ for cells assembled with or without an interlayer). The electrolyte used was a 50:50 v/v DOL to DME, with 1.0 mol dm^{-3} LiTFSI and 0.8 mol dm^{-3} $LiNO_3$. MWCNT films were added between the cathode and the separator. Cells were crimp-sealed within the dry box.

2.6. Characterisation

Materials characterisation: Raman spectroscopy was conducted on a Thermo Scientific DXR Raman microscope using a 532 nm excitation laser (6 mW power setting) and a grating with 900 lines mm^{-1} at 5 cm^{-1} resolution. Every spectrum was taken as an average of ten 1 s accumulations. All spectra were baseline corrected and normalised to the carbon G band at 1592 cm^{-1}. Spectral transformations used Spectragryph (version 1.2.16.1) software.

X-ray diffraction (XRD) profiles were taken with the use of a Malvern Panalytical X'pert powder diffractometer; the recording conditions were 2 h measurements between $10° \leq 2\theta \leq 70°$ at 40 mA and 45 kV with a Cu K_α radiation source. X-ray photoelectron spectroscopy (XPS) results were taken with the use of a Thermo Scientific K-Alpha$^+$ spectrometer (UK) using a monochromatic Al K_α (hν = 1486.6 eV, where 1 eV = 96.487 kJ mol^{-1}) X-ray source; the spot had an approximate radius of 400 μm. All survey spectra used a 200 eV pass energy, and each core level spectrum used a 50 eV pass energy. Core level spectra were corrected for charging effects by charge referencing relative to the C1s peak at 285.0 eV.

Infrared (IR) spectra were recorded using a PerkinElmer Spectrum Two FT-IR spectrometer (UK) with the use of a diamond/ZnSe UATR accessory. Each spectrum consisted of an average of 32 Fourier transform (FT) scans, at 2 cm^{-1} resolution between 4000–400 cm^{-1}. All scanning electron microscopy (SEM) images studied samples with a 3 nm gold coating were recorded on a Thermo Scientific Apreo S electron microscope (UK) that used a secondary electron detector, at a 1.6 nA current and 5 kV energy. Elemental mapping utilised an energy-dispersive X-ray (EDX) detector (UK). Nitrogen sorptiometry for porosity and specific surface area studies was undertaken at 77 K using a Belsorp mini II (UK) soptiometer (UK). Elemental microanalysis (CHNS studies) was performed with a Thermo Scientific Thermo FlashEA 1112 in a helium environment following dynamic flash combustion (DFC) gas at 1800 °C; separation and quantification of the vaporised products was performed with a gas chromatography (GC) column and a thermal conductivity detector (TCD), respectively.

Electrochemical studies: Galvanostatic discharge–charge (GDC) cycling at 0.05 C was operated within 1.8–2.8 V window using a Solartron SI 1470E potentiostat (UK). All currents were solely based on the active material in the cathode; any added interlayer bore no influence on the applied GDC currents. Electrochemical impedance spectroscopy (EIS) was achieved on a Solartron SI Cell (UK) Test 1400 unit (1455A frequency response analyser channels) working in union with the Solartron SI 1470E potentiostat (UK). Potentiostatic EIS was carried out over the frequency range 1 MHz–10 MHz; the perturbation was 5 mV_{rms} at the open circuit voltage (OCV).

3. Results and Discussion

3.1. Powder X-Ray Diffractometry (XRD)

A lithium phenylsulfonate functionalised MWCNT film was studied as an LSB interlayer placed between the cathode and separator. A control unfunctionalised MWCNT film was also fabricated. Figure 2 shows XRD diffraction profiles of both films and shows no discernible difference between their profiles. The two key features in both are broad peaks at 26° and 44° of 2θ (signified by the aromatic rings), corresponding to the (002) and (100) crystal planes of elemental carbon, respectively [27]. Instead of the sharp peaks given by high-crystallinity elemental carbon, the peaks are broad due to the nanoscale dimensions of the MWCNTs [28].

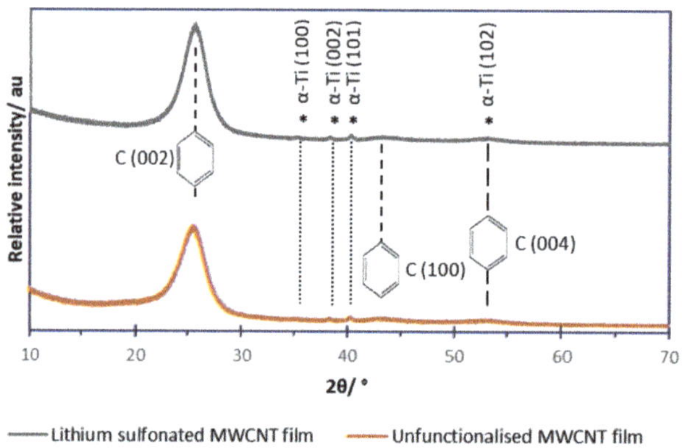

Figure 2. The powder XRD profiles of the MWCNT films. The crystal planes of carbon (shown by the aromatic ring) and hexagonal α-Ti (shown by the asterisk) are overlaid.

The profiles in Figure 2 both show very low-intensity peaks at 35.2°, 38.3°, 40.2° and 53.2° of 2θ (signified by the asterisks); these peaks are also present for the commercially available MWCNTs (Nanocyl) before functionalisation, and the peaks are consistent with a low content of elemental titanium metal (α-phase, hexagonal). Titanium surfaces are frequently used in the synthesis of CNTs through anodising a titanium growth substrate for CNTs, which is subsequently etched away [29].

3.2. Nitrogen Sorptiometry

Figure 3 shows the N_2 sorption–desorption isotherms at 77 K for the MWCNT films; the films themselves were studied because the parent nanomaterial powders would not accurately reflect the film's surface area or porosity. The isotherms of both films are Type IV with H1 hysteresis, which characterises both films as mesoporous (with pores with 2 < pore diameter (nm) < 50) and with pores of cylindrical shape. The modelled porosity data are given in Figure S1 (Supplementary Information).

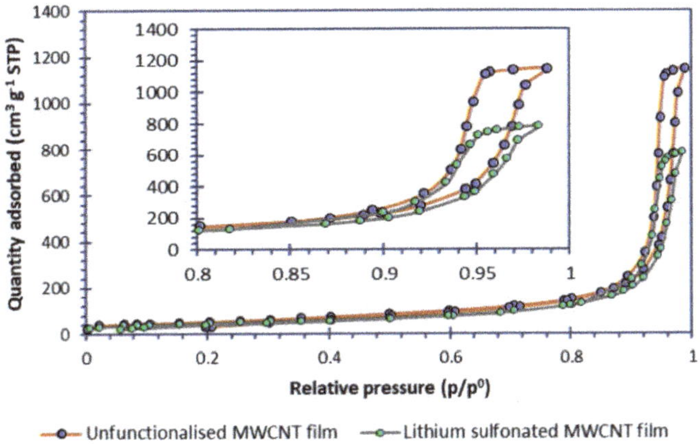

Figure 3. The nitrogen sorption isotherms for the MWCNT film.

Modelling from the literature suggests that micropores are effective at polysulfide adsorption relative to mesopores, yet, for this interlayer, application pores that have small diameters would theoretically be ineffective [30,31]. When the interlayer is placed between the cathode and separator of an LSB, the film ideally should provide large enough mesopores to permit migration of the solvent (both DOL and DME) and solvated lithium ions.

Using a model based on Brunauer–Emmett–Teller (BET) theory, the specific surface areas (computed using the typical mesoporous range of 0.05–0.35 P/P°) are shown in Table 2. The functionalised MWCNT film showed a lower specific surface area relative to the unfunctionalised MWCNT film. This provides indirect evidence of the functionalisation (grafted polar lithium phenylsulfonate groups), which improved MWCNTs' dispersibility in water without the use of surfactant (that additive was needed in the case of the dispersion of unfunctionalised MWCNTs). The better dispersed solution produced a film with a lower specific surface area because of less interstitial large mesoporosity and less microporosity [32]. The lower specific surface area observed for the lithium sulfonated MWCNT film was attributed to more agglomerations and bundles of MWCNTs in dispersion in aqueous media, relative to the unfunctionalised film, which used a Triton X-100 surfactant in its synthesis [33].

Table 2. The specific surface areas of the unfunctionalised MWCNT and lithium phenylsulfonated MWCNT films, which were themselves computed from the respective isotherm data (Figure 3).

Sample	Specific Surface Area # (m^2 g^{-1})
Unfunctionalised MWCNT film	197.4 ± 2.2
Functionalised MWCNT film	163.2 ± 0.4

Calculated using the Brunauer–Emmett–Teller (BET) model and computed using the Micromeritics® 3Flex MicroActive (version 5.02) software within the relative pressure (P/P°) range of between 0.05 and 0.35.

3.3. Scanning Electron Microscopy (SEM)

Figure 4 shows SEM micrographs relating to the unfunctionalised MWCNT film (a and b) and those relating to the lithium phenylsulfonated MWCNT film (c and d). Since both MWCNT films consist of carbon nanotubes, the predominant element detected is carbon (given by blue in the EDX maps), and the carbon presence is shown as a widely distributed expanse of carbon. The red and yellow false-colourings of EDX mapping images in Figure 4b,d are due to oxygen and sulfur colocations on the surface of the functionalised film that are not present on the unfunctionalised film; oxygen and sulfur are within the functional groups of the lithium sulfonated MWCNT films. Figure 4c,d show the functionalised film with a dark blue (carbon) heterogeneity defect formed during sample preparation for SEM. The defect exposes the lower carbon tape that was used to fix the sample to the SEM sample stub. This shows and contrasts the extent of functionalisation.

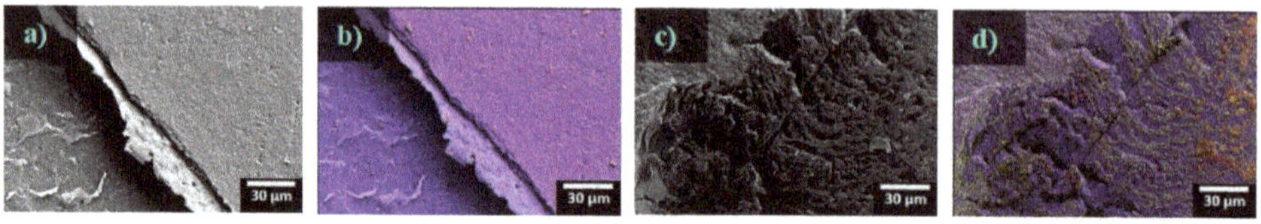

Figure 4. Scanning electron micrographs of the (**a**) unfunctionalised and (**c**) lithium sulfonated MWCNT films and their false-coloured EDX elemental maps of the (**b**) unfunctionalised film and (**d**) lithium sulfonated films. The false colours correspond to carbon (blue), oxygen (red) and sulfur (yellow).

3.4. *X-Ray Photoelectron Spectroscopy (XPS)*

Figure 5 shows the XPS spectra for each film type. Figure 5a is the survey spectrum. Changes occur in element-specific spectra before and after functionalisation, especially the sulfur and oxygen high-resolution spectra present in Figures 5d and 5e, respectively.

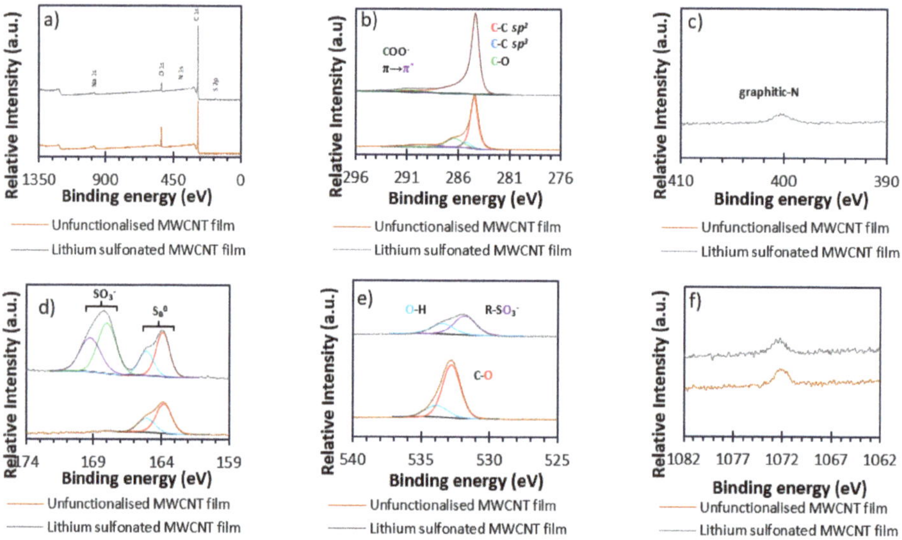

Figure 5. The high-resolution XPS spectra of the unfunctionalised and lithium sulfonated MWCNT films of (**a**) survey spectra, (**b**) C 1s, (**c**) N 1s, (**d**) S 2p, (**e**) O 1s and (**f**) Na 1s.

3.4.1. Carbon (Figure 5b)

The spectrum for the unfunctionalised MWCNT film gives a major peak at 284.5 eV and a minor peak at 286.6 eV for the film of unfunctionalised MWCNTs; those peaks correspond to aromatic sp^2-type carbon (expected for CNTs) and C-O bonds, respectively [34,35]. After functionalisation, the minor peak is no longer visible, and the same major peak persists, which may suggest that the added functionality eclipses the previously detected C-O bonds. The fact that, following the surface functionalisation, the major carbon peak associated with the surface is unchanged at 284.5 eV (corresponding to aromatic sp^2-type carbon), rather than shifting to a binding energy closer to 285.0 eV (corresponding to sp^3-type carbon), could suggest a very light degree of functionalisation. However, this may also reflect that the added lithium sulfonate functional groups also contain a phenyl (sp^2-carbon/aromatic) ring. Closer inspection of the fitted peaks for carbon XPS spectra are shown for both MWCNT films in Figure S3.

3.4.2. Nitrogen (Figure 5c)

The unfunctionalised MWCNT film shows no detectable nitrogen presence, but, following the diazotisation reaction, the functionalised MWCNT film has a nitrogen peak–the MWCNT film has been azotised, introducing C-N bonds. The peak at 401.0 eV (assigned to graphitic C-N bonds) was thought to be added as a by-product during the multi-stage functionalisation [36].

3.4.3. Sulfur (Figure 5d)

Prior to surface functionalisation, the sulfur-specific peaks are consistent with bridging/elemental sulfur's ($S_B{}^0$) $2p_{1/2}$ and $2p_{3/2}$ peaks at 165.2 and 164.0 eV [37]. Yet, following the functionalisation, these peaks coexist with another set of peaks at 168.9 and 167.7 eV, corresponding to $2p_{1/2}$ and $2p_{3/2}$ of added sulfonate groups on the interlayer's surface [38].

3.4.4. Oxygen (Figure 5e)

The oxygen-specific peak of the unfunctionalised film gave a single peak at 532.5 eV with a shoulder at 533.6 eV, which were assigned to carbonyl/carboxyl C-O groups and carboxyl and alcohol O-H groups (also shown in the carbon-specific peaks), respectively [39]. However, once functionalised, these peaks are replaced by a major peak corresponding to sulfonate groups' S-O bonds at 531.6 eV and a minor shoulder at 533.6 eV, consistent with carboxyl and alcohol O-H groups [40].

3.4.5. Sodium (Figure 5f)

The high-resolution sodium spectra also show peaks at 1072.3 eV as Na^+ impurities in the unfunctionalised MWCNT, as well as the MWCNTs following functionalisation, implying that this impurity predates functionalisation using sodium nitrite to form nitrous acid for the formation of the diazonium ion in situ [41]. Elemental surface concentrations are given in Table 3.

Table 3. The elemental surface concentrations of the unfunctionalised and lithium phenylsulfonated MWCNT films as determined through XPS.

Sample	C ($\%_{atom}$)	N ($\%_{atom}$)	S ($\%_{atom}$)	O ($\%_{atom}$)	Na ($\%_{atom}$)
Unfunctionalised MWCNT film	87.3	<LOD	0.4	12.1	0.1
Fuctionalised MWCNT film	92.7	0.6	1.1	5.5	0.1

3.5. *Infrared (IR) Spectroscopy*

The two MWCNT film types show different infrared spectra (Figure 6), due to surface functionalisation of the carbon that is fully consistent with the XPS data. The unfunctionalised film shows little-to-no surface functionality (no distinct sharp and/or broad peaks) in the range 4000–400 cm^{-1}. However, following lithium phenylsulfonate functionalisation, several peaks appear at frequencies <1600 cm^{-1}.

Narrow peaks are present at 464, 700, 854 and 1057 cm^{-1}, and broad peaks between 1200–1600 cm^{-1}, with maxima at 1362 and 1450 cm^{-1}. The sharp narrow peaks are assigned to vibration modes involving coordinated Li^+ (464 cm^{-1}), C=C and C-H bending (700 and 854 cm^{-1}, respectively) of a para-substituted phenyl ring. The broad peak at 1600–1400 cm^{-1} (maximum at 1450 cm^{-1}) is assigned as para-substituted phenyl ring C=C stretching. The two remaining peaks are assigned to vibrations shared by sulfonate groups at 1057 (SO_3) and 1362 cm^{-1} (S=O) [42]; those groups are assigned to sulfonate rather than sulfonic acid groups, as there is no broad peak about 3300 cm^{-1} that would correspond to an O-H stretch. This implies total conversion of the arylsulfonic acid groups to arylsulfonate groups through washing with lithium hydroxide; sulfonates also have subtle spectroscopic differences when compared to sulfonic acids; the SO_3 mode of a sulfonic acid group occurs at a lower frequency, approximately 1033 cm^{-1} [42,43].

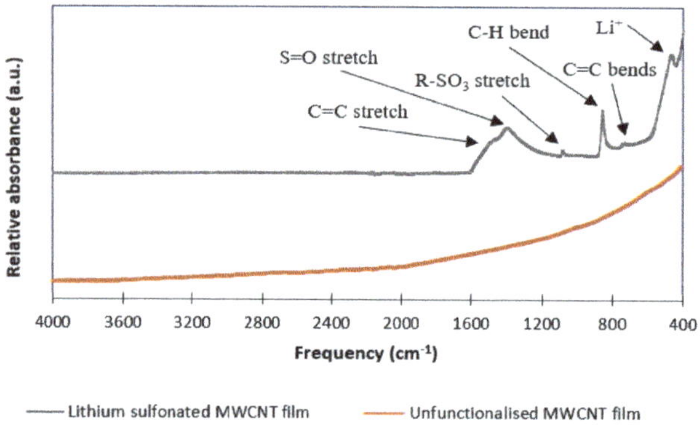

Figure 6. The FT-ATR-IR absorbance spectra of the unfunctionalised and lithium phenylsulfonate functionalised MWCNT films.

3.6. Raman Spectroscopy

The Raman spectra (Figure S3) show a very similar, unremarkable carbon spectrum for both non-functionalised and functionalised films. The two significant peaks in both spectra occur at 1343 and 1576 cm^{-1}, corresponding to the D-band and the G-band of the carbon nanotubes, respectively. The I_D/I_G ratio (being a measure of disorder/order) was found to vary by 0.09 (1.29 for the unfunctionalised and 1.20 for the lithium phenylsulfonated MWCNT films).

A carbon D′-band is also observed at 1615 cm^{-1} for both film types and is commonly seen for MWCNTs [44,45]. There are also two lower energy peaks present, at 2692 and 2957 cm^{-1}, which are the 2D and (D + G) combination-bands respectively. The only difference between the spectra of MWCNTs before and after phenylsulfonation is the addition of low-intensity peaks close to the baseline around the D- and G-bands for the sulfonated film. The peak at 1100–1200 cm^{-1} was assigned to a complex vibration ν(C, S, SO_3) due to functionalisation [46]. The peak at 1420–1520 cm^{-1} is consistent with an added phenyl ring which introduces aromatic ν(C-C) [46].

3.7. Elemental Microanalysis

The CHNS results in Table 4 show that both films were >85$\%_{mass}$ carbon (irrespective of functionalisation). Once functionalised, the carbon content reduced by 0.52$\%_{mass}$ C. The nitrogen and sulfur contents rose after functionalisation as a direct consequence of the addition route via diazotised functional groups. The sulfur content simultaneously increased by only 0.97$\%_{mass}$ S after functionalisation; this is believed to be (i) due to a phenyl ring being added during the addition of the sulfonate group (44$\%_{mass}$ of the added lithium phenylsulfonate moieties is carbon) and (ii) due to light decoration of MWCNTs with those sulfonate groups; CHNS microananalysis is a gravimetric average of the surface and subsurface/bulk of the sample. In comparison, the XPS elemental surface concentrations (shown in Table 3) show a large 12.1$\%_{atom}$ O surface concentration before lithium phenylsulfonate functionalisation of the MWCNTs (due to surface impurities), which reduced to 5.5$\%_{atom}$ O after functionalisation.

Table 4. CHNS analyses of the unfunctionalised and lithium phenylsulfonated MWCNT films.

Sample	C ($\%_{mass}$)	H ($\%_{mass}$)	N ($\%_{mass}$)	S ($\%_{mass}$)
Unfunctionalised MWCNT film	86.50	2.01	<LOD	0.66
Functionalised MWCNT film	85.98	2.10	0.62	1.63

Limit of detection = $0.10\%_{mass}$. Measurement error = $\pm 0.30\%_{mass}$.

3.8. In-Cell Electrochemical Investigations via GDC Cycling

The low E/S of 10 μL $mg^{-1}{}_{sulfur}$ did not factor in an interlayer's addition, which, therefore, precluded the fast cycling at C-rates of 1–2 C due to accelerated electrolyte depletion at higher discharge/charge rates [47]. Therefore, slow cycling at 0.05 C was used as the testing C-rate.

Figure 7 shows the GDC cycling data of three differing cells of the same cathode composition (where the sulfur has been deposited on, as opposed to melted in, the resole-type carbon via thiosulfate acidification [i.e., the composites are deposit-loaded]) but with the presence of (and if present, types of) carbon interlayers [26]. Figure 7a,d shows cycling of a cell over 40 cycles at 0.05 C for a cell without a carbonaceous interlayer. The cell showed great cycling stability but only displayed a specific capacity of ca. 600 mA h $g^{-1}{}_{sulfur}$ (the theoretical specific capacity of an LSB being 1675 mA h $g^{-1}{}_{sulfur}$). When studying the discharge–charge profile of that cell (Figure 7a), the poor performance was due to poor recovery of the lower discharge plateau (~2.10 V) associated with lower-order polysulfides (low n-value $S_n{}^{2-}$) [48].

Figure 7b,e show the cycling performance of a cell assembled including an unfunctionalised MWCNT film (using the same cathode as in Figure 7a); the specific capacities for the first few cycles were greater (1440 mA h $g^{-1}{}_{sulfur}$ at the first cycle, with a reduction of 474 mA h $g^{-1}{}_{sulfur}$ by the second cycle) relative to the specific capacities of the cell assembled without an interlayer but decayed to 700 mA h $g^{-1}{}_{sulfur}$ by the seventh cycle. Moreover, the coulombic efficiency of the first cycle was ~116% (off the scale shown in Figure 7e), supporting the hypothesis that unfunctionalised carbonaceous interlayers facilitate charge transfer yet expedite irreversible losses associated with side reactions. This result was repeatable with similar fresh cells. The cell's capacity then decayed slowly to 600 mA h $g^{-1}{}_{sulfur}$ at the fortieth cycle. After 40 cycles, and because of polysulfide shuttling, the cell performed closely and similarly to the cell assembled without the unfunctionalised interlayer. Therefore, adding an unfunctionalised MWCNT interlayer increased the conductive surface (thereby increasing sulfur utilisation), but as the interlayer surface was not optimised, since it was unfunctionalised (and therefore non-polar), it was unable to retain the soluble and polar polysulfide intermediates [10,31].

Figure 7c,f show the performance for a cell including a functionalised MWCNT interlayer, showing an initial specific capacity of 1100 mA h $g^{-1}{}_{sulfur}$ at 0.05 C (with a reduction of 20 mA h $g^{-1}{}_{sulfur}$ by the second cycle). The specific capacities varied slightly over 40 cycles, but the specific capacity remained high (800 mA h $g^{-1}{}_{sulfur}$ at cycle 40). That indicates that the functionalised interlayer improves the performance of the cell, producing a longer lower voltage plateau. However, the lower effective E/S, caused by the necessary wetting for the added interlayers, increased the polarisation voltages experienced by the cells assembled with interlayers, relative to the cells assembled without an interlayer [49]. All cells produced coulombic efficiencies of 98–100% from the second cycle onwards; this suggests that the added lithium nitrate in the electrolyte largely forestalled rapid polysulfide shuttling across the cells by forming a stable solid–electrolyte interface (SEI) on the anode surface [50,51]. However, the greatest extent of polysulfide shuttling was the cell assembled with an unfuctionalised MWCNT interlayer, which also corresponded to the lowest coulombic efficiencies that were displayed. What appears to be clear is that the interlayers initially improve the sulfur utilisation and polysulfide conversion kinetics within the LSBs. However, the rapidness of the capacity loss that eventually ensues can be delayed and limited in the initial cycles by adding functionality. The capacity reduction within the first 10 cycles were shown to be ~ 740 mA h g^{-1}_{sulfur} for a cell with an unfunctionalised interlayer and 160 mA h g^{-1}_{sulfur} for a cell with a lithium phenylsulfonated interlayer.

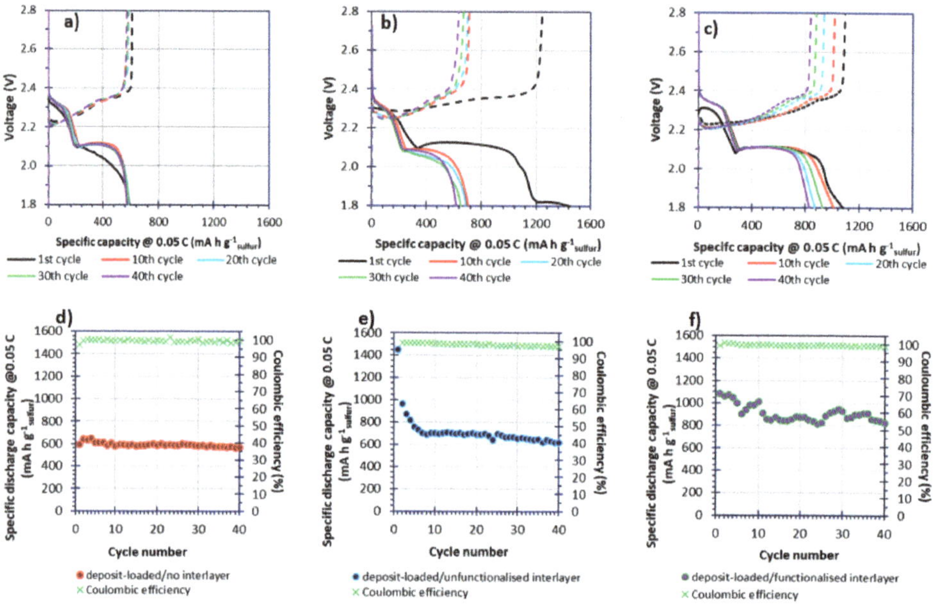

Figure 7. The 10-cycle interval discharge (solid line) and charge profiles (dash line) for a deposit-loaded cell with (**a**) no added interlayer, (**b**) an unfunctionalised MWCNT interlayer and (**c**) a lithium sulfonated MWCNT interlayer, as well as the relationship over 40 cycles at 0.05 C (**e**), (**d**) and (**f**) [respective for (**a–c**)].

Figure S4a,b show a cell with a cathode of melt-loaded sulfur on Reg-600 carbon and assembled with a functionalised MWCNT interlayer (to compare against the behaviour observed in Figure 7c,f with deposition-loaded sulfur composite on the cathode). The melt-loaded cathode led to a greatly increased initial specific capacity of ca. 1000 mA h g^{-1}_{sulfur} at 0.05 C, which slowly decayed to 600 mA h g^{-1}_{sulfur}; from the previous work featuring melt-loaded Reg-600, specific capacities were stable over the same cycling regime at 600 mA h g^{-1}_{sulfur} when assembled without an interlayer [26]. Thus, as sulfur-loadings were consistent between both works (being this work and the cited work), the deposit-loaded cell proved to have better capacity recovery, when an interlayer was used, over the same testing regime (relative to the melt-loaded cell).

The results show that, after only 40 cycles, the cell with the unfunctionalised interlayer performed as well as the cell without any interlayer. Furthermore, the cell with the functionalised interlayer was still ~200 mA h g^{-1}_{sulfur} greater than either of the other cell results at the same C-rate.

The rationale given for the improved upper and lower voltage plateaus (2.30 and 2.10 V, respectively) suggests that the lithium sulfonated interlayer's sulfonate group centres promote, and sustain, favourable conversions for both the higher- (S_n^{2-}, S_8^{2-} to S_4^{2-}) and lower-order polysulfides (S_n^{2-}, S_4^{2-} to S_2^{2-}). Mechanistically, this was thought to be caused by the intermediate polysulfides reacting with the sulfonate group centres, producing oxygenated-polysulfide species [52]. Literature sources have suggested that this effect might be consistent with the formation of intermediate side-products such as dithionite, thiosulfate and polythionate species [53,54].

Mirroring that of the previous study that featured the cells of Reg-600, the observed charge storage capacities were less than the theoretical maximum of 1675 mA h $g^{-1}{}_{sulfur}$ [26]. The reasons for the lower-than-theoretical charge storage abilities of those cathodes was attributed to (i) high-sulfur-content cathodes (a cathode coating of $70\%_{mass}$ sulfur), (ii) the low content of sulfur-host ($10\%_{mass}$) and (iii) the use of such a restrictive 10 μL $mg^{-1}{}_{sulfur}$ E/S ratio for batteries [26]. Yet, the E/S was unchanged from the previous study (10 μL $mg^{-1}{}_{sulfur}$) irrespective of an interlayer's addition in the coin cell [i.e., cells assembled with an interlayer received no more electrolyte relative to the previous study]. Thus, the E/S of 10 μL $mg^{-1}{}_{sulfur}$ was only based on the cathode; the requirement to wet an added interlayer means that the true/effective E/S was more restrictive.

With reference to Figure 8, the proposed mechanisms of the carbonaceous MWCNT interlayers used in this research are three-fold. Firstly, the added conductivity (that is largely neglected at the cathode surface/farthest from the electrode's current collector) greatly improves the charge-transfer mediation needed for sulfur/polysulfide/disulfide/sulfide species at the cathode surface [55]. Secondly, the lithium phenylsulfonate groups that adorn the MWCNT surfaces (if functionalised) act as a charged screen that largely blocks repulsively charged negative, but soluble, polysulfide species during cycling [31]. Finally, the sulfonate centres interacted with the polysulfide species, forming oxygenated polysulfide species which can immobilise other remaining polysulfide species [56,57].

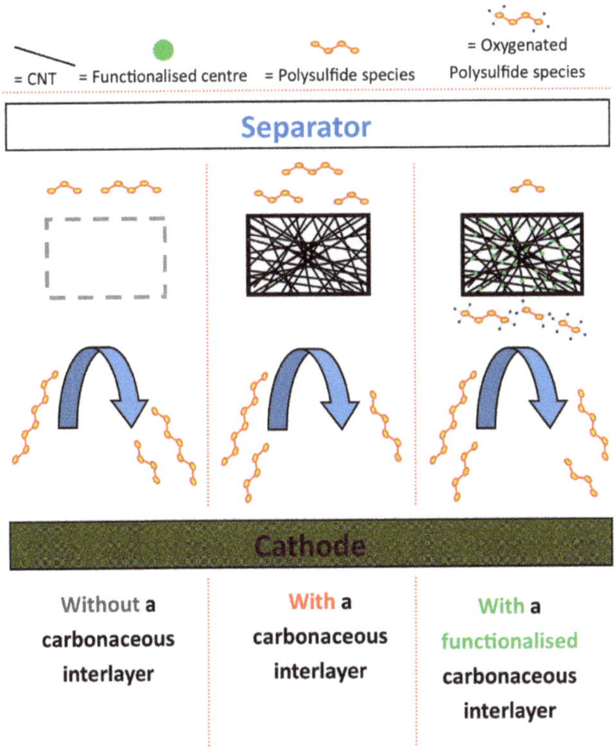

Figure 8. The proposed differing mechanisms for cells assembled without a carbonaceous MWCNT interlayer, cells assembled with an unfuctionalised carbonaceous MWCNT interlayer, and cells assembled with a lithium sulfonate fuctionalised carbonaceous MWCNT interlayer.

3.9. In-Cell Studies Using Electrochemical Impedance Spectroscopy (EIS)

Figure 9 shows EIS data recorded for cells before and after GDC cycling and presents that data in the form of Nyquist plots. The cell in Figure 9a has an unfunctionalised MWCNT interlayer, while the cell in Figure 9b differs by using a functionalised interlayer. Changing to the functionalised interlayer leads to significant changes in the impedance spectra, both before and after GDC cycling.

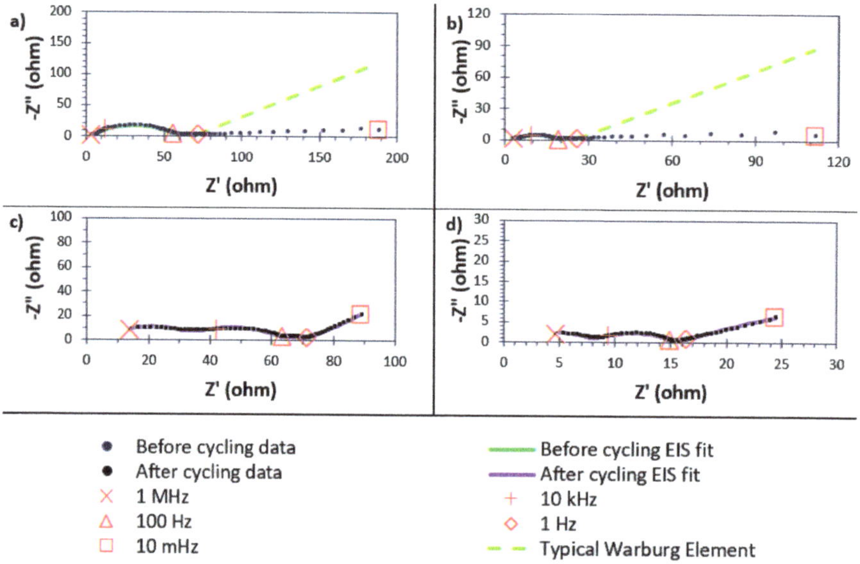

Figure 9. EIS spectra for cells before and after 40 GDC cycles at 0.05 C, showing data points and the output from fitted equivalent circuit. The cells were made with cathodes coated with deposition-loaded sulfur supported on carbon. The cells in (**a**,**c**) are of an unfunctionalised MWCNT interlayer before and after cycling (respectively), whereas the cells in (**b**,**d**) are of a lithium sulfonated MWCNT interlayer before and after cycling (respectively). Blue data points were obtained before cycling, and black data points after cycling. The green line series shows the EIS fitted model before cycling and the red line series shows the EIS fitted model after cycling. The red data points display the frequency of each measured data point every 2 decades. The lime dashed line indicates a typical Warburg Element that would be expected by the resistance increase observed.

At the lowest frequencies (10 Hz and lower), lithium-ion diffusion within porous structures typically leads to analysis in terms of a Warburg element (at approximately 45° from the Z'/real impedance axis) in the Nyquist plot. Before cycling, the cells in this study produce low-frequency slopes at ca. 10° and which cannot be fitted as a Warburg element. Moreover, this effect of a heavily "compressed" charge-transfer semicircle occurred for both cells that were assembled with an interlayer; this suggests that the effect is independent of interlayer fuctionalisation and is instead a function of cell architecture and assembly. This effect was thought to be caused by a combination of (i) the interlayers' increased surface areas required to wet, as well as the cathode, under lean electrolyte conditions (E/S >> 10 μL $mg^{-1}{}_{sulfur}$), and (ii) the increased electrolyte viscosity of the polysulfide species caused by an increase in sulfur utilisation because of the interlayers' added conductivities [58,59]. All of the described reasons could lead to a reduction of ionic and particle diffusion, caused by heterogenous cathode/interlayer make-ups, manifesting as a collapse of the charge-transfer capacitance behaviour.

Similar effects have been observed when studying solid-state (SS) lithium-cobalt-oxide-type LIBs (LCO cells) using the SS electrolyte $Li_{10}GeP_2S_{12}$ (LGPS); the associated paper describes a synthesis procedure for modification of the anode surface [60]. For the LGPS cell, the low-frequency behaviour is attributed to deterioration of the electrolyte–anode interface (of an unfunctionalised anode) and associated lowering of lithium-ion migration to the anode surface [60]. In relation to liquid electrolyte cells with interlayers from this study, the interlayer before cycling behaves like a sponge for electrolytes (as the E/S remained at a constant 10 $\mu L\ mg^{-1}{}_{sulfur}$ [only relative to the cathode]) and therefore struggles to properly wet the electrode surfaces prior to cycling. Once current is applied (on cycling), the electrolyte wets the electrode surfaces in a suitable manner, to result in a Warburg-like element. In the current study, wetting is improved by cycling. Furthermore, after polysulfide shuttling, as shown by the reduction in cycling performances, fewer polysulfides are present at the cathode/interlayer, and this would reduce the electrolyte viscosity. This is supported by the formation of an SEI and increased series resistance once cycled [59,61].

Before cycling, the spectrum for the uncycled cell with an unfunctionalised interlayer produces two arcs/semicircles (Figure 9a, black data), the larger of which is asymmetrical, implying overlap of a model equivalent circuit with more than one RC pair. Before cycling, the behaviour can be described by three RC pairs that are attributed to (i) a positive bulk contribution of the carbon cathode, (ii) charge transfer of the polysulfide species and (iii) the presence of a Li_2S/Li_2S_2 film on the electrode surfaces (in order of increasing capacitance). The low-frequency data points were attributed to the electrolyte's problematic transfer of charge to and from the electrode surfaces [60]. After cycling, the in-cell resistance decreases, and the detectable processes were SEI formation (left-hand semicircle), charge transfer by the polysulfide species (right-hand semicircle), and the presence of a Li_2S/Li_2S_2 film (in order of increasing capacitance), as well as typical Warburg-like diffusion of the lithium ions into the pores of the cathode. The spectra shown in Figure 9a,b show the complete EIS data between 1 MHz and 10 MHz. However, this has also given the unfortunate effect of compressing the display of the higher-frequency electronic data (1 MHz to 10 Hz). Therefore, for completeness and clarity, a restriction of the higher frequencies for Figure 9a,b is shown in Figure S5.

The impedance spectrum of the cell assembled using the functionalised (lithium phenylsulfonated) interlayer is also two arcs/semicircles, the larger of which is again asymmetrical, implying overlap of more than one RC pair, and was described by three RC pairs, but with smaller resistances than with the unfunctionalised interlayer; the processes were again consistent with a positive bulk contribution of the carbon cathode, charge transfer of the polysulfide species, and the presence of a Li_2S/Li_2S_2 film (in order of increasing capacitance). As with the functionalised interlayer, the low-frequency points that were not fitted were attributed to issues with wetting the anode surface [60]. After cycling, the cell resistance again decreases, and the processes detected are SEI formation, charge transfer of the polysulfide species and diffusional impedance (in order of increasing capacitance). No presence of a Li_2S/Li_2S_2 film was detected, and this could suggest a catalytic effect from the phenylsulfonate groups from the interlayer redistributing the sulfur more optimally [61,62]. Figure S6 and Table S1 describe the equivalent circuits and the fitted component values, respectively.

4. Conclusions

Adding interlayer films of MWCNTs improves utilisation of the sulfur active material, as shown by providing larger specific capacities. One factor driving this effect is the improved conductivity across the surface of the cathode coating. However, if the MWCNT film interlayer is unfunctionalized, this enhanced initial performance quickly dies away during cycling; that rapid fade is indicative of polysulfide shuttling which also seems enhanced relative to not adding an interlayer to the LSB. In contrast, when a functionalised interlayer is used, the cycle lifetime of the cell improves. These effects are not mutually exclusive, as functionalisation will convert some sp^2 carbons (associated with MWCNT electronic conductivity) to sp^3 carbons, thereby lowering the electronic conductivity of the fabricated film as polysulfide-shuttle-suppressing functional groups are added. However, the results clearly indicate the necessity of these functional groups to reduce polysulfide shuttling. XPS, SEM-EDX and FTIR results demonstrate surface functionalisation with grafted phenylsulfonate functionalities and consequent colocation of sulfur and oxygen at the atomic level. CHNS and powder XRD results indicate a low level of surface functionalisation of MWCNTs with no observable difference in the crystallinity relative to unfunctionalised MWCNTs.

The limited performance displayed by the cells, leading to the limited storage capacities, was because a commercially relevant (if less than ideal) and restrictive 10 $\mu L\ mg^{-1}{}_{sulfur}$ E/S ratio (based solely on the cathode) was used. All LSB cells that were tested in this paper are demonstrable proofs of concept that further the scientific discussion for the beneficial use of polar/functionalised carbonaceous interlayers and their abilities to act as secondary current collectors for the cathodes of LSBs.

Supplementary Materials: The following supporting information can be downloaded at: https://www.mdpi.com/article/10.3390/batteries11070266/s1, Figure S1: The pore size distributions (left) and cumulative pore volumes (right) of the unfunctionalised and lithium sulfonated MWCNT films, calculated using the respective nitrogen sorption isotherms present in Figure 3 and computed using a non-localised density functional theory model within the 3Flex MicroActive (Micromeritics®) software package; Figure S2: Peak fittings of the C 1s XPS peaks given in Figure 5: (a) unfunctionalised MWCNT film and (b) lithium sulfonated MWCNT films; Figure S3: Raman spectra of the unfunctionalised and lithium sulfonated MWCNT films; Figure S4: (a) The 10-cycle interval discharge and charge profiles for a melt-loaded Reg-600 cell assembled without an interlayer film, as well as (b) the progression over 40 cycles at 0.05 C; Figure S5: EIS spectra for cells before cycling, showing data points and the output from fitted equivalent circuit restricted to the higher frequency datapoints (caused by electronic effects) as previously displayed in Figure 9. The cells in Figure 9a have the unfunctionalised MWCNT interlayer, whereas the cells in Figure 9b have the lithium sulfonated MWCNT interlayer. Blue data points were obtained before cycling. The green line series shows the EIS fitted model before cycling. The red data points display the frequency of each measured data point every 2 decades. The lime dashed line indicates a typical Warburg element that would be expected by the resistance increase observed; Figure S6: The equivalent circuit diagrams of the two cells under study by EIS: (a) and (c) deposit-loaded cell assembled with an unfunctionalised MWCNT interlayer film, and (b) and (d) deposit-loaded cell assembled with lithium sulfonated MWCNT interlayer film. The equivalent circuits of the cells before cycling are given as (a) and (c), and the equivalent circuits of the cells after cycling are given as (b) and (d); Table S1: Values for the constructed equivalent circuit models used to fit to the impedance data of deposit-loaded cells assembled with either lithium sulfonated or unfunctionalised MWCNT film (taken after 2+ h at OCV before and after cycling).

Author Contributions: Conceptualisation, R.C.T.S. and C.C.; methodology, L.D.J.B. and C.C.; investigation, L.D.J.B. and S.J.H.; writing, original draft preparation, L.D.J.B.; writing, review, and editing, R.C.T.S., J.F.W. and C.C.; XPS assignment review, S.J.H. and J.F.W.; Project administration, R.C.T.S. and C.C. All authors have read and agreed to the published version of the manuscript.

Funding: This research was funded by The Faraday Institution grant number FIRG014 (LiSTAR project). The Article Processing Charges (APCs) were funded under the Institutional Open Access Program (IOAP) through the University of Surrey and MDPI.

Data Availability Statement: Data for this paper are available at the University of Surrey Open Research repository at https://doi.org/10.15126/thesis.901070. The data pertinent to this research is, however, currently under Embargo Status (until 1 May 2026) and so are not available at the time of submission for confidentiality reasons.

Acknowledgments: We thank EPSRC for a strategic equipment grant for Raman microscopy (EP/M022749/1). We thank Medac Ltd. for CHNS analyses and the University of Surrey for access to the electron microscopes at its central MicroStructural Studies Unit (MSSU). The authors have reviewed and edited the output and take full responsibility for the content of this publication.

Conflicts of Interest: The authors declare no conflicts of interest. The funders had no role in the design of the study; in the collection, analyses, or interpretation of data; in the writing of the manuscript; or in the decision to publish the results.

Abbreviations

The following abbreviations are used in this manuscript:

MWCNT	multiwalled carbon nanotubes
LSB	Lithium–sulfur battery
LIB	Lithium-ion battery
$S_B{}^0$	Bridging sulfur
$S_T{}^{1-}$	Polysulfide termini
CNT	Carbon nanotubes
E/S	Electrolyte-to-sulfur
ACS	American chemical society
CTAB	Cetyltrimethyl ammonium bromide
NMP	N-methyl 2-pyrolidone
PVDF	Polyvinylidene difluoride
MW	Molecular weight
GPC	Gel permeation chromatography
DME	1,2-Dimethoxyethane
DOL	1,3-Dioxolane
BHT	Butylated hydroxytoluene
LiTFSI	Lithium bis(trifluromethanesulfonyl)imide
HEPA	High-efficiency particulate air
XRD	X-ray diffraction
XPS	X-ray photoelectron spectroscopy
IR	Infrared
UATR	Universal attenuated total reflectance
FT	Fourier transform
SEM	Scanning electron microscopy
EDX	Energy-dispersive X-ray
CHNS	Carbon, hydrogen, nitrogen and sulfur
DFC	Dynamic flash combustion
TDC	Thermal conductivity detector
GDC	Galvanic discharge–charge
EIS	Electrochemical impedance spectroscopy
OCV	Open-circuit voltage
SEI	Solid–electrolyte interface
Bet	Brunauer–Emmett–Teller
LCO	Lithium cobalt oxide
LGPS	$Li_{10}GeP_2S_{12}$

References

1. Zhu, J.; Zou, J.; Cheng, H.; Gu, Y.; Lu, Z. High Energy Batteries Based on Sulfur Cathode. *Green Energy Environ.* **2019**, *4*, 345–359. [CrossRef]
2. Shi, C.; Takeuchi, S.; Alexander, G.V.; Hamann, T.; O'Neill, J.; Dura, J.A.; Wachsman, E.D. High Sulfur Loading and Capacity Retention in Bilayer Garnet Sulfurized-Polyacrylonitrile/Lithium-Metal Batteries with Gel Polymer Electrolytes. *Adv. Energy Mater.* **2023**, *13*, 2301656. [CrossRef]
3. Nakamura, N.; Ahn, S.; Momma, T.; Osaka, T. Future Potential for Lithium-Sulfur Batteries. *J. Power Sources* **2023**, *558*, 232566. [CrossRef]
4. Danner, T.; Latz, A. On the Influence of Nucleation and Growth of S8 and Li2S in Lithium-Sulfur Batteries. *Electrochim. Acta* **2019**, *322*, 134719. [CrossRef]
5. Saedi, S.; Shokri, M.; Rhim, J.-W. Antimicrobial Activity of Sulfur Nanoparticles: Effect of Preparation Methods. *Arab. J. Chem.* **2020**, *13*, 6580–6588. [CrossRef]
6. Xiao, R.; Yu, T.; Yang, S.; Chen, K.; Li, Z.; Liu, Z.; Hu, T.; Hu, G.; Li, J.; Cheng, H.M.; et al. Electronic Structure Adjustment of Lithium Sulfide by a Single-Atom Copper Catalyst toward High-Rate Lithium-Sulfur Batteries. *Energy Storage Mater.* **2022**, *51*, 890–899. [CrossRef]
7. Feng, Z.; Kim, C.; Vijh, A.; Armand, M.; Bevan, K.H.; Zaghib, K. Unravelling the Role of Li_2S_2 in Lithium-Sulfur Batteries: A First Principles Study of Its Energetic and Electronic Properties. *J. Power Sources* **2014**, *272*, 518–521. [CrossRef]
8. Wang, H.; Zhang, W.; Xu, J.; Guo, Z. Advances in Polar Materials for Lithium–Sulfur Batteries. *Adv. Funct. Mater.* **2018**, *28*, 1707520. [CrossRef]
9. Lin, Y.; Huang, S.; Zhong, L.; Wang, S.; Han, D.; Ren, S.; Xiao, M.; Meng, Y. Organic Liquid Electrolytes in Li-S Batteries: Actualities and Perspectives. *Energy Storage Mater.* **2021**, *34*, 128–147. [CrossRef]
10. Shutthanandan, V.; Nandasiri, M.; Zheng, J.; Engelhard, M.H.; Xu, W.; Thevuthasan, S.; Murugesan, V. Applications of XPS in the Characterization of Battery Materials. *J. Electron Spectros. Relat. Phenom.* **2019**, *231*, 2–10. [CrossRef]
11. Fan, Y.; Niu, Z.; Zhang, F.; Zhang, R.; Zhao, Y.; Lu, G. Suppressing the Shuttle Effect in Lithium-Sulfur Batteries by a UiO-66-Modified Polypropylene Separator. *ACS Omega* **2019**, *4*, 10328–10335. [CrossRef]
12. Fan, L.; Li, M.; Li, X.; Xiao, W.; Chen, Z.; Lu, J. Interlayer Material Selection for Lithium-Sulfur Batteries. *Joule* **2019**, *3*, 361–386. [CrossRef]
13. Zhang, X.; Bai, X.; Wei, C.; Wang, Z.; Xi, B.; Xiong, S.; Feng, J. Triggering the Electronic Microenvironment of Extraordinary Nitrogen-Bridged Atomic Iron Coordinated with in-Plane Nitrogen by Manipulating Phase-Reconfigured 2D Vanadium Nitride MXenes toward Invigorated Lithium–Sulfur Batteries. *Energy Environ. Sci.* **2024**, *17*, 7403–7415. [CrossRef]
14. Lee, B.-J.; Zhao, C.; Yu, J.-H.; Kang, T.-H.; Park, H.-Y.; Kang, J.; Jung, Y.; Liu, X.; Li, T.; Xu, W.; et al. Development of High-Energy Non-Aqueous Lithium-Sulfur Batteries via Redox-Active Interlayer Strategy. *Nat. Commun.* **2022**, *13*, 4629. [CrossRef] [PubMed]
15. Kim, J.H.; Ko, Y.i.; Kim, Y.A.; Kim, K.S.; Yang, C.M. Sulfur-Doped Carbon Nanotubes as a Conducting Agent in Supercapacitor Electrodes. *J. Alloys Compd.* **2021**, *855*, 157282. [CrossRef]
16. Salice, P.; Fabris, E.; Sartorio, C.; Fenaroli, D.; Figà, V.; Casaletto, M.P.; Cataldo, S.; Pignataro, B.; Menna, E. An Insight into the Functionalisation of Carbon Nanotubes by Diazonium Chemistry: Towards a Controlled Decoration. *Carbon* **2014**, *74*, 73–82. [CrossRef]
17. Fretz, S.J.; Pal, U.; Girard, G.M.A.; Howlett, P.C.; Palmqvist, A.E.C. Lithium Sulfonate Functionalization of Carbon Cathodes as a Substitute for Lithium Nitrate in the Electrolyte of Lithium–Sulfur Batteries. *Adv. Funct. Mater.* **2020**, *30*, 2002485. [CrossRef]
18. Guthrie, J.P. Hydrolysis of Esters of Oxy Acids: PKa Values for Strong Acids; Brønsted Relationship for Attack of Water at Methyl; Free Energies of Hydrolysis of Esters of Oxy Acids; and a Linear Relationship between Free Energy of Hydrolysis and PKa Holding over a Ran. *Can. J. Chem.* **1978**, *56*, 2342–2354. [CrossRef]
19. Ma, L.; Zhuang, H.L.; Wei, S.; Hendrickson, K.E.; Kim, M.S.; Cohn, G.; Hennig, R.G.; Archer, L.A. Enhanced Li-S Batteries Using Amine-Functionalized Carbon Nanotubes in the Cathode. *ACS Nano* **2016**, *10*, 1050–1059. [CrossRef]
20. Park, S.K.; Lee, J.K.; Kang, Y.C. Yolk–Shell Structured Assembly of Bamboo-Like Nitrogen-Doped Carbon Nanotubes Embedded with Co Nanocrystals and Their Application as Cathode Material for Li–S Batteries. *Adv. Funct. Mater.* **2018**, *28*, 1705264. [CrossRef]
21. Gaikwad, M.M.; Sarode, K.K.; Pathak, A.D.; Sharma, C.S. Ultrahigh Rate and High-Performance Lithium-Sulfur Batteries with Resorcinol-Formaldehyde Xerogel Derived Highly Porous Carbon Matrix as Sulfur Cathode Host. *Chem. Eng. J.* **2021**, *425*, 131521. [CrossRef]
22. Kim, P.J.H.; Kim, K.; Pol, V.G. Towards Highly Stable Lithium Sulfur Batteries: Surface Functionalization of Carbon Nanotube Scaffold. *Carbon* **2018**, *131*, 175–183. [CrossRef]
23. Lu, Y.; Gu, S.; Guo, J.; Rui, K.; Chen, C.; Zhang, S.; Jin, J.; Yang, J.; Wen, Z. Sulfonic Groups Originated Dual-Functional Interlayer for High Performance Lithium-Sulfur Battery. *ACS Appl. Mater. Interfaces* **2017**, *9*, 14878–14888. [CrossRef] [PubMed]

24. Lee, J.H.; Kang, J.; Kim, S.W.; Halim, W.; Frey, M.W.; Joo, Y.L. Effective Suppression of the Polysulfide Shuttle Effect in Lithium-Sulfur Batteries by Implementing RGO-PEDOT:PSS-Coated Separators via Air-Controlled Electrospray. *ACS Omega* **2018**, *3*, 16465–16471. [CrossRef]
25. Chang, C.H.; Chung, S.H.; Manthiram, A. Ultra-Lightweight PANiNF/MWCNT-Functionalized Separators with Synergistic Suppression of Polysulfide Migration for Li-S Batteries with Pure Sulfur Cathodes. *J. Mater. Chem. A* **2015**, *3*, 18829–18834. [CrossRef]
26. Barter, L.D.J.; Mohammad, I.; Hinder, S.J.; Watts, J.F.; Slade, R.C.T.; Crean, C. Carbons Derived from Resole-Type Phenolic Resins for Use in Lithium–Sulfur Batteries: Templating the Resins with Sulfur Leads to Enhanced Cell Performance. *Energy Adv.* **2024**, 3, 471–481. [CrossRef]
27. Renda, C.G.; Bertholdo, R. Study of Phenolic Resin and Their Tendency for Carbon Graphitization. *J. Polym. Res.* **2018**, *25*, 241. [CrossRef]
28. Ungár, T. Microstructural Parameters from X-Ray Diffraction Peak Broadening. *Scr. Mater.* **2004**, *51*, 777–781. [CrossRef]
29. Sirivisoot, S.; Webster, T.J. 3.32 Carbon Nanotubes: Applications for In Situ Implant Sensors. In *Comprehensive Biomaterials II*; Ducheyne, P., Ed.; Elsevier: Oxford, UK, 2017; pp. 703–715. ISBN 978-0-08-100692-4.
30. Azman, N.H.N.; Sulaiman, Y. Hierarchical Porous Materials for Supercapacitors. In *Encyclopedia of Energy Storage*; Cabeza, L.F., Ed.; Elsevier: Oxford, UK, 2022; pp. 622–637. ISBN 978-0-12-819730-1.
31. Hippauf, F.; Nickel, W.; Hao, G.P.; Schwedtmann, K.; Giebeler, L.; Oswald, S.; Borchardt, L.; Doerfler, S.; Weigand, J.J.; Kaskel, S. The Importance of Pore Size and Surface Polarity for Polysulfide Adsorption in Lithium Sulfur Batteries. *Adv. Mater. Interfaces* **2016**, *3*, 1600508. [CrossRef]
32. Chen, X.; Gross, A.J.; Giroud, F.; Holzinger, M.; Cosnier, S. Comparison of Commercial and Lab-Made MWCNT Buckypaper: Physicochemical Properties and Bioelectrocatalytic O_2 Reduction. *Electroanalysis* **2018**, *30*, 1511–1520. [CrossRef]
33. Khan, Z.U.; Kausar, A.; Ullah, H. A Review on Composite Papers of Graphene Oxide, Carbon Nanotube, Polymer/GO, and Polymer/CNT: Processing Strategies, Properties, and Relevance. *Polym. Plast. Technol. Eng.* **2016**, *55*, 559–581. [CrossRef]
34. Ensling, D.; Stjerndahl, M.; Nytén, A.; Gustafsson, T.; Thomas, J.O. A Comparative XPS Surface Study of Li_2FeSiO_4/C Cycled with LiTFSI- and $LiPF_6$-Based Electrolytes. *J. Mater. Chem.* **2009**, *19*, 82–88. [CrossRef]
35. Liu, L.; Chen, X.; Wang, Z.; Wang, X.; Lin, S. The Removal Mechanism and Performance of Tetrabromobisphenol A with a Novel Multi-Group Activated Carbon from Recycling Long-Root: Eichhornia Crassipes Plants. *RSC Adv.* **2019**, *9*, 24760–24769. [CrossRef] [PubMed]
36. Jia, R.; Chen, J.; Zhao, J.; Zheng, J.; Song, C.; Li, L.; Zhu, Z. Synthesis of Highly Nitrogen-Doped Hollow Carbon Nanoparticles and Their Excellent Electrocatalytic Properties in Dye-Sensitized Solar Cells. *J. Mater. Chem.* **2010**, *20*, 10829–10834. [CrossRef]
37. Lee, M.-T.; Liu, H.; Brandell, D. The Surface Chemistry of Thin Lithium Metal Electrodes in Lithium-Sulfur Cells. *Batter. Supercaps* **2020**, *3*, 1370–1376. [CrossRef]
38. Shanthi, P.M.; Hanumantha, P.J.; Ramalinga, K.; Gattu, B.; Datta, M.K.; Kumta, P.N. Sulfonic Acid Based Complex Framework Materials (CFM): Nanostructured Polysulfide Immobilization Systems for Rechargeable Lithium–Sulfur Battery. *J. Electrochem. Soc.* **2019**, *166*, A1827–A1835. [CrossRef]
39. Rojas, J.V.; Toro-Gonzalez, M.; Molina-Higgins, M.C.; Castano, C.E. Facile Radiolytic Synthesis of Ruthenium Nanoparticles on Graphene Oxide and Carbon Nanotubes. *Mater. Sci. Eng. B* **2016**, *205*, 28–35. [CrossRef]
40. Koushik, D.; Verhees, W.J.H.; Zhang, D.; Kuang, Y.; Veenstra, S.; Creatore, M.; Schropp, R.E.I. Atomic Layer Deposition Enabled Perovskite/PEDOT Solar Cells in a Regular n–i–p Architectural Design. *Adv. Mater. Interfaces* **2017**, *4*, 1700043. [CrossRef]
41. Dubina, E.; Plank, J.; Black, L. Impact of Water Vapour and Carbon Dioxide on Surface Composition of C3A Polymorphs Studied by X-Ray Photoelectron Spectroscopy. *Cem. Concr. Res.* **2015**, *73*, 36–41. [CrossRef]
42. Parker, S.F.; Revill-Hivet, E.J.; Nye, D.W.; Gutmann, M.J. Structure and Vibrational Spectroscopy of Lithium and Potassium Methanesulfonates. *R. Soc. Open Sci.* **2020**, *7*, 200776. [CrossRef]
43. Jiang, K.; Gao, S.; Wang, R.; Jiang, M.; Han, J.; Gu, T.; Liu, M.; Cheng, S.; Wang, K. Lithium Sulfonate/Carboxylate-Anchored Polyvinyl Alcohol Separators for Lithium Sulfur Batteries. *ACS Appl. Mater. Interfaces* **2018**, *10*, 18310–18315. [CrossRef]
44. Childres, I.; Jauregui, L.A.; Park, W.; Caoa, H.; Chena, Y.P. Raman Spectroscopy of Graphene and Related Materials. In *New Developments in Photon and Materials Research*; Jang, J.I., Ed.; Nova Science Publishers, Inc.: Hauppauge, NY, USA, 2013; pp. 403–418. ISBN 9781626183391.
45. Mohan, S.; Oluwafemi, O.S.; Songca, S.P.; Rouxel, D.; Miska, P.; Lewu, F.B.; Kalarikkal, N.; Thomas, S. Completely Green Synthesis of Silver Nanoparticle Decorated MWCNT and Its Antibacterial and Catalytic Properties. *Pure Appl. Chem.* **2016**, *88*, 71–81. [CrossRef]
46. Alía, J.M.; Edwards, H.G.M.; Kiernan, B.M. Raman Spectroscopy of Benzenesulfonic and 4-Toluenesulfonic Acids Dissolved in Dimethylsulfoxide. *Spectrochim. Acta Part A Mol. Biomol. Spectrosc.* **2004**, *60*, 1533–1542. [CrossRef] [PubMed]
47. Zerrin, T.; Shang, R.; Dong, B.; Aguilar, E.C.; Malvin, J.; Ozkan, M.; Ozkan, C.S. An Overlooked Parameter in Li-S Batteries: The Impact of Electrolyte-to-Sulfur Ratio on Capacity Fading. *Nano Energy* **2022**, *104*, 107913. [CrossRef]

48. Wang, M.; Xia, X.; Zhong, Y.; Wu, J.; Xu, R.; Yao, Z.; Wang, D.; Tang, W.; Wang, X.; Tu, J. Porous Carbon Hosts for Lithium–Sulfur Batteries. *Chem. A Eur. J.* **2019**, *25*, 3710–3725. [CrossRef] [PubMed]
49. Zhao, M.; Li, B.-Q.; Zhang, X.-Q.; Huang, J.-Q.; Zhang, Q. A Perspective toward Practical Lithium–Sulfur Batteries. *ACS Cent. Sci.* **2020**, *6*, 1095–1104. [CrossRef]
50. Yang, X.; Li, X.; Adair, K.R.; Zhang, H.; Sun, X. Structural Design of Lithium–Sulfur Batteries: From Fundamental Research to Practical Application. *Electrochem. Energy Rev.* **2018**, *1*, 239–293. [CrossRef]
51. Jozwiuk, A.; Berkes, B.B.; Weiß, T.; Sommer, H.; Janek, J.; Brezesinski, T. The Critical Role of Lithium Nitrate in the Gas Evolution of Lithium-Sulfur Batteries. *Energy Environ. Sci.* **2016**, *9*, 2603–2608. [CrossRef]
52. Blanchard, D.; Slagter, M. In Operando Raman and Optical Study of Lithium Polysulfides Dissolution in Lithium–Sulfur Cells with Carrageenan Binder. *J. Phys. Energy* **2021**, *3*, 044003. [CrossRef]
53. Bondy, A.L.; Craig, R.L.; Zhang, Z.; Gold, A.; Surratt, J.D.; Ault, A.P. Isoprene-Derived Organosulfates: Vibrational Mode Analysis by Raman Spectroscopy, Acidity-Dependent Spectral Modes, and Observation in Individual Atmospheric Particles. *J. Phys. Chem. A* **2018**, *122*, 303–315. [CrossRef]
54. Xu, L.; Csekő, G.; Petz, A.; Horváth, A.K. Kinetics and Mechanism of the Oxidation of Pentathionate Ion by Chlorine Dioxide in a Slightly Acidic Medium. *J. Phys. Chem. A* **2014**, *118*, 1293–1299. [CrossRef]
55. Borah, R.; Hughson, F.R.; Johnston, J.; Nann, T. On Battery Materials and Methods. *Mater. Today Adv.* **2020**, *6*, 100046. [CrossRef]
56. Li, J.; Dai, L.; Wang, Z.; Wang, H.; Xie, L.; Chen, J.; Yan, C.; Yuan, H.; Wang, H.; Chen, C. Cellulose Nanofiber Separator for Suppressing Shuttle Effect and Li Dendrite Formation in Lithium-Sulfur Batteries. *J. Energy Chem.* **2022**, *67*, 736–744. [CrossRef]
57. Zhe, R.; Zhu, T.; Wei, X.; Ren, Y.; Qing, C.; Li, N.; Wang, H.E. Graphene Oxide Wrapped Hollow Mesoporous Carbon Spheres as a Dynamically Bipolar Host for Lithium-Sulfur Batteries. *J. Mater. Chem. A* **2022**, *10*, 24422–24433. [CrossRef]
58. Pathak, R.; Chen, K.; Gurung, A.; Reza, K.M.; Bahrami, B.; Pokharel, J.; Baniya, A.; He, W.; Wu, F.; Zhou, Y.; et al. Fluorinated Hybrid Solid-Electrolyte-Interphase for Dendrite-Free Lithium Deposition. *Nat. Commun.* **2020**, *11*, 93. [CrossRef] [PubMed]
59. Shen, C.; Andrei, P.; Zheng, J.P. Stable Cycling of Lithium-Sulfur Batteries by Optimizing the Cycle Condition. *Electrochim. Acta* **2019**, *326*, 134948. [CrossRef]
60. Zhang, Z.; Chen, S.; Yang, J.; Wang, J.; Yao, L.; Yao, X.; Cui, P.; Xu, X. Interface Re-Engineering of $Li_{10}GeP_2S_{12}$ Electrolyte and Lithium Anode for All-Solid-State Lithium Batteries with Ultralong Cycle Life. *ACS Appl. Mater. Interfaces* **2018**, *10*, 2556–2565. [CrossRef]
61. Guo, W.; Zhang, W.; Si, Y.; Wang, D.; Fu, Y.; Manthiram, A. Artificial Dual Solid-Electrolyte Interfaces Based on in Situ Organothiol Transformation in Lithium Sulfur Battery. *Nat. Commun.* **2021**, *12*, 3031. [CrossRef]
62. Yan, J.; Liu, X.; Li, B. Capacity Fade Analysis of Sulfur Cathodes in Lithium–Sulfur Batteries. *Adv. Sci.* **2016**, *3*, 1600101. [CrossRef]

batteries

MDPI

Review

Carbon/High-Entropy Alloy Nanocomposites: Synergistic Innovations and Breakthrough Challenges for Electrochemical Energy Storage

Li Sun [1], Hangyu Li [1], Yu Dong [1], Wan Rong [1], Na Zhou [1], Rui Dang [1], Jianle Xu [2], Qigao Cao [1,*] and Chunxu Pan [3,*]

1 State Key Laboratory of Porous Metal Materials, Northwest Institute for Nonferrous Metal Research, Xi'an 710016, China
2 College of Physics and Electronic Information, Jiangsu Second Normal University, Nanjing 210013, China
3 MOE Key Laboratory of Artificial Micro- and Nano-Structures, School of Physics and Technology, Wuhan University, Wuhan 430072, China
* Correspondence: caoqigao@c-nin.com (Q.C.); cxpan@whu.edu.cn (C.P.)

Abstract

Against the backdrop of accelerating global energy transition, developing high-performance energy-storage systems is crucial for achieving carbon neutrality. Traditional electrode materials are limited by a single densification storage mechanism and low conductivity, struggling to meet demands for high energy/power density and a long cycle life. Carbon/high-entropy alloy nanocomposites provide an innovative solution through multi-component synergistic effects and cross-scale structural design: the "cocktail effect" of high-entropy alloys confers excellent redox activity and structural stability, while the three-dimensional conductive network of the carbon skeleton enhances charge transfer efficiency. Together, they achieve synergistic enhancement via interfacial electron coupling, stress buffering, and dual storage mechanisms. This review systematically analyzes the charge storage/attenuation mechanisms and performance advantages of this composite material in diverse energy-storage devices (lithium-ion batteries, lithium-sulfur batteries, etc.), evaluates the characteristics and limitations of preparation techniques such as mechanical alloying and chemical vapor deposition, identifies five major challenges (including complex and costly synthesis, ambiguous interfacial interaction mechanisms, lagging theoretical research, performance-cost trade-offs, and slow industrialization processes), and prospectively proposes eight research directions (including multi-scale structural regulation and sustainable preparation technologies, etc.). Through interdisciplinary perspectives, this review aims to provide a theoretical foundation for deepening the understanding of carbon/high-entropy alloy composite energy-storage mechanisms and guiding industrial applications, thereby advancing breakthroughs in electrochemical energy-storage technology under the energy transition.

Keywords: batteries; carbon/high-entropy alloys; composites; supercapacitors

1. Introduction

Since the 21st century, the global energy structure has been undergoing a profound transformation from fossil fuels to renewable energy sources (such as solar and wind power) [1]. Against this backdrop, electrochemical energy storage (EES) systems, by virtue of their high energy conversion efficiency, flexible deployment characteristics, and rapid response capabilities, have emerged as a core enabling technology for building new power systems [2,3]. Energy-storage devices such as lithium-ion batteries (LIBs), sodium-ion batteries (SIBs), supercapacitors (SCs), and sodium/potassium-ion batteries (SIBs/PIBs) play indispensable roles in fields including electric vehicles, smart grids, and portable electronics [4–9]. However, existing energy-storage technologies still face critical bottlenecks, such as relatively low energy density (LIBs typically below 300 Wh/kg), insufficient cycle life (SIBs < 2000 cycles), and safety hazards (e.g., lithium dendrite growth). These limitations hinder their ability to meet the demands of emerging fields like 5G communications and aerospace for energy storage devices operating under extreme conditions [10–12]. Consequently, developing electrode materials with high capacity, long cycle stability, and fast kinetics has become a major research focus in the EES field.

Electrode materials play a decisive, core role in the performance of energy-storage devices. Taking LIBs as an example, their commercialization began in 1991 when Sony Corporation adopted a system with graphite anodes and lithium cobalt oxide cathodes. After over 30 years of development, conventional materials are gradually approaching their theoretical limits: the specific capacity of graphite anodes is only 372 mAh/g, while lithium cobalt oxide faces challenges due to the scarcity of cobalt resources (global reserves ~7.1 million tons) and structural instability (layered structure prone to collapse under high voltage), making it difficult to meet the demands for next-generation high energy density [13–15]. Similarly, Prussian blue analogue cathodes for SIBs are limited by side reactions caused by crystalline water, and the energy density enhancement of activated carbon electrodes in SCs is constrained by low specific surface area utilization (<30%) [16,17]. More critically, electrode materials suffer from issues like volume expansion during charge/discharge (silicon anodes exhibit up to 300% expansion), active material dissolution (e.g., polysulfide shuttling in lithium–sulfur batteries), and interfacial side reactions. These lead to accelerated capacity decay and an increased risk of thermal runaway [18–20]. Although some researchers have achieved effective improvements in battery performance by designing intermediate layers that can prevent polysulfide dissolution and accelerate redox reaction kinetics, the root causes of these defects lie in the single-component nature, rigid electronic structure, and uncontrollable microstructure of traditional material systems, which struggle to simultaneously meet the multiple requirements of high conductivity, strong mechanical stability, and efficient ion transport [21,22].

The concept of a high-entropy alloy (HEA) was first proposed by Yeh et al. in 2004 [23,24]. They are defined as single-phase solid solutions formed by five or more principal elements (each with a molar fraction between 5% and 35%) [25,26]. The composition selection of the HEA should follow three principles: prioritizing thermodynamic stability, ensuring kinetic and structural compatibility, and implementing function-oriented collaborative design [21]. Firstly, the high-entropy effect (with at least five principal elements and a mixing entropy > 1.5 R) must be satisfied to suppress intermetallic compounds and promote the formation of solid solutions [22,27,28]. Meanwhile, the mixing enthalpy (ΔHmix < 15 kJ/mol) needs to be controlled to avoid segregation or precipitation of ordered phases; the competitive relationship between ΔHmix and ΔSmix determines phase stability. Secondly, the difference in atomic radii among elements (δ < 6.5%) is required to alleviate lattice distortion. Refractory metals with similar diffusion behaviors should be selected to slow down segregation [23,29,30]. Finally, 3D transition metals should be optimally chosen in line with application requirements to regulate catalytic performance, while refractory elements should be prioritized to enhance high-temperature oxidation resistance [31,32]. This design paradigm breaks from the traditional alloy approach dominated by 1–2 elements, leveraging the high configurational entropy effect (ΔSconfig $\geq$ 1.5 R) of multiple components to suppress element segregation and phase transformations, thereby endowing the materials with four unique advantages:

- **Structural Stability:** The high-entropy effect lowers the Gibbs free energy, enabling the alloy to maintain a single-phase structure under high-temperature, corrosive, or irradiation environments [32,33]. For instance, refractory HEA like $Nb_{25}Mo_{25}Ta_{25}W_{25}$ and $V_{20}Nb_{20}Mo_{20}Ta_{20}W_{20}$ remain stable and disordered, retaining a single-phase body-centered cubic structure without superlattice reflections even after exposure to 1400 °C [34].
- **Tunable Mechanical Properties:** By adjusting element types and ratios, the synergistic optimization of hardness and ductility can be achieved, surpassing traditional stainless steels and nickel-based alloys [34,35].
- **Enhanced Catalytic Activity:** Multi-element synergy tunes the d-band center position, optimizing the adsorption free energy of reaction intermediates. For example, the design of PtFeCoNiCu high-entropy alloys adheres to three core principles: atomic size compatibility (radius 1.24–1.39 Å), high mixing entropy (1.576 R), and electronic synergistic effects [36,37]. The comparable atomic sizes facilitate the formation of face-centered cubic (fcc) solid solutions, while the high mixing entropy suppresses elemental segregation through positive dissolution energies confirmed based on DFT calculations, thereby enhancing structural stability. The electronic effects manifest as follows: electronegativity differences induce electron transfer, resulting in a negative shift of 0.3 eV in Pt 4f binding energy and a downward shift of 0.36 eV in the d-band center, which optimizes the -OOH adsorption energy (ΔG = 0.369–0.428 eV). Specifically, Fe/Co/Ni synergistically activates the O-O bond with low activation barriers (0.291–0.326 eV), while Cu modulates electron density to prevent overoxidation. This multi-active-site synergistic mechanism endows the material with exceptional ORR performance, achieving a mass activity of 1.738 A/mg Pt (surpassing commercial Pt/C by 15.8 times) and unprecedented stability with only 3 mV decay after 10,000 cycles.
- **Improved Ion Storage Capacity:** Some HEAs possess open crystal structures, exhibiting lithium-ion diffusion coefficients significantly superior to graphite [38].

These properties have propelled HEAs to prominence in fields like electrocatalysis, battery anodes, and electromagnetic shielding [39]. However, HEA nanoparticles are prone to agglomeration due to high surface energy (especially under volume changes during cycling), and their intrinsic conductivity remains lower than carbon-based materials, limiting their application under high-current conditions [40].

Carbon materials, owing to their rich allotropes (graphite, diamond, graphene, etc.) and tunable microstructures, consistently occupy a central position in EES research [41]. Taking graphene as an example, its theoretical specific surface area is as high as 2630 m^2/g, electron mobility exceeds 200,000 $cm^2/(V \cdot s)$, and chemical modification can introduce functional groups (-OH, -COOH, etc.) to enhance interfacial reactivity [42]. The core functionalities of carbon materials can be summarized into four aspects:

- **Conductive Network Construction:** Three-dimensional conductive frameworks formed by carbon nanotubes (CNTs) or reduced graphene oxide can reduce overall electrode resistance below 10 Ω/sq [43].
- **Mechanical Stress Buffering:** Carbon coatings (e.g., amorphous carbon) absorb volume changes of active materials through elastic deformation, enabling silicon anodes to achieve double the capacity after 800 cycles [44].
- **Multi-scale Mass Transport Optimization:** Hierarchically porous carbon (micro-meso-macro pores) provides rapid ion transport channels, allowing SCs to achieve power densities exceeding 512 kW/kg [45].

- **Interfacial Chemistry Regulation:** Heteroatom doping (N, S, P) alters the electron distribution of the carbon skeleton, enhancing the chemical anchoring of polysulfides (capacity decay rate < 0.032% per cycle over 500 cycles in Li-S batteries) [46].

However, the energy storage capacity of single-component carbon materials is constrained by the electric double-layer mechanism or lithium intercalation chemistry (graphite), making it difficult to surpass theoretical ceilings [17]. Therefore, compositing carbon materials with high-capacity active materials has become an essential strategy for balancing energy density and cycle stability.

The integration of carbon materials (graphene, CNTs, porous carbon) with HEA to form composites demonstrates core advantages in the EES field, primarily through the synergistic enhancement of conductivity and structural stability [47]. Carbon materials (especially graphene and CNTs) provide exceptional intrinsic conductivity and mechanical flexibility, while porous carbon contributes a high specific surface area and ion transport channels, collectively establishing an efficient 3D electron conduction network and skeleton [48,49]. Although HEAs possess relatively weaker conductivity, their unique solid–solution structure grants them excellent structural stability and resistance to pulverization, particularly during the volume changes inherent to repeated charge/discharge cycles [50,51]. The composite achieves crucial complementarity: the carbon network significantly improves the electron transport efficiency of HEAs and buffers volume change stresses, inhibiting particle fragmentation and agglomeration to maintain electrode integrity; simultaneously, the stability of the HEA protects the carbon structure, enhancing its long-term cycling durability [52].

Furthermore, this composite significantly enhances the utilization efficiency of active materials and reaction kinetics. The high specific surface area and pore structure of carbon materials provide an ideal dispersion matrix for HEA nanoparticles, effectively preventing agglomeration and fully exposing HEA active sites [53]. HEA, leveraging their multi-principal element nature which induces the "cocktail effect" and severe lattice distortion, potentially generate more highly active catalytic sites (crucial for reactions like sulfur conversion in Li-S batteries), optimize the adsorption energy for reaction intermediates (e.g., polysulfides), and provide richer redox reactions (pseudocapacitive behavior) [54]. The synergistic interplay ensures that the carbon carrier maximizes the exposure of HEA active sites and enables rapid ion/electron contact; meanwhile, the unique electronic structure and catalytic activity of HEA significantly accelerate interfacial reaction rates (e.g., ion insertion/extraction, polysulfide conversion, Faradaic reactions), thereby enhancing rate capability and reversible capacity [55].

Moreover, the composite offers targeted solutions and enables multifunctional integration:

- In **Li-S batteries**, carbon materials (especially porous carbon) provide physical confinement and some chemical adsorption to capture polysulfides, while HEAs exert strong chemical adsorption and efficient catalytic action to accelerate polysulfide conversion. This synergy forms a "physical confinement + chemical adsorption + efficient catalysis" triple mechanism, effectively suppressing the shuttle effect [56].
- In **SIBs**, the composite leverages the sodium-storage capacity and buffering effect of carbon (e.g., hard carbon) combined with the potentially suitable open structure of HEAs for Na^+ storage, jointly improving rate performance and cycle stability [57].

Additionally, the diversity of carbon material dimensions, the flexibility to tune HEA components for precise performance control, and the controllability of composite structures (e.g., core-shell, porous networks) provide a vast scope for optimizing the performance of various energy storage devices.

This review systematically summarizes the recent progress in carbon/HEA nanocomposites for EES in Figure 1. Firstly, the intrinsic properties of HEAs and carbon materials, along with their synergistic mechanisms, are thoroughly analyzed. Secondly, material preparation strategies and structure control methods are summarized. Thirdly, their performance breakthroughs in LIBs, SIBs, SCs, and SIBs/PIBs are critically reviewed. Finally, technical bottlenecks for scalable application and future research directions are explored. Through an interdisciplinary perspective (encompassing materials science, electrochemistry, computational science, etc.), this review aims to provide theoretical guidance and a technical roadmap for developing next-generation high-performance energy storage devices.

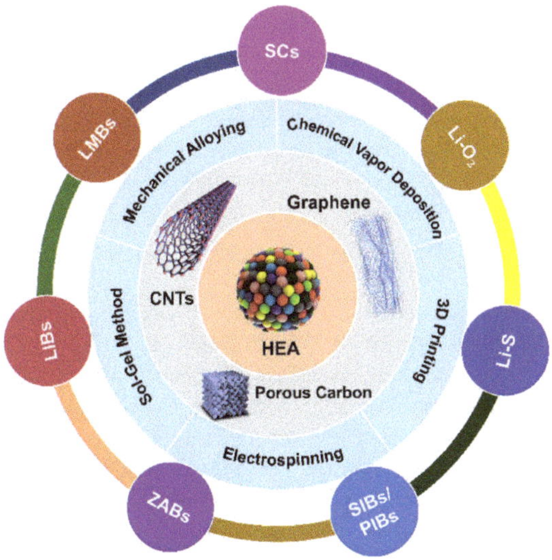

Figure 1. Schematic illustration of carbon/HEA nanocomposites for EES.

2. Synergistic Mechanisms of Carbon/HEA Nanocomposites

Carbon/HEA nanocomposites significantly enhance the performance of EES devices through multiple synergistic mechanisms. The core mechanisms include the following: interfacial electronic structure reconstruction (enhancing conductivity and catalytic activity), nano-confinement and steric hindrance effects (suppressing particle growth/agglomeration, anchoring active materials, improving structural stability), bicontinuous network and biomimetic channel design (promoting uniform ion/electron transport), and strain-adaptive characteristics (alleviating cycling stress) [58]. These mechanisms work synergistically to optimize reaction kinetics, interfacial stability, and electrode structural integrity, thereby achieving high specific capacity, excellent rate capability, a long cycle life, and high Coulombic efficiency.

2.1. Interfacial Electronic Reconstruction and Band Structure Modulation Mechanism

Carbon/HEA nanocomposites enable interfacial charge reconstruction and band structure modulation through electronic structure coupling [59]. At the interface between the carbon matrix (e.g., graphene, CNTs) and HEA nanoparticles within the composite, unique electronic structure coupling occurs. This coupling primarily functions via the following mechanisms: the high conductivity and electron-rich nature of carbon materials combine with the multi-element synergistic effects of HEAs, leading to charge transfer and redistribution (charge reconstruction) at the interface, thereby optimizing the electronic density of states at active sites. Simultaneously, band alignment between the two components modulates the overall band structure of the composite, for instance, by forming metal-semiconductor heterojunctions or adjusting the Fermi level position, thus enhancing carrier separation efficiency or catalytic activity. This electronic-level synergy provides new insights for designing high-performance electrocatalysts, photocatalysts, and energy-storage materials. The interfacial electronic coupling between carbon materials and HEA extends beyond simple charge transfer; it achieves the synergistic optimization of catalytic activity and ion transport through atomic-level hybrid orbital reconstruction [38,58]. The locally distorted lattice induced by the high-entropy effect on the HEA surface can form hybrid sp^2-d orbitals with the π-electron cloud of graphene, significantly increasing the density of states near the Fermi level [38,60].

For example, HEA nucleates at defect sites of activated carbonized wood (ACW) while catalyzing the deposition of adjacent carbon atoms to form a few-layer graphitic carbon shell. This physical encapsulation effectively isolates the HEAs from electrolyte corrosion and mitigates the nanoparticle detachment caused by bubble impact. PtNiCoFeCu@ACW exhibits a high electronic density of states near the Fermi level, enabling superior electron transport capability. The d-orbitals of Ni, Co, and Fe dominate intermediate adsorption near the Femi level, while Pt and Cu provide an electron-rich environment, synergistically accelerating charge transfer. The self-encapsulated structure, high-entropy synergy, and interfacial stability collectively enable highly efficient and durable HER performance (no degradation observed during a 500 h stability test) [61]. HEA nanoparticles are uniformly embedded onto the rGO surface, forming tight interfacial bonding with stable M-C bonds via defect sites. The multi-metallic synergy within the HEAs potentially optimizes the electronic structure of the M-C bonds, enhancing interfacial stability. Consequently, the HEA/rGO interface exhibits high reversibility during charge/discharge, suppressing side reactions and structural degradation. The strong chemical adsorption and catalytic conversion of lithium polysulfides (LiPSs) by HEA nanoparticles effectively inhibit the LiPS shuttle effect. An HEA/rGO@PP battery maintains a Coulombic efficiency close to 100% after 200 cycles, with the shuttle effect significantly suppressed [56].

2.2. *Nano-Confinement and Steric Hindrance Stabilization Mechanism*

The nanoscale confinement effect in carbon/high-entropy alloy (HEA) nanocomposites refers to the physical spatial constraints imposed by the carbon matrix (e.g., graphene or carbon nanotubes) on HEA components at the nanoscale [58]. This effect enhances material performance through two mechanisms: first, kinetic stability enhancement—the nanoscale spatial confinement mediated by the carbon matrix significantly suppresses the atomic diffusion and lattice migration of HEA components. Under high temperatures or stress conditions, this effectively delays phase separation and grain coarsening, maintaining structural integrity under thermodynamic metastability over extended periods. Second, electronic structure modulation—the confined environment of the carbon matrix tunes the electronic states of HEA components (e.g., curvature-induced electron density redistribution in carbon nanotubes), optimizing the intrinsic reactivity of catalytic active sites. The steric hindrance effect originates from the geometric arrangement of surface atoms/groups of HEA components, functioning through two mechanisms: first, selective exposure of active sites—the steric hindrance from specific HEA surface configurations (e.g., step edges, kink sites) precisely controls reactant molecule access to active sites, suppressing side reactions. Second, volume expansion suppression—during charge–discharge cycles, the rigid structure of HEA components resists lattice distortion via atomic-scale steric hindrance, mitigating powdering caused by volumetric changes. The synergistic interaction between nanoscale confinement and steric hindrance effects collectively enhances the catalytic activity and energy-storage stability of the composite material [62].

This enhancement of kinetic stability and optimization of steric hindrance through nano-confinement provides an important theoretical basis and technical pathway for developing novel multifunctional composites with both high stability and high activity, showing broad application prospects in frontier fields like energy storage and catalytic conversion. For example, a multi-layer graphene shell (~3–5 layers) suppresses HEA nanoparticle growth (average size < 5 nm) and enhances kinetic stability via the confinement effect, enabling the material to maintain structural stability even at 800 °C. A 3D porous laser-induced graphene (LIG) substrate restricts nanoparticle migration through steric hindrance, achieving uniform HEA NP distribution (density > 10^5 particles/μm^2). The HEA NPs/LIG electrode exhibits low overpotential (268 mV@10 mA/cm^2) and high stability (>100 h), attributed to the graphene shell preventing active component detachment (only a 0.8 nm particle size increase after cycling) and the porous structure facilitating mass transport (porosity > 80%) [63]. HEA nanocrystallites (40–60 nm) uniformly embedded in a nitrogen-doped carbon (NC) matrix form nano-confined spaces that suppress grain aggregation and provide abundant active sites. The 3D porous structure of the NC matrix (specific surface area 198.1 m^2/g) anchors polysulfides through dual mechanisms: physical confinement and chemical adsorption (via nitrogen-doped defect sites). HEAs and NC synergistically construct steric hindrance, limiting polysulfide diffusion paths. The highly conductive NC matrix accelerates electron transport, while HEA catalyzes polysulfide conversion, achieving a sulfur utilization of 89.4% (1079.5 mAh/g) at 0.1 C. The 3D porous structure promotes Li^+ diffusion (validated by GITT showing increased diffusion coefficient), delivering 440.5 mAh/g even at 5 C. The confinement effect and chemical adsorption synergistically suppress the shuttle effect, resulting in a 99.0% capacity retention after 100 cycles. Thus, nano-confinement optimizes the reaction kinetics, steric hindrance inhibits side reactions, and nitrogen doping enhances interfacial stability, ultimately achieving a high specific capacity, long cycle life, and excellent rate performance [64].

2.3. Multiscale Structure Synergy and Strain-Adaptive Mechanism

Carbon/HEA nanocomposites exhibit multiscale structure synergy through bicontinuous networks and biomimetic transport channels. Within these composites, the bicontinuous network structure manifests as a 3D interpenetrating framework of the carbon matrix (e.g., 3D porous carbon or graphene aerogel) and HEA nanoparticles. This structure constructs efficient mass/charge transport channels via biomimetic design (e.g., hierarchical channels resembling plant vascular bundles) [65]. The continuous conductive skeleton of the carbon network ensures rapid electron conduction, while the continuous active interface of the HEA network promotes ion/molecule transport. In electrochemical systems, pore structures across different scales play critical and complementary roles: micropores (<2 nm), with sizes comparable to electrolyte ion diameters, provide high-density charge storage sites. However, single-phase microporous systems inherently increase ion transport resistance due to narrow channels, limiting rate performance. At this stage, mesopores (2–50 nm) act as "ion highways" to accelerate ion diffusion toward microporous regions, shortening transport pathways. Experimental evidence confirms that materials with a high mesopore content maintain superior capacitance retention at high current densities. Moreover, mesopore wall structures reduce ion migration energy barriers, synergistically enhancing kinetics alongside micropores. Macropores (>50 nm) function as macroscopic buffers and guiding channels. As "ion reservoirs," they shorten diffusion distances to active regions (mesopores/micropores), particularly preventing local concentration polarization in thick electrodes. In reactions involving gaseous products, graded macroporous structures (e.g., 200–300 nm) facilitate directional bubble release, avoiding active site blockage. Ultimately, macropores, mesopores, and micropores form an efficient three-tier synergistic mechanism: macropores buffer ions → mesopores enable rapid transport → micropores store charges [66]. This structural synergy endows the material with both high activity and high kinetic performance in applications like electrocatalysis and battery electrodes.

For example, CuInNiSnCd HEA nanoparticles (HEA-NPs) are uniformly loaded onto a 3D carbon fiber (CF) framework, forming a bicontinuous network structure. This structure combines a high specific surface area with low local current density, promoting uniform lithium deposition. The lithiophilic nature of HEA-NPs reduces the Li nucleation barrier, and biomimetic transport channels accelerate ion transport. This multiscale synergy significantly enhances battery performance, enabling stable cycling for over 3000 h at 10 mA cm^{-2}/10 mAh cm^{-2}. A full cell exhibits 93.3% capacity retention after 160 cycles at 1 C, with Coulombic efficiency reaching 99.2% [67]. Combining strain-adaptive characteristics and a potential auxetic-like (negative Poisson's ratio) effect, HEAs/graphite composites (HEAs/C) significantly optimize lithium-ion battery electrode performance. The strain-adaptive property allows the material to self-adjust stress during charge/discharge, maintaining structural stability; the auxetic effect promotes isotropic expansion, suppressing volumetric deformation. After 1000 cycles, the electrode thickness of HEAs/C increased from 14.78 μm to 20.25 μm, yet it still delivered a discharge specific capacity of 1196.5 mAh g^{-1} at 0.5 A g^{-1} with nearly 100% Coulombic efficiency. Compared to pure HEA electrodes, HEAs/C exhibit significantly enhanced cycling stability and drastically reduced capacity decay. The synergistic effect of dispersed HEA multi-components and the graphite buffer alleviates volume expansion and promotes electron transport, reducing the electrode volume change rate by 84% (37% vs. 235%) and extending the cycle life [68].

3. Preparation Methods

In recent years, the nanoscale composite technology integrating carbon materials (graphene, CNTs, porous carbon, etc.) with HEAs has advanced rapidly, focusing on interface regulation and multiscale structural design. Preparation techniques have expanded from traditional mechanical alloying and chemical vapor deposition (CVD) to sol-gel methods, electrospinning, 3D printing, and other additive manufacturing technologies, and has different advantages and disadvantages in Table 1. Concurrently, Atomic Layer Deposition (ALD) is emerging due to its advantages in atomic-level precision and biomimetic structures. Current research hotspots include low-temperature/green processes (e.g., plasma-assisted CVD, bio-templated low-temperature synthesis), heterointerface optimization (enhancing interfacial bonding strength via ALD or gradient ball milling), and function-oriented design (e.g., 3D printed customized porous electrodes, electrospun flexible devices).

Table 1. Comparative Analysis of Preparation Technologies.

Technology	Advantages	Disadvantages	Application Scenarios
Mechanical Alloying	Low cost, easy to scale up	Easy to introduce impurities, uneven particle size	Bulk materials, energy storage electrodes
CVD	Strong interface bonding, controllable structure	Expensive equipment, high energy consumption at a high temperature	Thin films, catalytic materials
Sol-Gel	Porous structure, high specific surface area	Calcination shrinkage, low mechanical strength	Porous electrodes, catalyst carriers
Electrospinning	Flexible fibers, high specific surface area	Low fiber strength, difficult mass production	Flexible devices, sensors
3D Printing	Customized complex structures, high precision	Limited material selection, complex post-processing	Bionic structures, functional devices

3.1. Mechanical Alloying

Mechanical alloying combined with Spark Plasma Sintering, Hot Pressing (HP), or Hot Isostatic Pressing is a mainstream process enabling the uniform dispersion of carbon materials and strong interfacial bonding. This primarily involves the mechanical mixing of carbon materials and HEA metal powders via high-energy ball milling, achieving atomic-level compounding through repeated collisions [69,70]. Strengthening mechanisms include load transfer, grain refinement, dislocation pinning, and interfacial reactions (e.g., carbide formation) [71]. Carbon materials significantly enhance the strength, hardness, and wear resistance of HEAs, while some systems can retain good ductility [72]. Key parameters include the milling time (10–50 h), rotation speed (200–500 rpm), and atmosphere control (Ar/N_2). Liu et al. fabricated graphene nanoplatelet-reinforced layered CoCrFeNiMn HEA matrix composites (GNPs/CoCrFeNiMn HEAs) via mechanical ball milling and flake powder metallurgy, followed by vacuum hot-press sintering, forming a unique nacre-like structure in Figure 2. The microstructure comprised a FCC matrix phase and $Cr_{23}C_6$ and $CrMn_{1.5}O_4$ precipitate phases, along with numerous dislocations and twins. Isothermal oxidation testing at 1000 °C for 100 h revealed that the composite's mass increased over time, initially linearly and later exponentially, indicating superior long-term oxidation resistance. The layered structure's anisotropy endowed the composite with excellent resistance to high-temperature steam oxidation perpendicular to the lamellae. The oxide layer consisted mainly of outer $(Mn, Cr)_3O_4$ and Mn_3O_4, with an inner Cr_2O_3 layer. Elemental depletion zones of Mn and Cr appeared in the matrix near the oxide scale, and the diffusion of these elements effectively enhanced the material's oxidation resistance [72]. Singh et al. prepared a FeCoCrNiCu HEA powder via mechanical alloying (MA), with functionalized CNTs (0.1–7.0 wt.%) uniformly dispersed within the HEA matrix via ball milling. Subsequent densification was achieved via Spark Plasma Sintering at 800 °C and 50 MPa, optimizing the process to balance grain growth and CNT structural integrity. The uniform CNT dispersion and interface regulation addressed issues of microstructural inhomogeneity and insufficient corrosion performance in conventional HEAs. The synergistic combination of the HEA multi-principal element effects and CNT reinforcement optimized material performance. The optimal CNT content (2 wt.%) significantly promoted HEA microstructural homogenization, forming a single FCC phase; excessive CNT (>2 wt.%) triggered $Cr_{23}C_6$ carbide precipitation, leading to renewed phase separation. CNTs inhibited grain coarsening during sintering, reduced lattice strain, and increased interface density. The composite with 2 wt.% CNTs exhibited an 88.6% reduction in the corrosion rate compared to pure HEA (0.52 vs. 4.58 $mil/yr/cm^2$) in a 3.5% NaCl solution, with a significant positive shift in corrosion potential. However, an excessive CNT content (7 wt.%) caused a sharp 58% increase in the corrosion rate, demonstrating a clear threshold effect. The corrosion performance improvement stemmed from CNTs promoting elemental diffusion uniformity and reducing galvanic corrosion risk. At a high CNT content, $Cr_{23}C_6$ precipitation induced intergranular corrosion, disrupting chemical homogeneity. CNTs synergistically enhanced corrosion resistance through grain refinement, increased dislocation density, and interfacial barrier effects; the bimodal phase distribution and carbide control were key factors in the performance transition [73].

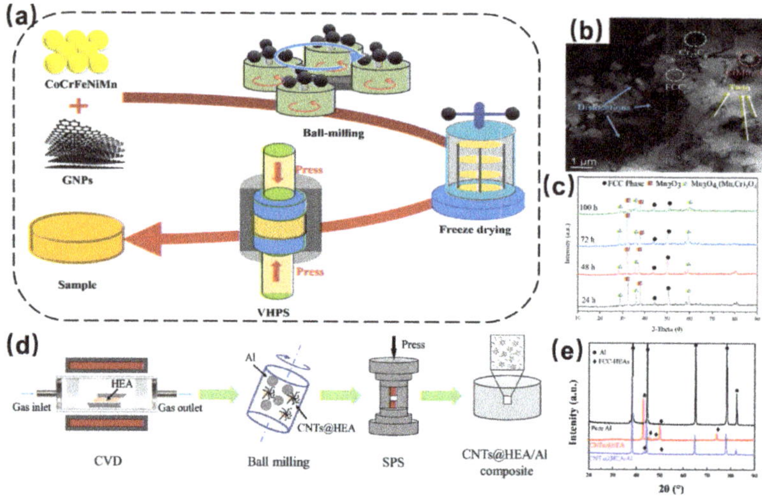

Figure 2. Mechanical alloying and CVD method for the fabrication of carbon/HEAs nanocomposites: (**a**) Schematic diagram, (**b**) TEM and (**c**) XRD of GNPs/CoCrFeNiMn HEAs [72], copyright 2023, Elsevier; (**d**) sSchematic diagram, and (**e**) XRD of CNTs/HEAs [74], copyright 2024, Elsevier.

3.2. Chemical Vapor Deposition (CVD)

CVD is primarily used to grow carbon materials (CNTs, graphene, etc.) on or within HEAs. Key aspects include the catalyst design, temperature control, and suppression of interfacial reactions. This involves decomposing carbon sources (e.g., CH_4) and metal precursors in a high-temperature reaction chamber to co-deposit carbon materials and HEA nanoparticles onto a substrate [75,76]. Carbon materials enhance strength and toughness through load transfer, dislocation pinning, and grain refinement. For instance, graphene/HEA heterojunctions can optimize conductivity, catalytic activity, and oxidation resistance [74]. Peng et al. demonstrated the in situ growth of CNTs on the surface of HEA particles via CVD in Figure 2. Subsequently, CNTs/HEAs-particle synergistically reinforced aluminum matrix composites were fabricated using powder metallurgy (ball milling and Spark Plasma Sintering). In situ synthesis of the CNTs@HEAs hybrid reinforcement ensured high crystallinity and structural integrity of the CNTs, resolving issues of poor dispersion and weak interfacial bonding in the aluminum matrix. The composite exhibited a bimodal grain distribution (coarse and fine grains), with CNTs uniformly dispersed in the aluminum matrix. Interfacial Al_4C_3 and Cr_7C_3 carbides enhanced the load-bearing capacity of the CNTs. Furthermore, the presence of high-density stacking faults further improved material performance. Compared to pure aluminum and HEA-only reinforced composites, the CNTs@HEAs/Al composite demonstrated significantly improved tensile strength (230 MPa) and elongation, achieving a synergy of strength and plasticity superior to similar studies. Strengthening mechanisms included load transfer, back stress strengthening, Orowan strengthening, grain refinement, and stacking fault strengthening. The bimodal grain structure and interfacial carbides jointly promoted strain hardening capability, while the strong HEAs-CNTs-Al interfacial bonding and the presence of SFs significantly enhanced ductility [74]. Hassan et al. successfully synthesized $FeCoNiCuMn_x/C$ nanoparticles with excellent electromagnetic wave absorption performance via an innovative core-shell structure design, composition tuning, and Metal-Organic Chemical Vapor Deposition. Encapsulating FeCoNiCuMn HEA nanoparticles with a graphitic carbon shell effectively solved the agglomeration problem of magnetic nanoparticles, reduced the material density, and improved impedance matching. The fine-tuning of electromagnetic properties was achieved by adjusting the Mn content, further optimizing the absorption performance. The nanoparticles featured a uniform core-shell structure, with a metal core diameter of ~14 nm and a carbon shell thickness of ~3 nm. The carbon shell possessed a high degree of graphitization, beneficial for conductive loss and polarization loss. The metal core exhibited an FCC structure, with slight lattice constant changes as the Mn content increased. The core-shell design and Mn content tuning effectively optimized impedance matching, reducing electromagnetic wave reflection and enhancing absorption efficiency. The heterogeneous interface between the carbon shell and metal core promoted interfacial polarization, while defects in the carbon shell enhanced defect polarization. Concurrently, magnetic loss mechanisms (natural resonance, exchange resonance, eddy current loss) in the HEA nanoparticles further strengthened the absorption capability. $FeCoNiCuMn_{0.5}/C$ nanoparticles achieved a reflection loss of −52.3 dB at a thickness of 2.35 mm, with an effective absorption bandwidth covering 5.52 GHz [77].

3.3. Sol-Gel Method

The sol-gel method is a wet-chemical process for synthesizing materials via liquid-phase reactions. It utilizes highly reactive metal salts or organic precursors as raw materials, forming a stable sol through hydrolysis and condensation polymerization, followed by gelation, drying, and thermal treatment to obtain nanostructured materials [78]. Carbon/HEA nanocomposites are fabricated by combining a carbon matrix (e.g., CNTs, graphene, or polymer-derived carbon) with multi-component HEA nanoparticles via this method, forming functional materials with combined dielectric and magnetic loss characteristics [79]. The sol stage achieves atomic/molecular-level mixing, ensuring a uniform distribution of the carbon matrix and HEA components, enabling precise control of the porosity and nanoparticle size [80]. The gel network forms nanopores, increasing the material's specific surface area and enhancing multiple reflections and interfacial polarization effects in electromagnetic wave absorption [79]. By combining the molecular controllability of sol-gel with the multi-element synergy of HEAs, this method provides efficient and designable nanocomposite solutions for electromagnetic shielding, new energy storage, and other fields. Zhang et al. synthesized FeCoNiCuAl HEAs via the sol-gel method. Subsequently, a heterojunction between HEAs and nitrogen-doped carbon (NC) was constructed through the self-polymerization of dopamine hydrochloride and subsequent annealing in Figure 3 [81]. This approach leveraged the uniformity of the sol-gel and the flexibility of self-polymerization to successfully prepare a composite with a core-shell structure. Introducing nitrogen-doped carbon nanospheres to form a heterojunction with HEAs not only addressed the impedance mismatch issue of single-phase HEA materials in EM absorption but also enhanced dielectric loss through defects and interface effects in NC, achieving synergistic magnetic and dielectric loss for improved absorption performance. The HEAs/NC composite featured tightly bonded heterojunction interfaces, with NC exhibiting the (002) crystal plane of graphitic carbon, forming a graphitic carbon network. Nitrogen doping introduced defects and interface effects that significantly enhanced dielectric loss. The heterojunction interface promoted interfacial polarization. Furthermore, the graphitic network improved material conductivity, enhancing conductive loss. Finally, the inherent magnetic loss mechanisms of the HEAs (natural resonance, exchange resonance, eddy current loss) contributed to absorption. This synergistic effect made HEAs/NC an efficient EM absorber. The prepared HEAs/NC2 composite achieved a minimum reflection loss of −56.38 dB at a thickness of 1.80 mm, with a maximum effective absorption bandwidth of 5.69 GHz. Yu et al. prepared $Pt_{18}Ni_{26}Fe_{15}Co_{14}Cu_{27}$ HEA nanocrystals via a colloidal synthesis strategy, with reduced graphene oxide (rGO) synthesized simultaneously during nanocrystal formation [82]. $Pt_{18}Ni_{26}Fe_{15}Co_{14}Cu_{27}$ nanocrystals grew uniformly on the rGO surface, forming a $Pt_{18}Ni_{26}Fe_{15}Co_{14}Cu_{27}$/rGO composite. The nanocrystals were uniformly dispersed on rGO with high crystallinity. Combining HEA nanocrystals with rGO leveraged the 2D structure and superior properties of rGO to enhance the composite's EM absorption. Monodisperse HEA nanocrystals (~3.3 nm) were successfully synthesized, providing more polarization-active sites. Multiple new polarization interfaces formed between the HEA nanocrystals and rGO; charge rearrangement at these interfaces generated strong polarization effects, enhancing absorption. Compared to pure $Pt_{18}Ni_{26}Fe_{15}Co_{14}Cu_{27}$ nanocrystals, the composite exhibited superior EM absorption performance in the 2–18 GHz range. At a 4 mm thickness, the minimum reflection loss reached −41.8 dB, with an effective absorption bandwidth of 2.5 GHz in the 9.4–11.9 GHz band.

3.4. Electrospinning

Electrospinning is a technique that utilizes a high-voltage electric field to induce the jetting and stretching of a polymer solution/melt to form nanofibers [83]. Carbon/HEA nanocomposites fabricated via this technology combine multi-component HEA nanoparticles with a carbon matrix (e.g., carbon nanofibers, graphene composite fibers), forming functional materials characterized by porous structures, a high specific surface area, and multi-element synergistic effects [84]. Electrospinning enables the atomic-level uniform distribution of HEA nanoparticles (<10 nm) within the carbon matrix, reducing agglomeration. By precisely controlling fiber microstructure and component distribution, this method provides a design paradigm for high-performance nanocomposites in fields like new energy and electromagnetic protection [85]. Liu et al. used a bottom-up synthesis approach combining electrospinning and calcination to prepare FeCoNiMnCu HEAs/carbon nanofiber (HEAs/CNF) composites [86]. The calcination temperature was tuned to regulate the composite's crystallinity, graphitization degree, and electromagnetic parameters. FeCoNiMnCu HEA nanoparticles were uniformly embedded within carbon nanofibers, forming partial core-shell structures and abundant interfaces. The fiber diameter was uniform (170–230 nm), with alloy particles averaging ~20 nm in diameter. HEA particles had an FCC structure, and the carbon matrix was highly crystalline. The 1D carbon nanofiber network promoted electron migration and transition, enhancing conductive loss. Abundant interfaces between alloy particles and the carbon matrix induced multiple interfacial polarizations, increasing polarization loss. Natural resonance and eddy current loss from the HEAs contributed to magnetic loss. This design yielded a composite achieving strong absorption (−64.4 dB) and a broad bandwidth (4.1 GHz) at low filler loading (10 wt%) and thin thickness (1.9 mm). Wang et al. prepared FeCoNiMnRu-HCB composites via electrospinning, activation, and carbonization, embedding HEA catalysts within a highly mesoporous carbon material to significantly increase the specific surface area and catalytic activity in Figure 3 [87]. Benzooxazine was added to the electrospinning solution; its thermal polymerization and subsequent activation process created rich mesopores, increasing the specific surface area and catalytic sites. The prepared nanofibers had high aspect ratios and uniform diameters (50–150 nm), forming a 3D porous structure. HEA particles (FeCoNiMnRu) were uniformly dispersed in the carbon nanofiber matrix with high crystallinity (lattice spacings of 2.1 Å and 1.8 Å, corresponding to (111) and (200) planes). The abundant mesoporous structure provided more active sites, facilitating electrolyte penetration and gas product release, thereby boosting catalytic performance. The unique composition and structural features of the HEAs (high configurational entropy, lattice distortion, sluggish diffusion, synergistic "cocktail effect") endowed the catalyst with excellent electrocatalytic activity and stability. Nitrogen from the residual PBZ resin formed metal–nitrogen bonds, stabilizing the crystal structure and enhancing catalytic efficiency. The FeCoNiMnRu-$HCB_{0.5}$ electrode exhibited outstanding electrocatalytic performance for alkaline water electrolysis: overpotentials for hydrogen evolution reaction (HER) and oxygen evolution reaction (OER) at 10 mA cm^{-2} were 42 mV and 229 mV, respectively, lower than commercial noble metal catalysts. Both HER and OER current densities remained stable after 20 h of continuous electrolysis, demonstrating good long-term stability.

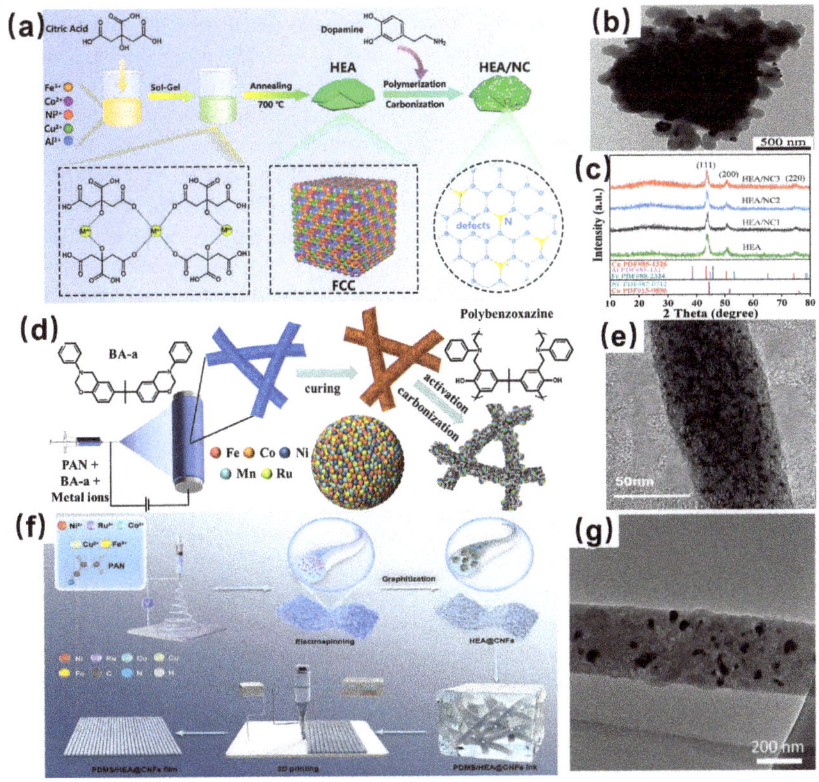

Figure 3. Sol-gel and electrospinning method for the fabrication of carbon/HEA nanocomposites: (**a**) schematic diagram, (**b**) TEM, and (**c**) XRD of HEAs/NC [82], copyright 2025, Royal Society of Chemistry; (**d**) schematic diagram, and (**e**) TEM of FeCoNiMnRu-HCB [87], copyright 2025, Elsevier; (**f**) schematic diagram, and (**g**) TEM of HEAs@CNFs [88], copyright 2024, Springer.

3.5. 3D Printing

The 3D printing of carbon/HEA nanocomposites refers to the use of additive manufacturing processes (e.g., Laser Powder Bed Fusion (LPBF), Fused Deposition Modeling (FDM)) to combine multi-principal element HEAs with carbon-based materials (e.g., CNTs, graphene), forming advanced functional materials with nanoscale multiphase composite structures [89,90]. Carbon-based materials (e.g., CNTs) are uniformly dispersed within the alloy matrix through shear forces and high temperatures during printing, forming strong "alloy-carbon" interfacial bonding. Rapid cooling suppresses elemental segregation, promoting single-phase solid solution formation in the HEAs, while the nanoconfinement effect of the carbon matrix stabilizes alloy lattice distortion [90]. These materials combine the strength and toughness of HEAs with the lightweight, high thermal/electrical conductivity of carbon materials, showing potential in electromagnetic shielding, thermal management, catalysis, etc. [91]. By precisely controlling the microstructure and multiphase synergy, this method provides a new paradigm for designing and scaling high-performance composites. Yan et al. prepared HEOs@C-GR/PLA composites via high-temperature carbonization and 3D printing, achieving uniform dispersion and structural densification [92]. The composite exhibited a distinct double core-shell structure: HEO cores were uniformly coated by a carbon shell, while graphene (GR) dispersed within the PLA matrix formed a continuous conductive network. This double core-shell design (HEOs@C), with an inner high-entropy multi-metal oxide phase and an outer carbon shell, combined with the GR hierarchical conductive network. The "cocktail effect" of the high-entropy material modulated lattice distortion and oxygen vacancies, synergizing with GR's conductivity to achieve coupled dielectric-magnetic loss enhancement, breaking the limitations of single materials. The 3D printing enabled precise shaping of complex structures. Double core-shell interfacial polarization and the GR network induced multiple relaxations. Lattice distortion, oxygen vacancies, and carbon defects formed dipole polarization centers, enhancing energy dissipation. The ferromagnetic components (Co, Fe, Ni) in HEOs converted EM energy via natural resonance and eddy current effects; the high-entropy effect optimized permeability, improving low-frequency absorption. GR modulated material conductivity, while the HEOs@C core-shell structure reduced surface reflection. Thickness-frequency synergistic matching based on $\lambda/4$ theory broadened the absorption bandwidth. The coupling of multilevel heterogeneous interfaces and the conductive network enabled multiple reflection-scattering-absorption of EM waves, significantly boosting the overall absorption efficiency. The composite achieved a minimum reflection loss of −51.36 dB and an effective absorption bandwidth of 5.20 GHz covering the C to Ku bands within 2–18 GHz. Compared to pure HEO systems, filler usage was reduced by 40%, while achieving light weight (density < 2.0 g/cm^3) and high mechanical strength (tensile strength > 45 MPa). Wu et al. prepared the flexible films by mixing HEAs@CNFs or pure CNFs with PDMS resin using electrospinning, pyrolysis, and 3D printing, followed by curing to form composites with different filler contents (10 wt%, 15 wt%, 20 wt%) in Figure 3 [88]. HEAs@CNFs were nanoscale-diameter fibers embedding uniformly distributed HEA nanoparticles (~10 nm), with five metals (Fe, Co, Ni, Cu, Ru) uniformly distributed within the alloy particles (atomic ratio 5–35%). Embedding HEA nanoparticles within carbon nanofibers (CNFs) significantly reduced the electron delocalization capability, tuning carrier concentration, and effective electron mass, achieving epsilon-near-zero (ENZ) performance in the radio frequency (RF) range (21 MHz) for the first time. Combined with the flexibility and biocompatibility of PDMS, 3D printing enabled the construction of wearable electronic devices suitable for human motion detection and medical monitoring. HEA introduction reduced the delocalization ability of electrons around carbon atoms (decreased ELF value), increased the work function (3.438 eV $\rightarrow$ 4.831 eV), and reduced the carrier concentration. HEAs flattened the CNF band structure (enhanced non-parabolicity), increasing the effective electron mass and further lowering the plasma frequency. Charge accumulation at the PDMS/CNFs interface enhanced positive permittivity, while conductive network formation led to negative permittivity. Consequently, at 20 wt% HEAs@CNFs, the PDMS/HEAs@CNFs film achieved ENZ (permittivity transition from negative to positive) at 21 MHz, with plasma frequency significantly reduced to the RF range. The high filler content formed a continuous conductive network, endowing the film with high conductivity and excellent flexibility (bendable, foldable). Cell experiments (H&E staining, CCK-8) showed no cytotoxicity (cell viability >95%). This study innovatively introduced HEAs into a CNF system via electrospinning and 3D printing, achieving breakthroughs in both RF ENZ performance and biocompatibility, providing new material solutions for wearable medical electronics and flexible sensing.

4. EES Applications

4.1. Lithium-Ion Batteries (LIBs)

Nanocomposites of carbon materials (e.g., graphene, CNTs, porous carbon) and HEAs, leveraging their unique synergistic effects (high conductivity, multiple active sites, excellent mechanical stability), have demonstrated broad application prospects in LIBs in recent years. The carbon framework provides continuous electron/ion transport pathways and mitigates volume expansion, while the multi-element synergistic effect of HEA nanoparticles enhances structural stability, significantly extending the cycle life. The abundant active sites in HEAs contribute to high theoretical capacity (>500 mAh/g), and the carbon carrier accelerates reaction kinetics, supporting high-rate charging/discharging. HEA surfaces can efficiently catalyze electrode reactions (e.g., sulfur conversion or oxygen reduction), suppressing the shuttle effect and improving energy efficiency.

In terms of anode materials, carbon matrices buffer the volume expansion of HEAs, enhancing cycling stability. HEA nanoparticles composed of Ge, Sn, Sb, Si, Fe, Cu, and P elements were synthesized with carbon materials via high-energy mechanical ball milling for LIB anodes by Wei et al. in Figure 4 [38]. The rational selection of elements with complementary electrochemical properties (e.g., high-capacity Si, Ge, Sn, Sb, and P and high-conductivity Cu and Fe) enabled the construction of a novel HEA/C composite anode featuring a unique dragon-fruit-like dense structure. This composite process ensured the uniform dispersion and encapsulation of HEA nanoparticles within the carbon matrix, which functions as a multifunctional framework: (1) providing mechanical support that prevents nanoparticle agglomeration and pulverization during cycling; (2) buffering volume expansion during lithiation (reduced from 34% to 25%) to maintain electrode integrity; (3) facilitating rapid electron transport through its high conductivity (2.14×10^1 S/m), flexibility, and chemical stability. The multi-elemental synergy within HEA nanoparticles generated distinct electrochemical behaviors during lithiation, enhancing the lithium storage capacity and reaction activity while alleviating volume stress. This synergistic effect—combined with the carbon matrix's defective structure (D/G band ratio $\approx$1:1), increasing reaction sites and uniform element distribution (e.g., Ge, Sn, Sb) and exposing active sites—was further reinforced by P-C/P-O-C bonds at the HEA–carbon interface (confirmed by XPS), stabilizing the framework and promoting electron/ion diffusion. Consequently, the anode achieved a high initial Coulombic efficiency of 91%, stable cycling exceeding 1600 h, and 63% capacity retention at 2000 mA/g, significantly outperforming conventional alloy anodes. Xiao et al. synthesized Sn-Si-Co-Cu-P high-entropy alloy/graphite (HEA/C) composite anode materials via high-energy ball milling, forming an amorphous composite structure with HEA nanoparticles uniformly dispersed at the atomic scale within the graphite matrix [68]. This architecture delivers threefold synergistic advantages: (1) the graphite matrix provides high conductivity (significantly reducing charge transfer resistance R_{ct}) and layered mechanical support, effectively suppressing phase separation and particle agglomeration while buffering volume expansion (experimental value 37% vs. theoretical 235%); (2) its porous structure (specific surface area: 13.93 m^2/g) increases lithium-ion adsorption/diffusion sites. Elemental mapping confirms the homogeneous distribution of Sn/Si/Co/Cu/P/C without segregation, where high-capacity elements (Sn/Si/P) synergistically optimize the electronic environment of active sites with conductive copper and structurally stabilizing cobalt. The multi-component synergy enhances the lithium storage capacity and reaction activity, coupled with ion transport facilitation through graphite interlayer channels and lithium storage enhancement at HEA active sites. Consequently, the material delivers an initial discharge capacity of 1881 mAh/g, retains 1196.5 mAh/g after 1000 cycles, achieves 47.2% capacity retention at a 10 A/g high current density, and maintains structural integrity over 1000 cycles.

In terms of cathode materials, the multi-element synergy of HEAs optimizes structural stability at high voltages. Yi et al. synthesized an FeCoNiMnCuAl@C material with a hierarchical structure via carbothermal reduction using recycled metal ions from spent LIBs, serving as a cathode catalyst for rechargeable Li–CO_2 batteries in Figure 4 [93]. The hierarchical nanosheet architecture, derived from MOF-pyrolyzed carbon material, integrates high electrical conductivity with a large specific surface area (221.1483 m^2/g). This unique structure ensures the uniform distribution of HEA nanoparticles across carbon nanosheets, simultaneously providing the following: (1) high porosity enhancing mass transport and CO_2 diffusion; (2) abundant ion transport channels boosting catalytic activity (discharge capacity: 27,664 mAh/g at 100 mA/g). The carbon substrate facilitates rapid electron transport while buffering volume changes, with its flexibility preventing HEA structural collapse during cycling. DFT calculations confirm that multi-element sites (e.g., Co, Mn, Al) in the HEA exhibit optimal CO_2 adsorption energy. Through high-entropy effects and carbon–HEA synergy, inert CO_2 molecules are activated to drive reversible Li_2CO_3 formation/decomposition, reducing overpotential to 1.05 V and maintaining a high discharge voltage plateau (2.77 V). These mechanisms collectively enable stable cycling over 134 cycles. Wen et al. fabricated FeCoNiCuRu high-entropy alloy/carbon nanofiber (HEA/CNF) composites as Li-CO_2 battery cathodes via electrospinning and thermal treatment, achieving tunable configurational entropy [94]. The 3D conductive CNF network (specific surface area: 295.6 m^2/g) uniformly anchors sub-10 nm HEA nanoparticles, concurrently providing mechanical support to prevent agglomeration, establishing rapid electron transport pathways, and facilitating CO_2/electrolyte diffusion through its flexible framework. The HEA nanoparticles drive efficient CO_2 adsorption/activation and reversible Li_2CO_3 conversion via triple synergistic mechanisms: (1) high-entropy effects optimizing active site distribution; (2) electron transfer from low-electronegativity elements (Fe, Co) to Ru generating electron-rich sites (verified by XPS/XANES); (3) defective carbon matrix structures increasing catalytic sites. This synergy enhances CO_2 adsorption (evidenced by negative adsorption energy) and reduces Li_2CO_3 decomposition barriers to 0.70 eV, ultimately delivering a discharge capacity of 6160 mAh/g at 200 mA/g with a low charge voltage plateau (<4.0 V), enhanced rate capability, and ultralong cycling stability over 550 cycles (5500 h).

Despite the outstanding advantages of carbon materials (such as graphene, CNTs, porous carbon, etc.) and HEA nanocomposites in the field of LIBs, this material system still faces some key challenges. The precise control of HEA element ratios (e.g., equiatomic) and uniform dispersion are difficult; current methods (e.g., ball milling, sputtering) are costly, energy-intensive, and hard to scale. Insufficient carbon/HEA interfacial bonding strength can lead to particle detachment during cycling. Active metals in HEAs (e.g., Fe, Cr) may dissolve in the electrolyte, causing continuous SEI growth (>50% increase in interfacial impedance after cycling), increasing resistance, and reducing the Coulombic efficiency. Incorporating precious metals (e.g., Pt, Ir) increases the cost, and the high density of HEAs (>8 g/cm^3) limits gravimetric energy density, hindering EV applications. Future research must focus on developing low-cost synthesis processes, optimizing interfacial coating techniques (e.g., ALD of Al_2O_3), and exploring precious-metal-free HEA systems to advance practicality.

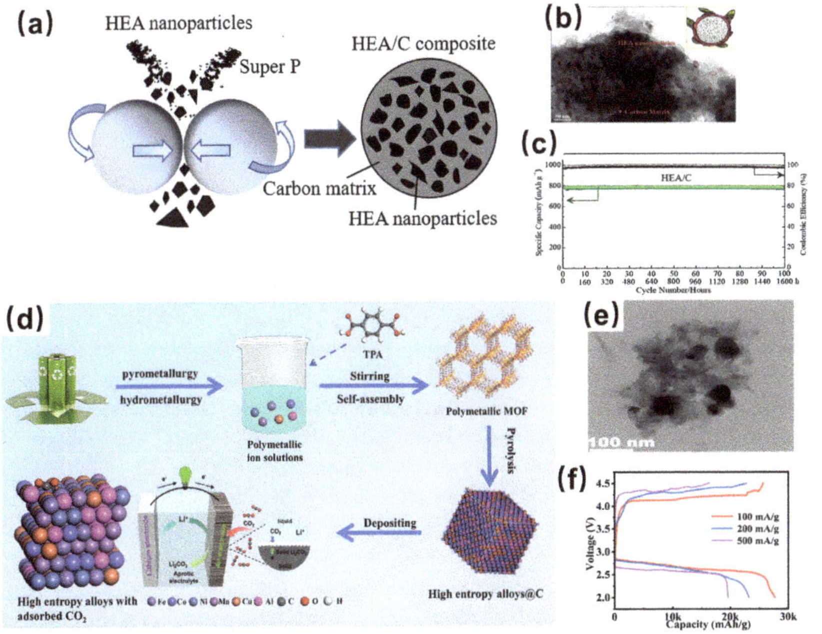

Figure 4. Carbon/HEA nanocomposites for LIBs: (**a**) schematic diagram, (**b**) TEM, and (**c**) cycle stability of HEAs/C [38], copyright 2022, Elsevier; (**d**) schematic diagram, (**e**) TEM, and (**f**) discharge and charge profiles of FeCoNiMnCuAl@C [93], copyright 2024, Wiley-VCH GmbH.

4.2. Lithium Metal Batteries (LMBs)

Carbon/HEA nanocomposites significantly optimize LMB performance by synergizing carbon-based carriers (e.g., graphene, CNTs) with multi-element HEA nanoparticles. HEA nanoparticle-decorated carbon skeletons homogenize Li^+ flux, suppress dendrite growth, and enhance cycling stability. Multi-element active sites in HEAs accelerate the reversible conversion of polysulfides/lithium peroxide, mitigating the shuttle effect and improving energy efficiency. The carbon/HEA composite interface provides high mechanical strength and 3D ion/electron channels, reducing the local current density and delaying electrode pulverization.

Wang's group developed three types of high-entropy alloy/carbon (HEA/C) composite materials through innovative methods to enhance lithium metal battery performance, including an AgCuInCdZn HEA nanoparticle-embedded porous carbon fiber (HEA/PCF) system synthesized via a fusion method [55]. This design utilizes a rigid 3D porous PCF skeleton constructed via Ag^+ etching (calcined at 1000 °C to form a single-phase alloy), synergistically integrated with molten-state embedded HEA nanoparticles. The composite preserves the porosity while enhancing ion/electron conduction through HEA lattice distortion. Key synergistic mechanisms include the following: (1) the PCF skeleton buffers volume expansion during lithium deposition and regulates mass transfer under high currents; (2) HEA nanoparticles uniformly distributed within pores (verified via molecular skeleton-level analysis) provide abundant lithiophilic sites, significantly reducing lithium nucleation overpotential (~5.3 mV); (3) the highly conductive network lowers the local current density, cooperatively suppressing dendrite formation and "dead lithium". Through coordinated pore engineering and catalytic site optimization, this HEA/PCF architecture achieves homogeneous lithium deposition and volume change regulation, ultimately enabling stable symmetric cell cycling > 1200 h at 60 mA/cm^2 and full cells (NCM-811 cathode) maintaining > 99.5% Coulombic efficiency over 200 cycles.

Secondly, CuInNiSnCd high-entropy alloy/carbon fiber (HEA/CF) composites were synthesized via a high-temperature redox strategy with calcination at 1100 °C, forming a single-phase face-centered cubic (FCC) structure featuring uniformly distributed HEA nanoparticles on the CF surface [67]. This 3D framework delivers triple synergistic functions: (1) the CF skeleton structurally regulates lithium deposition by reducing the local current density, buffering volume expansion, and providing spatial accommodation; (2) HEA nanoparticles electrochemically optimize nucleation through abundant lithiophilic sites with ultralow overpotential (3.4 mV, surpassing pure carbon's 14 mV and copper's 40 mV), enabling uniform Li deposition/stripping; (3) multi-element synergy kinetically enhances stability, as confirmed via dynamic electrochemical impedance spectroscopy (DEIS) demonstrating improved Li^+ transport kinetics and cycling reversibility. These mechanisms collectively contribute to dendrite-free symmetric cell operation exceeding 3000 h at 10 mA cm^{-2}, while maintaining Coulombic efficiencies > 99.2% over 160 cycles in full cells and 99.5% over 800 cycles at 2 mA cm^{-2} in half-cells.

Finally, a NiCdCuInZn high-entropy alloy/carbon fiber (HEAs/C) composite was fabricated via a thermodynamically driven phase transition method (Figure 5) [95]. This material features an ultrathin (~20 nm), uniformly distributed high-entropy alloy (HEA) layer comprising nanoparticles (NPs) and nanosheets on the carbon fiber (CF). This structural design provides a high-performance solution for anode-free lithium metal batteries (LMBs): the carbon fiber skeleton ensures efficient ion/electron transport and mechanical flexibility, while the single-phase face-centered cubic (FCC) structured HEA forms a strong bond with CF at 1100 °C, enhancing interfacial stability through lattice distortion. Leveraging the "cocktail effect" of its components, the HEA generates a gradient adsorption energy (−3.18 eV to −2.03 eV; e.g., Ni/Cu sites exhibit strong adsorption), enabling selective lithium-ion adsorption and rapid diffusion. This effectively reduces nucleation overpotential and guides uniform deposition. The ultrathin HEA nanostructure (nanosheets and NPs) provides continuous Li^+ transport pathways, abundant active sites for lithium deposition, and shortened Li^+ diffusion paths, thereby significantly suppressing dendrite growth. The exceptional performance is demonstrated as follows: asymmetric cells maintain > 99.6% Coulombic efficiency (CE) after 2000 cycles; symmetric cells achieve > 7200 h of dendrite-free stable cycling even at a high current density of 60 mA cm^{-2}; and full cells with a high-loading NCM-811 cathode deliver an average CE of 99.5% after 160 cycles at a 1C rate.

Although carbon/HEA nanocomposite-based lithium-metal batteries present excellent potential for application, they still face bottlenecks. The high catalytic activity of HEA surfaces can exacerbate electrolyte decomposition (e.g., oxidation of carbonate solvents), leading to thick, unstable CEI layers (>100 nm), increased impedance, and reduced Coulombic efficiency (<95% after cycling). The uniform dispersion of HEA NPs relies on high-energy ball milling or sputtering; composite processes (e.g., CVD growth) are complex and costly for mass production. Transition metals in HEAs (e.g., Mn, Co) may dissolve at high voltages (>4.3 V), causing active material loss and electrode passivation (>30% capacity fade after 100 cycles). The high density of HEAs hinders battery lightweighting. Future efforts need to develop in situ coating strategies (e.g., ALD of Al_2O_3 interlayers), design precious-metal-free HEA systems (e.g., TiZrNbMo), and explore solid-state electrolyte-compatible interfaces to drive practical application.

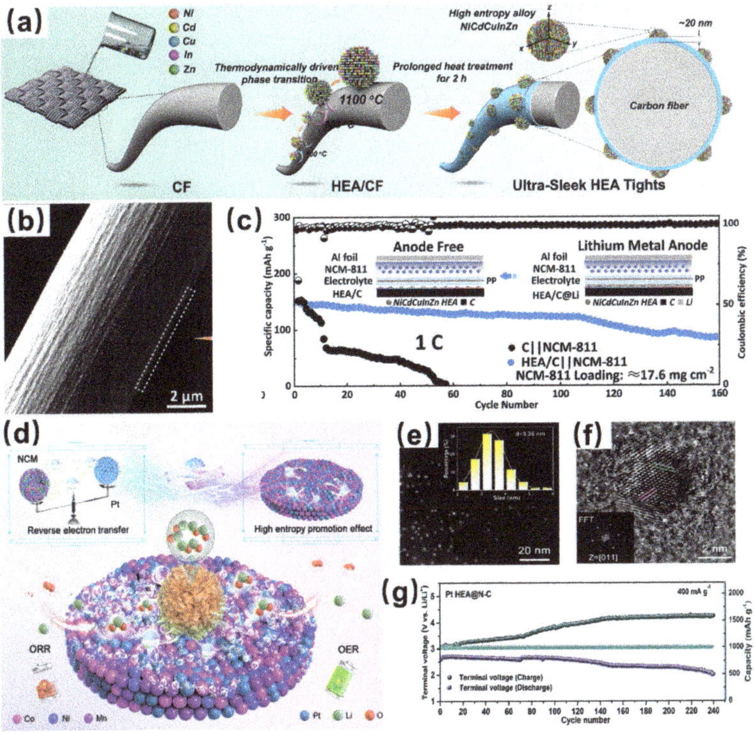

Figure 5. HEAs/C for LMBs: (**a**) schematic diagram, (**b**) SEM, and (**c**) cycle stability [95], copyright 2023, Wiley-VCH GmbH. Pt HEAs@N-C for Li-O_2: (**d**) schematic diagram, (**e**,**f**) TEM, and (**g**) cycle stability [96], copyright 2025, American Chemical Society.

4.3. Lithium-Oxygen Batteries (Li-O_2)

Carbon/HEA nanocomposites demonstrate unique advantages in Li-O_2 batteries. These materials combine HEA nanoparticles with carbon-based carriers (e.g., porous carbon fibers, graphene) to construct electrode systems integrating high catalytic activity and conductivity. The multi-element synergistic effect of HEAs significantly enhances the catalytic activity for the ORR and OER, while the carbon matrix provides a conductive network and reaction interface, collectively optimizing the battery's discharge capacity and cycling stability. The tunable composition of HEAs allows precise regulation of the electronic structure of catalytic sites. The porous structure of the carbon matrix facilitates oxygen/electrolyte transport. Synergistic interfacial effects within the composite suppress side reactions.

Peng Wang et al. synthesized Pt HEAs@N-C via a Joule-heating strategy using recycled metals (Ni, Co, Mn) from spent $LiNi_{1/3}Mn_{1/3}Co_{1/3}O_2$ cathodes combined with Pt in Figure 5 [96]. The Pt HEAs@N-C catalyst features uniformly anchored HEA nanoparticles (~3.36 nm) on an N-doped carbon (NC) carrier, forming abundant active sites and efficient electron pathways. The NC carrier delivers three core functions: (1) a high specific surface area (261.36 m^2/g) with abundant mesopores (Type IV isotherm) facilitates electrolyte infiltration, rapid Li^+/O_2 transport, and Li_2O_2 deposition; (2) superior conductivity ensures efficient charge transfer and reduced interfacial resistance; (3) N-doping defects enhance interactions with Pt HEA, optimizing the electronic structure. Its molecular skeleton—composed of porous ultrathin nanosheets with a tortuous 2D architecture—confines HEA nanoparticles to prevent aggregation and ensures uniform dispersion, while the rigid framework stabilizes HEA loading during cycling. During synthesis, the nitrogen atoms/defects on NC electrostatically anchor Ni/Co/Mn/Pt ions, enabling the Joule heating-induced formation of single-phase solid-solution alloys. This process triggers reverse electron transfer (Pt $\rightarrow$ Ni/Co/Mn), optimizing the Pt d-band structure to enhance O-intermediate adsorption and accelerate ORR/OER kinetics. The NC/HEA interface establishes an efficient triple-phase boundary (catalyst/electrolyte/O_2), promoting LiO_2 adsorption and Li_2O_2 decomposition. Consequently, the catalyst achieves an ultra-low polarization voltage (0.27 V) at 200 mA/g, stable cycling over 240 cycles, and excellent rate performance even at 500 mA/g. A PtFeCoNiCu@rGO composite catalyst was prepared via high-temperature annealing by Wu et al. [97]. Constructed on a reduced graphene oxide (rGO) carrier, this material features a layered skeleton that retains the wrinkled texture of pristine GO, with flexible layers conformally encapsulating uniformly distributed PtFeCoNiCu high-entropy alloy (HEA) nanoparticles (~50 nm). The rGO carrier delivers three core functions: (1) an ultrahigh specific surface area (512.28 m^2/g) and hierarchical porosity accelerate Li^+/O_2 diffusion while providing ample space for Li_2O_2 storage; (2) exceptional conductivity synergizes with HEA to enhance charge transfer efficiency; (3) surface oxygen-containing functional groups strengthen interactions with the HEA, stabilizing the composite structure. The porous network formed via interlayer stacking effectively buffers volume fluctuations during Li_2O_2 deposition/decomposition. Active site analysis reveals the following: the high defect density (I_D/I_G = 1.21) cooperates with HEA metal sites to boost O_2/LiO_2 intermediate adsorption, while tight rGO-HEA interfacial contact establishes efficient electron pathways. The multi-element synergistic effect of HEA optimizes oxygen intermediate adsorption/conversion processes, significantly enhancing ORR/OER kinetics. Consequently, the catalyst achieves the following: a high initial discharge capacity (13,949 mAh/g at 100 mA/g), low overpotential (0.77 V), and stable cycling over 148 cycles (under 500 mAh/g capacity limitation), with the Li_2O_2-dominated formation pathway confirming its high selectivity and reversibility. Tian et al. designed an HEAs/C electrocatalyst featuring high-entropy alloy (HEA) nanoparticles, combining Pt/Ir with Fe/Co/Ni/Mn, uniformly dispersed within a carbon matrix [98]. This carbon matrix plays several pivotal roles in the catalyst. As a functional support, it provides high electrical conductivity for efficient charge exchange between the HEA particles and the external circuit, it maintains chemical stability in the harsh oxidizing environment, and its porous structure (e.g., carbon paper fiber networks) facilitates O_2 diffusion and reversible Li_2O_2 conversion. As a structural framework, the loose skeleton of amorphous carbon particles or fibrous structures effectively accommodates the HEA nanoparticles, preventing their agglomeration while preserving structural integrity; 3D carbon paper networks further enhance the mechanical strength of the electrode. At the material sites, defects and functional groups on the carbon surface anchor the HEA particles, enabling their in situ growth and uniform dispersion. Critically, the carbon–HEA interfacial interactions modulate the d-band center distribution of the alloy. This broad distribution, combined with the diverse local atomic environments within the HEAs, enhances catalytic flexibility and optimizes the adsorption/desorption balance strength of oxygen species (O_2/LiO_2 intermediates). Benefiting from this synergistic structural design, the catalyst achieved outstanding performance: an energy conversion efficiency exceeding 80%, stable cycling for 2000 h at a 4000 mAh/g capacity with 66.7% efficiency retention, and a discharge capacity of 39.1 Ah/g, significantly outperforming conventional catalysts.

However, key challenges remain for carbon/HEA nanocomposites in Li-O_2 batteries. HEA nanoparticles are susceptible to elemental segregation or oxidation during cycling, leading to catalytic activity decay. Carbon matrices may corrode at high potentials, impacting the long-term stability. The interfacial interaction mechanisms, particularly charge transfer and catalysis at the HEA–carbon interface, are not fully elucidated and require deeper investigation. Furthermore, current synthesis methods (e.g., high-temperature fusion, electrodeposition) are costly, hindering scalability and practical application.

4.4. Sodium/Potassium-Ion Batteries (SIBs/PIBs)

Driven by lithium resource scarcity and rising costs, SIBs/PIBs have gained significant attention as alternative energy-storage technologies. However, the larger ionic radii and sluggish diffusion kinetics of Na^+/K^+ ions lead to bottlenecks like low capacity and a short cycle life in traditional electrode materials. Composites of carbon materials (e.g., hard carbon, graphene) with HEAs, synergistically optimizing ion-storage sites, electron conduction, and mechanical stability, present a key solution to overcome these limitations. The composite design shows significant promise in SIBs/PIBs.

Bai et al. developed a lightweight, mechanically flexible sodium deposition substrate through a carbon-thermal encapsulation strategy (Figure 6), integrating zinc-based multi-element alloys (ranging from ternary to medium/high-entropy alloys) with carbon nanotubes (CNTs) [99]. Leveraging their lightweight nature (1.0–1.2 mg/cm^2), mechanical flexibility, and high conductivity, the CNTs act as conductive substrates for sodium deposition, reducing localized current density and enabling uniform Na^+ diffusion. Their interwoven network structure accommodates high-capacity sodium deposition (up to 10 mA h/cm^2) while suppressing dendrite formation. Using carbon-thermal encapsulation, the CNTs uniformly encapsulate multi-element alloy nanoparticles (FeCoNiAlZn), forming a stable "alloy@CNT" composite framework. This design prevents nanoparticle aggregation and direct contact with the electrolyte, avoiding side reactions and preserving the structural integrity. Simultaneously, the integration embeds high-entropy alloy (HEA) nanoparticles within the CNT network, creating a conductive and sodiumophilic composite material. The synergistic architecture enhances mechanical stability, conductivity, Na^+ adsorption/transport kinetics, and deposition/stripping efficiency: CNTs provide rapid electron transport and structural support, while HEAs enable efficient Na^+ adsorption, uniform distribution, and reduced nucleation overpotential. In an anode-free sodium-ion battery prototype, the FeCoNiAlZn@CNT composite paired with an $NaVPO_4F$ cathode achieved a high energy density of 351.6 Wh/kg and power density of 1335.5 W/kg, retaining 93.7% of its capacity after 200 cycles. The system demonstrated excellent rate capability and mechanical flexibility, maintaining stable performance even under flexible conditions. Zhang et al. synthesized the HEAs-NPs@NC material using a sacrificial template method, encapsulating high-entropy alloy (HEA) nanoparticles (Mn, Co, Ni, Cu, Zn) within a nitrogen-doped carbon (NC) matrix [100]. The NC matrix, derived from the pyrolysis of a metal-organic framework (Me-(CHX)-MOF), features high conductivity, a porous structure (BET specific surface area: 188.5 $m^2\ g^{-1}$), and abundant surface defects (particularly N-doping sites), which collectively facilitate K^+ transport, enhance potassium adsorption, mitigate volume expansion, and provide additional K^+-storage sites. The chelation effect of the flexible ligand CHX ensures the uniform distribution of multiple metals, suppressing phase separation. A robust "HEA-NPs@NC" framework is formed via metal-N bonding, where the NC matrix not only acts as a conductive substrate and mechanical support but also prevents the aggregation of HEA nanoparticles while enabling rapid electron transport. The uniformly embedded HEA nanoparticles contribute high surface area and porosity, forming a multi-path K^+ transport network through their homogeneous distribution. The HEAs' "cocktail effect" and configurational entropy synergistically optimize electrochemical activity and K^+-adsorption capacity, with the resulting potassium-intercalated metal solid solution further enhancing K^+-storage efficiency and stability. Benefiting from this synergistic design, the HEAs-NPs@NC anode exhibits exceptional electrochemical performance: a high specific capacity (513 mAh/g at 0.1 A/g), excellent rate capability (retaining 202 mAh/g at 5 A/g), and ultra-long cycling stability (96.6% capacity retention after 3000 cycles). The assembled potassium-ion full battery retains 82.8% of its capacity after 500 cycles, fully validating the material's advantages in high-capacity and long-term cycling stability.

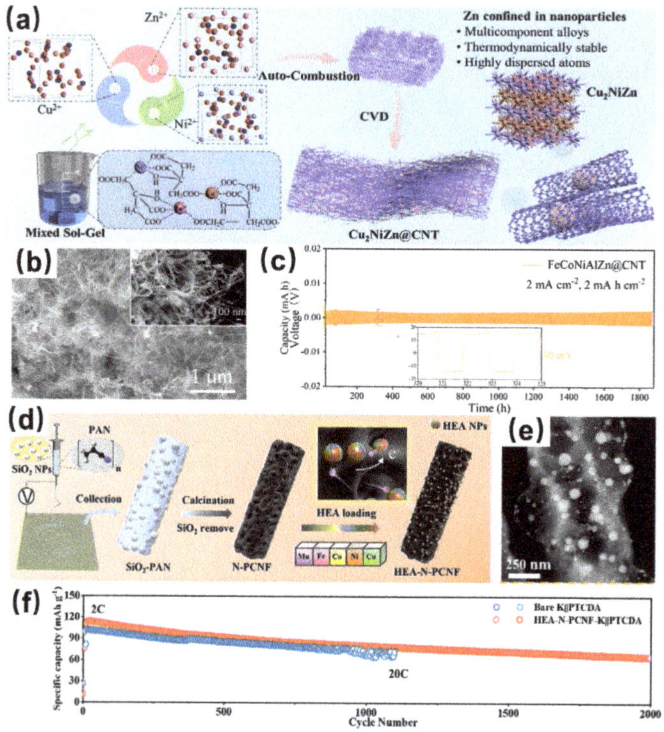

Figure 6. FeCoNiAlZn@CNT for SIBs: (**a**) schematic diagram, (**b**) SEM, and (**c**) cycle stability [99], copyright 2022, Royal Society of Chemistry. HEAs-N-PCNF for PIBs: (**d**) schematic diagram, (**e**) TEM, and (**f**) cycle stability [101], copyright 2024, Wiley-VCH GmbH.

Chang et al. fabricated an HEA-N-PCNF composite by integrating high-entropy alloy (HEA) nanoparticles (Mn, Fe, Co, Cu, Ni) with nitrogen-doped porous carbon nanofibers (N-PCNF) (Figure 6) [101]. The N-PCNF matrix, featuring a large specific surface area (122.44 m^2/g) and a porous structure, reduces localized current density while providing space for potassium deposition. Its fibrous network enhances electron/ion transport efficiency and rate capability. The three-dimensional porous N-PCNF network uniformly anchors HEA nanoparticles, forming a composite "HEA-N-PCNF" framework that offers conductivity, structural stability, and enhanced surface activity. The HEA nanoparticles distributed on N-PCNF pores/surfaces further increase the surface area, providing additional K^+-adsorption sites, improving electrolyte permeability and facilitating rapid ion/electron transport. Critically, the configurational entropy effect of HEAs and their induced local charge enhancement (notably the high electronegativity of Cu/Ni sites) significantly improve potassium metal affinity, optimizing the K^+-adsorption capacity and deposition uniformity. Electron transfer and lattice distortion within HEAs generate strong K^+-adsorption sites, inducing Frank–Van der Merwe-type layer-by-layer growth, effectively suppressing dendrite formation. This synergistic structural design regulates

K^+ flux and stabilizes the solid–electrolyte interphase layer. Benefiting from this design, the symmetric battery based on this material achieved an ultra-long cycling life of over 2350 h at 8 mA/cm^2. The assembled HEA-N-PCNF-K | | PTCDA full battery delivered a high energy density of 331 Wh/kg and retained 58% of its capacity after 2000 cycles, significantly outperforming conventional potassium batteries. This highlights the composite's efficiency in dendrite suppression and stability enhancement. Zheng et al. synthesized the HEA-CNF composite through an electrospinning technique, uniformly embedding HEA nanoparticles within CNFs [102]. The CNFs, derived from the pyrolysis of a polyacrylonitrile precursor after electrospinning, feature a three-dimensional network structure that provides high conductivity and prevents particle aggregation. Their porous nature (HEA-CNF specific surface area: 672 m^2/g) enhances electrolyte wettability, facilitates Li^+/K^+ diffusion, and buffers volume changes during cycling. HEA nanoparticles (composition: $Co_{0.2}Sb_{0.2}Fe_{0.2}Mn_{0.2}Ni_{0.2}$) are uniformly encapsulated within the CNFs, forming a fibrous "HEA-CNFs" composite framework: the tight connection between HEA nanoparticles and CNFs enables rapid electron/ion transport, while the CNF network offers mechanical support to accommodate volume deformation. The configurational entropy and synergistic multi-metal effects of HEAs significantly improve ion diffusion rates and structural stability, enhance Li^+/K^+-storage capacity via alloying reactions, and suppress aggregation and phase separation. Notably, this design ensures uniform elemental distribution—active metals (e.g., Co, Sb) participate in alloying reactions for Li^+/K^+ storage, while inert metals (e.g., Mn) maintain structural stability. Benefiting from this synergistic architecture, the HEA-CNF anode exhibits exceptional Li^+/K^+-storage performance: as a lithium-ion battery anode, it retains a capacity of 1400 mAh/g after 800 cycles at 0.5 A/g; as a potassium-ion battery anode, it delivers a reversible capacity of 280 mAh/g after 200 cycles at 0.2 A/g. Its performance surpasses that of the binary alloy counterpart (CoSb-CNFs), fully demonstrating the dual advantages of high capacity and long-term cycling stability offered by the composite structure.

Despite the promising performance of carbon/HEA nanocomposites in SIBs/PIBs, several bottlenecks persist. Active metals in HEAs (e.g., Sn, Sb) readily react with ester-based electrolytes, forming thick, inhomogeneous CEI layers, increasing impedance, and accelerating capacity fade (CE < 95% after 100 cycles). HEA nanoparticle synthesis relies on high-energy ball milling/sputtering, leading to high production energy consumption. The high density of HEAs constrains the battery gravimetric energy density. The Na^+/K^+-storage mechanism under multi-element synergy remains unclear. Transition metal dissolution causes active site loss, limiting the lifespan. Therefore, developing solid-state electrolyte-compatible interfaces (e.g., sulfide-coated HEAs), designing lightweight medium-entropy alloys (MEAs), and revealing multi-element dynamic evolution mechanisms via in situ characterization will be key future research focuses.

4.5. Lithium-Sulfur Batteries (Li-S)

Li-S batteries have attracted significant attention due to their high theoretical energy density (2600 Wh/kg), low cost, and environmental friendliness. However, their practical application is hindered by challenges such as the polysulfide "shuttle effect," low conductivity of the sulfur cathode, substantial volume expansion (~80%), and lithium anode dendrite growth. HEAs/carbon composites suppress the polysulfide shuttle effect through dual mechanisms of chemical adsorption and catalytic conversion.

Han et al. synthesized PtCuFeCoNi high-entropy alloy (PCFCN-HEA) nanoparticles via a hydrothermal method followed by subsequent annealing and loaded them onto hollow carbon spheres (HCSs) hybridized with mycelium-derived carbon nanobelt (HCNB) composites, serving as a sulfur host material for lithium-sulfur battery cathodes [103].

The carbon framework (HCSs and HCNBs) provides three core functionalities: (1) the high conductivity and porous structure enable rapid electron/ion transport, addressing the insulating nature of sulfur and its discharge products (Li_2S_2/Li_2S); (2) the synergistic physical adsorption and spatial confinement suppress the polysulfide (LiPSs) shuttle effect; (3) acting as a carrier for HEA nanoparticles, ensuring their uniform dispersion and maximizing catalytic site exposure. Specifically, the HCSs' hollow spherical structure, with a large surface area and internal voids, accommodates sulfur, while the HCNBs form an interwoven nanobelt network that enhances structural stability, mechanical strength, and electrolyte contact; this hierarchical architecture balances porosity with continuous conductive network connectivity. HEA nanoparticles (~5.6 nm) uniformly deposited on HCSs are interconnected via HCNBs. The HEA integrates five metals with distinct work functions (high: Pt/Ni; low: Fe/Co/Cu), achieving a d-band center shift upward (to −1.70 eV) through work function modulation. Electron transfer (from low- to high-work function metals) and multi-metal synergy precisely regulate the d-band center position, optimizing LiPSs adsorption and catalytic activity. Consequently, carbon sites physically anchor LiPSs, while HEA catalytic sites (synergistically enhanced by Pt/Cu/Fe) accelerate LiPS conversion, forming an integrated "adsorption-catalysis" synergy mechanism—carbon restricts LiPS diffusion, while HEA lowers reaction energy barriers—dramatically boosting sulfur reduction reaction (SRR) kinetics. Benefiting from this design, the material exhibits exceptional electrochemical performance: reversible capacity of 652 mAh/g at an ultrahigh 8 C rate; 40.9% capacity retention after 1500 cycles at 2 C (ultra-low decay rate: 0.039% per cycle); 81.5% capacity retention after 500 cycles at 6 C; and a specific activity of 2.58 mA/cm^2 for Li_2S_4-to-Li_2S conversion, 6.14 times higher than that of Pt catalysts. Ma et al. developed an efficient electrocatalyst composed of high-entropy alloy (HEA) nanoparticles, nitrogen-doped carbon (NC), and carbon nanotubes (CNTs) [104]. The material is based on NC derived from pyrolyzed metal–organic frameworks (MOFs), of which the porous structure is rich in nitrogen active sites (e.g., pyridinic nitrogen/pyrrolic nitrogen). These sites chemically anchor lithium polysulfides (LiPSs) through polar interactions, effectively suppressing the shuttle effect. The CNTs form an interconnected one-dimensional conductive network that permeates the NC, creating a three-dimensional framework. This not only prevents NC aggregation and enhances mechanical strength and conductivity, but also provides rapid electron/ion transport channels, reducing internal resistance. HEA nanoparticles (~13 nm) uniformly dispersed within the NC matrix are encapsulated by NC and interconnected with CNTs. This stable composite conductive framework integrates three functionalities synergistically: NC's nitrogen sites chemically immobilize LiPSs; CNTs' conductive sites ensure efficient electron transport; HEAs' catalytic sites (attributed to the oxidation states of Fe/Co/Ni) accelerate the bidirectional conversion reactions of LiPSs. This integrated "adsorption-transport-catalysis" system significantly improves the composite's conductivity, adsorption capacity, and structural stability, endowing the battery with excellent cycling stability and rate capability: a discharge capacity of 692.0 mAh/g after 300 cycles at a 1 C rate (ultra-low decay rate: 0.03% per cycle), and a retained capacity of 521.1 mAh/g even at a high 5 C rate. Wang et al. fabricated FeCoNiCuMn high-entropy alloy nanocrystals (HEA-NCs) supported on nitrogen-doped carbon (NC) using a reflux–chelation–confinement annealing strategy (Figure 7) [64]. The method employed 1,10-phenanthroline chelation, g-C_3N_4 templating, and polydopamine coating to achieve spatial confinement, effectively suppressing HEA aggregation and phase separation. The resulting HEA nanoparticles (40–60 nm) are uniformly dispersed on the NC surface, forming a single-face-centered cubic (FCC) phase. EDS mapping confirmed atomic-level mixing, while HRTEM revealed significant lattice distortion (lattice spacing of 0.28 nm > 0.20 nm for pure Ni). In this composite structure, ultrathin, highly crumpled

NC sheets with abundant pores and a large surface area act as a Lewis basic matrix, chemically anchoring lithium polysulfides (LiPSs) through acid–base interactions at pyridinic nitrogen/graphitic nitrogen sites (confirmed by UV-vis) and enhancing conductivity to mitigate sulfur's insulation. Meanwhile, HEA nanocrystals uniformly dispersed on NC synergistically leverage lattice distortion and the "cocktail effect," particularly their catalytic sites (Fe^{2+}/Co^{2+}/Ni^{2+}) collaborating with pyridinic nitrogen sites, to dramatically accelerate LiPS conversion (especially the solid-state $Li_2S_2 \rightarrow Li_2S$ step). This lowers the energy barrier (evidenced by increased P3-phase capacity in GDC curves), boosts reaction kinetics (exchange current density of 0.18 mA cm^{-2}), and induces reticular, loose Li_2S_2/Li_2S deposition (confirmed by SEM). This dual-site synergistic mechanism of NC anchoring and HEA catalysis enables exceptional practical performance under high sulfur loading and lean electrolyte conditions: a cathode capacity of 1079.5 mAh/g at 72.3% sulfur content (89.4% sulfur utilization); 807.8 mAh/g retention after 160 cycles at 4.4 mg/cm^2 sulfur loading and lean electrolyte (5 μL/mg) with 0.3% decay per cycle; 868.2 mAh/g (32.4 mAh/cm^2 area capacity) under ultra-high sulfur loading (27 mg/cm^2) and extreme lean electrolyte (3 μL/mg); and 95.2% capacity retention after 970 cycles at a 1 C rate with only 0.05% decay per cycle.

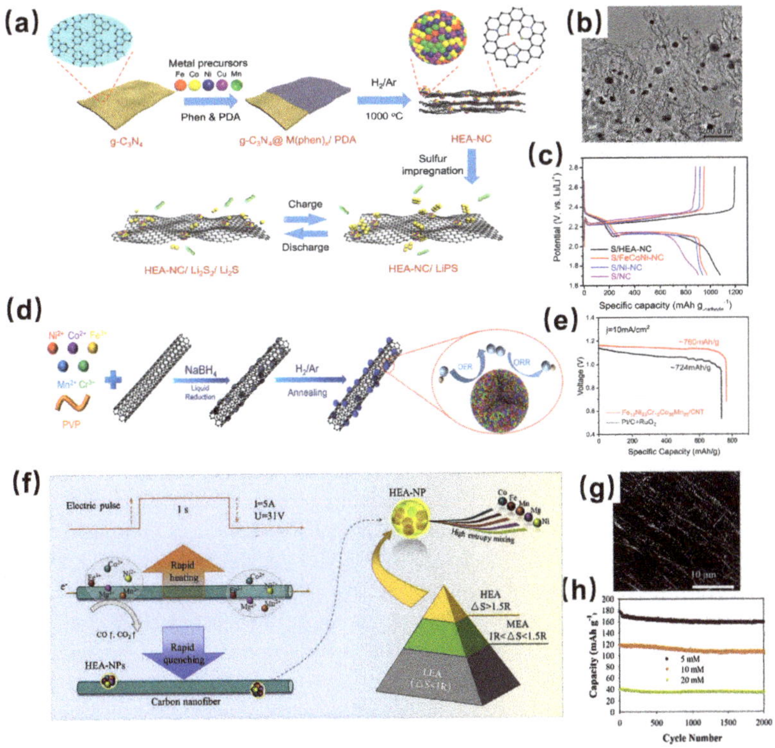

Figure 7. HEA-NC for Li-S: (**a**) sSchematic diagram, (**b**) TEM, and (**c**) charge-discharge curves [64], copyright 2022, Wiley-VCH GmbH. $Fe_{12}Ni_{23}Cr_{10}Co_{30}Mn_{25}$/CNT for ZABs: (**d**) sSchematic diagram and (**e**) specific capacity plots [105], copyright 2023, American Chemical Society. FeNiCoMnMg HEA-NPs/ACNFs for SCs: (**f**) sSchematic diagram, (**g**) SEM, and (**h**) cycle stability [54], copyright 2020, Elsevier.

Despite the promise of carbon/HEA composites in Li-S batteries, significant challenges remain. The dissolution and shuttle effect of LiPSs remain critical issues. The complex reaction mechanism (multi-step, multi-electron, various intermediate polysulfides) facilitates LiPS dissolution. While carbon offers some physical confinement, its adsorption capacity is limited, struggling to effectively suppress the shuttle and maintain capacity under lean electrolyte conditions. Regarding catalytic performance, simple-component catalysts lack sufficient activity for the complex 16-electron sulfur redox reactions. Some catalysts suffer from poor electrochemical stability, degrading cycling performance. Although HEAs show potential through multi-element synergy, their synthesis is complex, the precise control of element ratios/distributions is challenging, and cost is high, hindering large-scale adoption. Furthermore, under low-temperature and lean-electrolyte conditions, Li-S batteries face sluggish kinetics, electrolyte gelation, impeded ion transport, and LiPS agglomeration, severely impacting sulfur utilization and cycling stability.

4.6. Zinc–Air Batteries (ZABs)

ZABs rely on efficient ORR and OER reactions. Carbon/HEA composites are often employed as bifunctional catalysts for air electrodes. Carbon materials provide high conductivity, a porous structure for electron transport/reaction interfaces, and a 3D scaffold for dispersing HEA nanoparticles. HEAs leverage multi-element synergy to exhibit superior ORR/OER activity compared to traditional noble metal catalysts (e.g., Pt/C, IrO_2), with lower costs and better corrosion resistance. The synergy between carbon and HEAs optimizes the charge/discharge efficiency and cycling stability, showing significant potential for rechargeable ZABs.

Han et al. synthesized a PtFeCoNiMoY/CNT composite through a simple one-step vapor-phase synthesis strategy [106]. The method utilized pretreated CNTs as a support, where surface defects (functional groups/vacancies) provided uniform nucleation sites for HEA nanoparticles, effectively preventing aggregation. Pt-based hexanary HEA nanoparticles were tightly and uniformly loaded onto the CNTs, forming a single-phase solid solution (lattice spacing matching the (200) plane, with elemental homogeneity confirmed via XPS and TEM). The CNTs not only offer high conductivity, mechanical support, and a stable tubular framework but also enhance mass transport and active site utilization due to their low density, high specific surface area, and porous structure. The incorporation of Mo and Y synergistically modulated the alloy's d-band center and electronic structure: Mo improved the continuity of spin-up states near the Fermi level, reducing the intermediate adsorption energy; Y shifted the d-band center downward, optimizing the surface adsorption strength. This synergistic electronic structure optimization significantly enhanced catalytic activity, enabling excellent bifunctional oxygen electrocatalytic performance: an OER overpotential of only 238 mV (@10 mA/cm^2), an ORR half-wave potential of 0.75 V, and a bifunctional ΔE of 0.713 V. The assembled ZAB exhibited outstanding performance: an open-circuit voltage of 1.41 V, a peak power density of 128.4 mW/cm^2, a specific capacity of 797 mA·h/g, and stability exceeding 80 h, comprehensively outperforming the benchmark Pt/C + RuO_2 catalyst. Yao et al. developed a novel solid-state thermal reaction strategy to synthesize the HEA@N-GHCT composite [107]. This approach combines a solid-state reaction between multi-metal salts and 2-methylimidazole followed by carbonization, avoiding organic solvent use and offering eco-friendliness and scalability. The resulting material features FeCoNiMnCu high-entropy alloy (HEA) nanoparticles (<100 nm) encapsulated within nitrogen-doped graphitized hollow carbon tubes (GHCTs). Zinc volatilization during carbonization (at 908 K) effectively promotes pore formation, endowing the GHCT with a hollow, porous structure and high specific surface area (489.4 m^2/g). These graphitized hollow carbon tubes (GCTs) exhibit high conductivity (from graphitic structure) and chemical stability, with their robust tubular framework preventing HEA nanoparticle aggregation during catalytic reactions (e.g., ORR) and providing efficient electron transfer pathways; their hollow/porous architecture significantly enhances mass transport and reaction kinetics. HEA nanoparticles are tightly encapsulated within the GHCT (confirmed by TEM/EDS, showing uniform distribution of Fe, Co, Ni, Mn, and Cu elements), and the GHCT protection minimizes active site loss. The material's synergistic mechanism includes the following: graphitized layers (002 plane) and nitrogen doping (pyridinic, pyrrolic, and graphitic nitrogen) optimizing electron transfer and tuning the surface electron density to promote O_2 adsorption/activation; HEA's synergistic effects (notably Cu's FCC structure forming a solid solution) optimizing adsorption energies of ORR intermediates (-OH, -O), lowering reaction barriers. The assembled zinc–air battery (ZAB) demonstrates excellent performance: a peak power density of 81 mW/cm^2, open-circuit voltage (OCV) of 1.36 V, and over 200 h of cycling stability, outperforming the Pt/C-RuO_2 benchmark catalyst in overall performance. Cao et al. synthesized $Fe_{12}Ni_{23}Cr_{10}Co_{30}Mn_{25}$/CNT composites via liquid-phase reduction combined with high-temperature sintering in an H_2-Ar atmosphere, utilizing acid-treated CNTs as carriers and nucleation sites in Figure 7 [105]. The acid treatment generated surface defects (carboxyl/hydroxyl groups) and rich graphitic structures (002 plane), providing high-density, stable sites for the uniform deposition and reduction of metal ions (Fe, Ni, Cr, Co, Mn). Subsequent sintering resulted in the formation of single-phase FCC solid solution HEA nanoparticles (~30 nm) with uniform element distribution and lattice spacing characteristic of (111) and (200) planes. These HEA NPs were strongly anchored onto the CNTs via defect–particle interactions, forming a core-shell porous structure. The inherent tubular pores and high specific surface area (489.4 m^2/g) of the CNTs acted as efficient gas channels and nanoreactors, accelerating mass transport (e.g., O_2/OH^- diffusion) for oxygen catalysis while preventing HEA nanoparticle aggregation and exposing abundant active sites. Furthermore, the CNTs' graphitic structure and N-doping (pyridinic, pyrrolic N) facilitated electron transport. Critically, tuning the Co/Mn ratio (30:25) optimized the HEA electronic structure; the synergy between Co and Mn down-shifted the d-band center to −1.976 eV (confirmed via XPS and DFT calculations), lowering the -OOH adsorption energy (0.32 eV) to accelerate the ORR rate-determining step, while lattice distortion suppressed element dissolution. This "metal-carbon" synergy endowed the composite with superior bifunctional oxygen electrocatalysis: achieving an ORR half-wave potential ($E_1/_2$) of 0.81 V, an OER overpotential at 10 mA/cm^2 (E_{10}) of 284 mV, and a remarkably low potential gap (ΔE) of 0.7 V, outperforming the benchmark Pt/C + RuO_2 (ΔE = 0.72 V). Assembled zinc–air batteries demonstrated a peak power density of 128.6 mW/cm^2, a high specific capacity of 760 mA·h/g, an energy density of 865.5 Wh/kg, and exceptional stability exceeding 256 h with lower degradation than Pt/C + RuO_2.

Despite their advantages, practical applications face challenges. (1) Activity-Stability Trade-off: Selective dissolution of elements in strong alkaline electrolytes can cause activity decay; weak carbon/HEA interfacial bonding may lead to catalyst detachment during volume changes. (2) Complex Synthesis and Cost: Precise HEA composition control and nanoscale dispersion require complex processes (high-temperature sintering, sputtering), hindering scale-up; carbon surface modification to enhance bonding adds cost. (3) Operational Compatibility Issues: Electrolyte migration and Zn dendrite growth can block pores or damage the structure; repeated OER/ORR cycling may cause HEA lattice distortion, affecting durability. (4) Unclear Reaction Mechanisms: The interfacial catalysis mechanism between multi-element HEAs and carbon carriers is not fully understood, hindering theory-guided optimization for targeted performance enhancement.

4.7. Supercapacitors (SCs)

Carbon/HEA composites have demonstrated significant application potential in the field of SCs. Carbon materials, owing to their high specific surface area, excellent conductivity, and outstanding electrochemical stability, serve as ideal electrode materials for SCs. HEAs, benefiting from unique high-entropy effects, sluggish diffusion, and lattice distortion effects, exhibit superior properties in mechanics, electromagnetics, and other aspects, presenting new opportunities for enhancing supercapacitor electrode performance [51]. Research indicates that electrodes composed of HEA nanoparticles/carbon nanofibers, prepared via specific methods at certain precursor concentrations, exhibit a relatively high specific capacitance and specific energy density.

Shen et al. synthesized HEA-NP@MOL/HCPC composites through an adsorption–reductio—carbonization strategy: Fe^{2+}, Co^{2+}, Ni^{2+}, Cu^{2+}, and Sn^{2+} metal ions were adsorbed onto a hypercrosslinked polymer, in situ reduced by $NaBH_4$ to form high-entropy alloy nanoparticles (HEA-NPs), and subsequently carbonized at 800 °C [108]. This process yields a ternary "metal core-oxide shell-carbon carrier" structure. The carbonized hypercrosslinked polymer-derived carbon (HCPC) retains a stable 3D porous framework with abundant micropores/mesopores (pore size ~3–4 nm) and a high specific surface area (1330 m^2/g), providing ample adsorption sites for the uniform dispersion of HEA-NPs (8–135 nm) and rigid support to prevent their aggregation. Crucially, during carbonization, a nanoscale metal oxide layer (MOL, ~1–2 nm thick) spontaneously forms on the HEA-NP surface; this amorphous MOL, comprising multi-component oxides (Fe_2O_3, Co_3O_4, NiO, CuO, SnO_2), creates a "crystalline core (HEA)-amorphous shell (MOL)" interface that exposes numerous redox-active sites. The HCPC's graphitic carbon skeleton provides efficient electron conduction paths and contributes electric double-layer capacitance (EDLC) via ion adsorption/desorption. Simultaneously, the MOL delivers significant pseudocapacitance through rapid multivalent metal ion (e.g., Fe^{3+}/Fe^{2+}, Co^{3+}/Co^{2+}) redox reactions, enhanced by d-band center tuning for improved OH^- adsorption. The synergy between these components—the MOL (dominant pseudocapacitance), the HCPC porous structure (facilitating ion diffusion and EDLC), and the HEA-NP metallic core (ensuring fast electron transport)—drives exceptional performance. The optimized HEA-NP@MOL/HCPC-2.0 composite (with only 4.4 wt% MOL contributing ~420 F/g enhancement over pure HCPC) achieved a high specific capacitance of 495.4 F/g at 0.5 A/g in 1 M KOH, remarkable cycling stability (94.7% capacitance retention after 15,000 cycles), and fast kinetics (88.9% capacitance contribution at 200 mV/s). Mohanty et al. synthesized Fe-Co-Ni-Cr-Mn HEA/green carbon composites (FCNCM@Green Carbon) through induction melting, ball milling, and the pyrolysis of rice husk, utilizing a 50:50 mass ratio [109]. The rice husk-derived green carbon forms a stable porous framework with the optimized mesoporous structure (BET SSA = 25 m^2/g), preserving the natural biomass architecture to provide abundant loading sites and prevent HEA nanoparticle aggregation. Agglomerated FCNCM HEA nanoparticles (~520 nm) with multivalent oxide surfaces are uniformly dispersed within this carbon network, creating a synergistic "EDLC-pseudocapacitance" system. The green carbon contributes electric double-layer capacitance (EDLC) via its porous skeleton that facilitates electrolyte ion adsorption/diffusion, while the HEA nanoparticles deliver pseudocapacitance through rapid redox reactions of Fe/Co/Ni multivalent ions, with surface lattice distortion lowering the reaction overpotential. This synergy shortens ion diffusion paths, reduces electrode polarization, and enhances kinetics. Electrochemically, the composite achieved a specific capacitance of 450 F/g at 2 A/g (3 M KOH, three-electrode) with a 1.3 V voltage window. In symmetric aqueous supercapacitors, it delivered 78 F/g at 1 A/g, an energy density of 33.5 Wh/kg, a power density of 1800 W/kg (1.8 V), and 95.6% capacitance retention after 5000 cycles, significantly outperforming pure HEA or biochar electrodes.

The biomass-derived carbon's sustainable properties further amplify the composite's functional efficacy, demonstrating green carbon's critical role in dispersing HEA and boosting electrochemical activity. Xu et al. synthesized FeNiCoMnMg HEA-NPs/ACNFs composites via a carbothermal shock (CTS) strategy combined with self-designed aligned carbon nanofibers (ACNFs) derived from electrospun polyacrylonitrile (PAN) precursors in Figure 7 [54]. The ACNFs exhibit an oriented architecture, superior conductivity, and high specific surface area, providing ordered electron transport pathways and enhanced surface wettability to uniformly anchor ~30 nm HEA nanoparticles while preventing aggregation. Crucially, the CTS process—directing current along the fiber axis—converts metal chlorides into uniformly dispersed high-entropy solid solution NPs, with alloying-induced electron binding energy shifts ensuring tight HEA–carbon contact for efficient charge transfer. The aligned fiber structure simultaneously shortens ion diffusion paths and maintains structural integrity. Electrochemically, the composite leverages dual synergies: the ACNF skeleton delivers rapid electron conduction and stable mechanical support, while HEA-NPs provide rich redox-active sites through high-entropy effects and lattice distortion. At a 5 mM precursor concentration, the optimized FeNiCoMnMg/ACNFs achieved a specific capacitance of 203 F/g, an energy density of 21.7 Wh/kg, and 89.2% capacitance retention after 2000 cycles—significantly surpassing the performance of random CNF carriers. These gains are attributed to the oriented conductive network, homogeneous NP distribution via CTS, and enhanced interfacial kinetics, establishing ACNFs as an ideal substrate for maximizing HEA-NP activity.

However, the practical application of these composites still faces several challenges. Precise control of the HEA nanoparticle size remains challenging; achieving uniform dispersion and strong bonding with the substrate material is difficult. For instance, in constructing HEA/carbon nanotube (CNT) composite electrodes, existing bulk preparation techniques struggle to achieve effective combination. The wettability between different metals and CNTs varies significantly, and fundamental studies on HEA-CNT wettability are lacking, making it difficult to improve compatibility and achieve strong interfacial bonding. While composites show advantages, the further enhancement of energy density is needed to meet high-performance demands. Some materials suffer from capacity fading or structural degradation during long-term cycling, impacting the supercapacitor lifespan. Introducing electrochemically active sites into the carbon skeleton increases pseudocapacitance but often leads to sluggish reaction kinetics and compromises the structural robustness of the carbon framework. This results in the rapid decay of electrode activity and insufficient cycling stability. The performance comparison of different batteries is shown in Table 2.

Table 2. Comparison of different battery performances.

Type	Material	Capacity	Stability	Ref.
LIBs	HEAs/C	1462 mAh/g	1600 h	[38]
	HEA/C	1881 mAh/g	63.6% after 1000 cycles	[68]
	FeCoNiMnCuAl@C	27,664 mAh/g	134 cycles	[93]
	FeCoNiCuRu/CNFs	6160 mAh/g	550 cycles	[94]
LMBs	HEA/PCF	177.9 mAh/g	99.5% after 200 cycles	[55]
	HEA/CF	197.9 mAh/g	99.2% after 160 cycles	[67]
	HEA/C	166.3 mAh/g	99.5% after 160 cycles	[95]
$Li\text{-}O_2$	Pt HEAs@N-C	1000 mAh/g	240 cycles	[96]
	PtFeCoNiCu@rGO	13,949 mAh/g	148 cycles	[97]
	HEAs/C	4000 mAh/g	2000 h	[98]

Table 2. *Cont.*

Type	Material	Capacity	Stability	Ref.
SIBs/PIBs	FeCoNiAlZn@CNT	10 mAh/cm^2	93.7% after 200 cycles	[99]
	HEAs-NPs@NC	513 mAh/g	96.6% after 3000 cycles	[100]
	HEA-N-PCNF	120 mAh/g	2350 h	[101]
Li-S	HEA-CNFs	1164 mAh/g	1400 mAh/g after 1000 cycles	[102]
	PCFCN-HEA/HCS/HCNB	1221.6 mAh/g	82.9% after 200 cycles	[103]
	CNT/HEA-NC	622.5 mA h/g	only 0.03% per cycle after 300 cycles	[104]
	FeCoNiCuMn HEA-NC	1079.5 mAh/g	0.3% per cycle for 160 cycles	[64]
ZABs	PtFeCoNiMoY/CNT	797 mA·h/g	80 h	[106]
	FeCoNiMnCu HEA@N-GHCT	630.29 mA·h/g	>200 h	[107]
	$Fe_{12}Ni_{23}Cr_{10}Co_{30}Mn_{25}$/CNT	760 mA·h/g	256 h	[105]
SCs	HEA-NP@MOL/HCPC	495.4 F/g	94.7% after 15,000 cycles	[108]
	FCNCM@Green carbon	450 F/g	94.7% after 15,000 cycles	[109]
	FeNiCoMnMg HEA-NPs/ACNFs	203 F/g	89.2% after 2000 cycles	[54]

5. Summary and Outlook

This review systematically summarizes the research progress on carbon material/ HEA nanocomposites in the field of EES. Carbon materials, with their high conductivity, large specific surface area, and chemical stability, provide excellent electron transport pathways and structural frameworks for energy-storage systems. HEAs, leveraging high-entropy effects, lattice distortion effects, and sluggish diffusion effects, exhibit unique catalytic activity and structural stability. The synergistic effects arising from their combination significantly enhance material performance in EES systems, including LIBs, SIBs/PIBs, Li-S, and SCs through electronic structure optimization, enhanced structural stability, and functional complementarity. This represents a crucial pathway for overcoming the performance bottlenecks of traditional energy storage materials. However, carbon/HEA composites still face the following core challenges:

(1) **Complex and Costly Synthesis:** The multi-principal element nature of HEAs makes the precise control over synthesis difficult, requiring meticulous regulation of elemental ratios and reaction conditions. Traditional HEA preparation often relies on high temperatures, high pressures, and inert atmospheres, demanding stringent equipment specifications, significantly increasing energy consumption and production costs, and severely hindering large-scale production. Elemental segregation during multi-component mixing affects material homogeneity and reproducibility, leading to inconsistent product quality.

(2) **Ambiguous Interfacial Interaction Mechanisms:** The interfacial bonding strength and interaction mechanisms between carbon materials and HEAs remain unclear. During electrochemical cycling, differences in thermal expansion coefficients and chemical activity can cause stress concentration and structural mismatch at the interface, reducing interfacial stability. This can lead to composite structural failure, shortening the cycle life of energy storage devices and impacting long-term operational stability.

(3) **Lagging Theoretical Research:** While experiments demonstrate the excellent performance of carbon/HEA composites in EES, theoretical research providing a deep understanding of their intrinsic mechanisms at the atomic and electronic levels lags behind. The microscopic action mechanisms of high-entropy effects in complex composite systems are debated. The lack of comprehensive theoretical models hinders material design, performance optimization, and the full exploitation of material potential.

(4) **Performance-Cost Trade-off:** Pursuing high performance often necessitates the use of rare or expensive elements in HEAs, increasing material costs. However, energy-

storage devices are highly cost-sensitive in practical applications. Achieving a balance between high performance and low cost is a critical challenge for the large-scale adoption of these materials.

(5) **Slow Industrialization Process:** Transitioning from lab-scale R&D to industrial production presents numerous obstacles for carbon/HEA composites. Existing preparation processes struggle to meet large-scale demands, lacking standardized production workflows and quality control systems. Furthermore, the incomplete industry chain and insufficient upstream–downstream collaboration further delay industrialization.

Future Research Hotspots and Priorities:

(1) Precise Multi-scale Structural Design and Control:

- **Atomic Scale:** Utilize advanced computational simulations (e.g., Density Functional Theory—DFT) to precisely control the HEA elemental composition and ratio, optimize the electronic structure, and precisely regulate active sites.
- **Nanoscale:** Finely tune the HEA nanoparticle size, shape, and distribution, combined with the nanostructural features of carbon materials (e.g., graphene layer number, CNT diameter/length), to achieve efficient synergy.
- **Macroscale:** Construct composites with three-dimensionally ordered porous structures, optimizing the pore architecture and connectivity to promote ion transport and electrolyte wetting, enhancing the overall energy-storage performance. Examples include designing gradient-structured composites for efficient ion transport/storage across scales and boosting the energy density, power density, and cycling stability.

(2) **Cross-disciplinary Integration for New Material Systems:** Integrate knowledge from materials science, chemistry, physics, and computational science to develop novel HEA systems and carbon/HEA composites.

- Materials Science: Focus on synthesis, structural characterization, and performance testing.
- Chemistry: Deepen understanding of reaction mechanisms and surface chemistry.
- Physics: Provide theoretical support from perspectives of electronic structure and energy conversion.
- Computational Science: Employ machine learning, molecular dynamics simulations, etc., to accelerate material screening and design. Use machine learning algorithms to screen HEA components with specific properties, combined with experimental validation, to develop high-performance composites for novel energy-storage/conversion devices (e.g., all-solid-state batteries, fuel cells), expanding application fields.

(3) **Innovation and Development of Sustainable Synthesis Technologies:** Develop green, low-energy-consumption synthesis technologies for sustainable material production.

- Explore bio-templating methods, utilizing specific structures of biomacromolecules or microorganisms to synthesize complex-structured composites, reducing chemical usage and energy consumption.
- Optimize existing processes (e.g., improved Joule heating for precise control and higher energy efficiency, lowering costs).
- Enhance research on recycling and reuse technologies to achieve green practices throughout the material lifecycle, laying the foundation for large-scale application.

(4) Deep Application of Intelligent In situ Characterization Techniques: Employ intelligent in situ characterization techniques:

- In situ XRD: Real-time monitoring of crystal structure evolution during cycling.
- In situ TEM: Observation of nanostructural and interfacial changes.
- In situ XPS: Analysis of surface element chemical states and electronic structure evolution.

These techniques provide deep insights into performance change mechanisms under operating conditions, enabling the establishment of accurate structure–property relationship models to guide material optimization design and performance enhancement, accelerating R&D of novel composites.

(5) **Material Surface Engineering Optimization:** Utilize surface modification techniques (e.g., ALD, CVD) to introduce functional coatings or active sites onto composite surfaces.
    - Coatings can improve hydrophilicity/hydrophobicity, optimize electrolyte wettability, and promote fast ion transport.
    - Introducing active sites can enhance electrocatalytic activity and reduce reaction overpotentials (e.g., depositing ultra-thin metal oxide coatings to boost ORR catalytic performance in ZABs, improving efficiency and stability).
    - Surface engineering can also enhance corrosion resistance, preventing HEA oxidation/dissolution in electrolytes and extending device lifespan.

(6) **Research on Novel Electrolyte Compatibility:** Develop novel electrolytes highly compatible with carbon/HEA composites for different systems (e.g., LIBs, Li-S).
    - Create electrolytes with high ionic conductivity, wide electrochemical windows, and good thermal stability, suppressing polysulfide shuttling in Li-Ss.
    - Improve electrode–electrolyte interfacial compatibility using special additives or ionic liquid solvents, enhancing the cycle performance and Coulombic efficiency.
    - Deeply study interaction mechanisms between electrolytes and composite surfaces to optimize formulations, reduce side reactions, and improve device safety and reliability.

(7) **Research on Dynamic Response Mechanisms of Composite Structures:** Utilize advanced characterization and numerical simulations to deeply investigate the dynamic response mechanisms during cycling:
    - Structural evolution, stress distribution changes, and electrochemical reaction kinetics.
    - Molecular dynamic simulations of HEA nanoparticle diffusion within the carbon matrix and interaction changes during cycling.Understanding these mechanisms aids in optimizing material design, improving stability and reliability under complex operating conditions, and providing a theoretical basis for high-performance devices.

(8) **Device Integration and System Optimization:** Extend research from electrode materials to the level of complete energy-storage device systems for integration and optimization.
    - Study the compatibility and synergy between internal components (e.g., electrode-separator–electrolyte interfaces).
    - Optimize device structure design to increase energy/power density.
    - Tailor thermal management, safety management, and energy-management strategies to application scenarios (e.g., develop efficient heat dissipation structures, design safety mechanisms against overcharge/overdischarge/thermal runaway).

This will propel the practical application of composites in large-scale energy storage.

Carbon/HEA composites, through multi-element synergy and cross-scale structural design, provide revolutionary solutions for next-generation EES technologies. Future research must focus on core directions like intelligent design, dynamic interface regulation, and green manufacturing. Multi-disciplinary integration will drive their transition from the laboratory to industrialization, ultimately enabling the large-scale application of energy-storage systems characterized by high energy density, high safety, and environmental friendliness.

Author Contributions: Conceptualization, L.S. and C.P.; methodology, H.L. and Q.C.; formal analysis, W.R. and C.P.; investigation, L.S. and C.P.; re-sources, W.R., J.X., R.D. and Y.D.; data curation, N.Z. and Y.D.; writing—original draft preparation, L.S., H.L. and Y.D.; writing—review and editing, C.P. and Q.C.; visualization, W.R., N.Z. and R.D.; supervision, C.P. and Q.C.; project administration, L.S.; funding acquisition, L.S., J.X. and Y.D. All authors have read and agreed to the published version of the manuscript.

Funding: This work received funding from Shaanxi Province Qin Chuangyuan Cited High-Level Innovation and Entrepreneurship Talent Program (Approval No. QCYRCXM-2023-130 and QCYRCXM-2023-199), Science and Technology Project of Northwest Institute for Non-ferrous Metals Research (Approval No. 0901YK2316, 0901YK2411 and 0901YK2516), and a Basic Science (Natural Science) Research Project of Higher Education Institutions in Jiangsu Province (Approval No. KY24CZ05F05).

Conflicts of Interest: The authors declare no conflicts of interest.

References

1. Nandy, S.; Fortunato, E.; Martins, R. Green economy and waste management: An inevitable plan for materials science. *Prog. Nat. Sci.* **2022**, *32*, 1–9. [CrossRef]
2. Larcher, D.; Tarascon, J.M. Towards greener and more sustainable batteries for electrical energy storage. *Nat. Chem.* **2014**, *7*, 19–29. [CrossRef] [PubMed]
3. de la Peña, O.S.; Víctor, A.; Costa, R.D. Recent advances towards sustainable materials and processes for energy conversion and storage. *Adv. Energy Mater.* **2021**, *11*, 2102874. [CrossRef]
4. Zhang, W.; Lu, J.; Guo, Z. Challenges and future perspectives on sodium and potassium ion batteries for grid-scale energy storage. *Mater. Today* **2021**, *50*, 400–417. [CrossRef]
5. Hirsh, H.S.; Li, Y.; Tan, D.H.S.; Zhang, M.; Zhao, E.; Meng, Y.S. Sodium-ion batteries paving the way for grid energy storage. *Adv. Energy Mater.* **2020**, *10*, 2001274. [CrossRef]
6. Fagiolari, L.; Sampò, M.; Lamberti, A.; Amici, J.; Francia, C.; Bodoardo, S.; Bella, F. Integrated energy conversion and storage devices: Interfacing solar cells, batteries and supercapacitors. *Energy Storage Mater.* **2022**, *51*, 400–434. [CrossRef]
7. Baptista, A.C.; Martins, J.I.; Fortunato, E.; Martins, R.; Borges, J.P.; Ferreira, I. Thin and flexible bio-batteries made of electrospun cellulose-based membranes. *Biosens. Bioelectron.* **2011**, *26*, 2742–2745. [CrossRef]
8. Ferreira, I.; Bras, B.; Correia, N.; Barquinha, P.; Fortunato, E.; Martins, R. Performance and applications. *J. Disp. Technol.* **2010**, *6*, 332–335. [CrossRef]
9. Carvalho, J.T.; Correia, A.; Cordeiro, N.J.A.; Coelho, J.; Lourenço, S.A.; Fortunato, E.; Martins, R.; Pereira, L. MoS_2 decorated carbon fiber yarn hybrids for the development of freestanding flexible supercapacitors. *NPJ 2D Mater. Appl.* **2024**, *8*, 20. [CrossRef]
10. Goodenough, J.B.; Park, K.-S. The Li-ion rechargeable battery: A perspective. *J. Am. Chem. Soc.* **2013**, *135*, 1167–1176. [CrossRef]
11. Sada, K.; Darga, J.; Manthiram, A. Challenges and prospects of sodium-ion and potassium-ion batteries for mass production. *Adv. Energy Mater.* **2023**, *13*, 2302321. [CrossRef]
12. Wang, J.; Li, L.; Hu, H.; Hu, H.; Guan, Q.; Huang, M.; Jia, L.; Adenusi, H.; Tian, K.V.; Zhang, J.; et al. Toward dendrite-free metallic lithium anodes: From structural design to optimal electrochemical diffusion kinetics. *ACS Nano* **2022**, *16*, 17729–17760. [CrossRef]
13. Wu, F.; Maier, J.; Yu, Y. Guidelines and trends for next-generation rechargeable lithium and lithium-ion batteries. *Chem. Soc. Rev.* **2020**, *49*, 1569–1614. [CrossRef]
14. Fu, X.; Beatty, D.N.; Gaustad, G.G.; Ceder, G.; Roth, R.; Kirchain, R.E.; Bustamante, M.; Babbitt, C.; Olivetti, E.A. Perspectives on cobalt supply through 2030 in the face of changing demand. *Environ. Sci. Technol.* **2020**, *54*, 2985–2993. [CrossRef] [PubMed]
15. Wang, G.; Bi, Z.; Zhang, A.; Das, P.; Lin, H.; Wu, Z.-S. High-voltage and fast-charging lithium cobalt oxide cathodes: From key challenges and strategies to future perspectives. *Engineering* **2024**, *37*, 105–127. [CrossRef]
16. Xiao, Y.; Xiao, J.; Zhao, H.; Li, J.; Zhang, G.; Zhang, D.; Guo, X.; Gao, H.; Wang, Y.; Chen, J.; et al. Prussian blue analogues for sodium-ion battery cathodes: A review of mechanistic insights, current challenges, and future pathways. *Small* **2024**, *20*, 2401957. [CrossRef] [PubMed]

17. Olabi, A.G.; Abbas, Q.; Abdelkareem, M.A.; Alami, A.H.; Mirzaeian, M.; Sayed, E.T. Carbon-based materials for supercapacitors: Recent progress, challenges and barriers. *Batteries* **2022**, *9*, 19. [CrossRef]
18. Wang, W.; Wang, Y.; Yuan, L.; You, C.; Wu, J.; Liu, L.; Ye, J.; Wu, Y.; Fu, L. Recent advances in modification strategies of silicon-based lithium-ion batteries. *Nano Res.* **2022**, *16*, 3781–3803. [CrossRef]
19. Li, H.; Li, Y.; Zhang, L. Designing principles of advanced sulfur cathodes toward practical lithium-sulfur batterierus. *SusMat* **2022**, *2*, 34–64. [CrossRef]
20. Shen, Z.; Huang, J.; Xie, Y.; Wei, D.; Chen, J.; Shi, Z. Solid electrolyte interphase on lithium metal anodes. *ChemSusChem* **2024**, *17*, e202301777. [CrossRef]
21. Bruce, P.G.; Freunberger, S.A.; Hardwick, L.J.; Tarascon, J.-M. Li-O_2 and Li-S batteries with high energy storage. *Nat. Mater.* **2011**, *11*, 19–29. [CrossRef]
22. Versaci, D.; Canale, I.; Goswami, S.; Amici, J.; Francia, C.; Fortunato, E.; Martins, R.; Pereira, L.; Bodoardo, S. Molybdenum disulfide/polyaniline interlayer for lithium polysulphide trapping in lithium-sulphur batteries. *J. Power Source.* **2022**, *521*, 230945. [CrossRef]
23. Yeh, J.W.; Chen, S.K.; Lin, S.J.; Gan, J.Y.; Chin, T.S.; Shun, T.T.; Tsau, C.H.; Chang, S.Y. Nanostructured high-entropy alloys with multiple principal elements: Novel alloy design concepts and outcomes. *Adv. Eng. Mater.* **2004**, *6*, 299–303. [CrossRef]
24. Ma, Y.; Ma, Y.; Wang, Q.; Schweidler, S.; Botros, M.; Fu, T.; Hahn, H.; Brezesinski, T.; Breitung, B. High-entropy energy materials: Challenges and new opportunities. *Energ. Environ. Sci.* **2021**, *14*, 2883–2905. [CrossRef]
25. Zhou, Z.; Ma, Y.; Brezesinski, T.; Breitung, B.; Wu, Y.; Ma, Y. Improving upon rechargeable battery technologies: On the role of high-entropy effects. *Energ. Environ. Sci.* **2025**, *18*, 19–52. [CrossRef]
26. Ma, Y.; Du, H.; Zheng, S.; Zhou, Z.; Zhang, H.; Ma, Y.; Passerini, S.; Wu, Y. High-entropy approach vs. traditional doping strategy for layered oxide cathodes in alkali-metal-ion batteries: A comparative study. *Energy Storage Mater.* **2025**, *79*, 104295. [CrossRef]
27. Ma, Y.; Zhou, Z.; Brezesinski, T.; Ma, Y.; Wu, Y. Stabilizing Layered Cathodes by High-Entropy Doping. *Research* **2024**, *7*, 0503. [CrossRef]
28. Dong, Y.; Zhou, Z.; Ma, Y.; Zhang, H.; Meng, F.; Wu, Y.; Ma, Y. Layered-structured sodium-ion cathode materials: Advancements through high-entropy approaches. *ACS Energy Lett.* **2024**, *11*, 5096–5119. [CrossRef]
29. Wang, T.; Tan, X.; Ma, Y.; Ma, Y.; Wu, Y. High-entropy materials regulating lithium polysulfides for advanced lithium-sulfur batteries. *Chem. Eng. J.* **2025**, *519*, 165580. [CrossRef]
30. Oketola, A.M.; Adegbola, T.A.; Jamiru, T.; Ogunbiyi, O.; Salifu, S. Advances in high-entropy alloy research: Unraveling fabrication a techniques, microstructural transformations, and mechanical properties. *J. Bio-Tribo-Corrosio.* **2025**, *11*, 79. [CrossRef]
31. Barman, S.; Gupta, K.K.; Dey, S. Molecular dynamics-based explanation of the reinforcement geometry effects on CNT/graphene-reinforced $Al_{0.3}$ CoCrFeNi high-entropy alloys. *Sci. Rep.* **2025**, *15*, 24984. [CrossRef]
32. Kumar, D. Recent advances in tribology of high entropy alloys: A critical review. *Prog. Mater. Sci.* **2023**, *136*, 101106. [CrossRef]
33. Zhang, Y.; Zuo, T.T.; Tang, Z.; Gao, M.C.; Dahmen, K.A.; Liaw, P.K.; Lu, Z.P. Microstructures and properties of high-entropy alloys. *Prog. Mater. Sci.* **2014**, *61*, 1–93. [CrossRef]
34. Senkov, O.N.; Wilks, G.B.; Scott, J.M.; Miracle, D.B. Mechanical properties of $Nb_{25}Mo_{25}Ta_{25}W_{25}$ and $V_{20}Nb_{20}Mo_{20}Ta_{20}W_{20}$ refractory high entropy alloys. *Intermetallics* **2011**, *19*, 698–706. [CrossRef]
35. Gludovatz, B.; Hohenwarter, A.; Catoor, D.; Chang, E.H.; George, E.P.; Ritchie, R.O. A fracture-resistant high-entropy alloy for cryogenic applications. *Science* **2014**, *345*, 1153–1158. [CrossRef]
36. Ren, J.-T.; Chen, L.; Wang, H.-Y.; Yuan, Z.-Y. High-entropy alloys in electrocatalysis: From fundamentals to applications. *Chem. Soc. Rev.* **2023**, *52*, 8319–8373. [CrossRef] [PubMed]
37. Chen, T.; Ning, F.; Qi, J.; Feng, G.; Wang, Y.; Song, J.; Yang, T.; Liu, X.; Chen, L.; Xia, D. PtFeCoNiCu high-entropy solid solution alloy as highly efficient electrocatalyst for the oxygen reduction reaction. *iScience* **2023**, *26*, 105890. [CrossRef]
38. Wei, Y.; Liu, X.; Yao, R.; Qian, J.; Yin, Y.; Li, D.; Chen, Y. Embedding the high entropy alloy nanoparticles into carbon matrix toward high performance Li-ion batteries. *J. Alloys Compd.* **2023**, *938*, 168610. [CrossRef]
39. Zhao, J.; Wei, Z.; Chen, N.; Meng, F.; Tian, R.; Zeng, Y.; Du, F. High-entropy alloy anodes for low-strain and high-volumetric lithium-ion storage at ambient and subzero temperatures. *Energy Storage Mater.* **2024**, *65*, 103127. [CrossRef]
40. Wu, D.-H.; Haq, M.U.; Zhang, L.; Feng, J.-J.; Yang, F.; Wang, A.-J. Noble metal-free FeCoNiMnV high entropy alloy anchored on N-doped carbon nanotubes with prominent activity and durability for oxygen reduction and zinc–air batteries. *J. Colloid Interf. Sci.* **2024**, *662*, 149–159. [CrossRef]
41. Wang, X.; Yang, C.; Li, J.; Chen, X.A.; Yang, K.; Yu, X.; Lin, D.; Zhang, Q.; Wang, S.; Wang, J.; et al. Insights of heteroatoms doping-enhanced bifunctionalities on carbon based energy storage and conversion. *Adv. Funct. Mater.* **2020**, *31*, 2009109. [CrossRef]
42. Sun, Y.; Sun, J.; Sanchez, J.S.; Xia, Z.; Xiao, L.; Chen, R.; Palermo, V. Surface chemistry and structure manipulation of graphene-related materials to address the challenges of electrochemical energy storage. *Chem. Commun.* **2023**, *59*, 2571–2583. [CrossRef]

43. Li, X.; Zhang, G.; Bai, X.; Sun, X.; Wang, X.; Wang, E.; Dai, H. Highly conducting graphene sheets and Langmuir-Blodgett films. *Nat. Nanotechnol.* **2008**, *3*, 538–542. [CrossRef] [PubMed]
44. Qi, C.; Li, S.; Yang, Z.; Xiao, Z.; Zhao, L.; Yang, F.; Ning, G.; Ma, X.; Wang, C.; Xu, J.; et al. Suitable thickness of carbon coating layers for silicon anode. *Carbon* **2022**, *86*, 530–538. [CrossRef]
45. Xu, J.; Tan, Z.; Zeng, W.; Chen, G.; Wu, S.; Zhao, Y.; Ni, K.; Tao, Z.; Ikram, M.; Ji, H.; et al. A hierarchical carbon derived from sponge-templated activation of graphene oxide for high-performance supercapacitor electrodes. *Adv. Mater.* **2016**, *28*, 5222–5228. [CrossRef]
46. Song, Z.; Lu, X.; Hu, Q.; Ren, J.; Zhang, W.; Zheng, Q.; Lin, D. Synergistic confining polysulfides by rational design a N/P co-doped carbon as sulfur host and functional interlayer for high-performance lithium-sulfur batteries. *J. Power Source* **2019**, *421*, 23–31. [CrossRef]
47. Başgöz, Ö.; Güngör, A.; Güler, Ö.; Erdem, E. High-entropy alloys and oxides as supercapacitor electrodes: A structural and electrochemical perspective for energy storage. *Adv. Sustain. Syst.* **2025**, *9*, 2500201. [CrossRef]
48. Yang, Z.; Tian, J.; Yin, Z.; Cui, C.; Qian, W.; Wei, F. Carbon nanotube- and graphene-based nanomaterials and applications in high-voltage supercapacitor: A review. *Carbon* **2019**, *141*, 467–480. [CrossRef]
49. Zhang, J.; Jiang, J.; Li, H.; Zhao, X.S. A high-performance asymmetric supercapacitor fabricated with graphene-based electrodes. *Energ. Environ. Sci.* **2011**, *4*, 4009–4015. [CrossRef]
50. Hussain, I.; Lamiel, C.; Ahmad, M.; Chen, Y.; Shuang, S.; Javed, M.S.; Yang, Y.; Zhang, K. High entropy alloys as electrode material for supercapacitors: A review. *J. Energy Storage* **2021**, *44*, 103405. [CrossRef]
51. Wang, Y.; Wang, Y. High-entropy alloys in catalyses and supercapacitors: Progress, prospects. *Nano Energy* **2022**, *104*, 107958. [CrossRef]
52. Feng, D.; Dong, Y.; Nie, P.; Zhang, L.; Qiao, Z.-A. CoNiCuMgZn high entropy alloy nanoparticles embedded onto graphene sheets via anchoring and alloying strategy as efficient electrocatalysts for hydrogen evolution reaction. *Chem. Eng. J.* **2022**, *430*, 132883. [CrossRef]
53. Ashwini, R.; Kumar, M.K.P.; Rekha, M.Y.; Santosh, M.S.; Srivastava, C. Optimization of NiFeCrCoCu high entropy alloy nanoparticle-graphene (HEA-G) composite for the enhanced electrochemical sensitivity towards urea oxidation. *J. Alloys Compd.* **2022**, *903*, 163846. [CrossRef]
54. Xu, X.; Du, Y.; Wang, C.; Guo, Y.; Zou, J.; Zhou, K.; Zeng, Z.; Liu, Y.; Li, L. High-entropy alloy nanoparticles on aligned electronspun carbon nanofibers for supercapacitors. *J. Alloys Compd.* **2020**, *822*, 153642. [CrossRef]
55. Wang, J.; Liu, W.; Wang, Y.; Guo, Y.; Liu, M.; Ye, C.; Wang, S.; Zou, Q. AgCuInCdZn high-entropy alloy nanoparticles-embedded in porous carbon fibers for long-cycling lithium metal anodes. *Chem. Eng. J.* **2023**, *477*, 146884. [CrossRef]
56. Yao, Y.; Zhao, Z.; Niu, R.; Chen, J.; Wang, X. Rationally designed FeCoNiCrMn high-entropy alloys nanoparticles hybridized rGO modified separator for highly efficient lithium-sulfur batteries. *Chem. Phys. Lett.* **2024**, *839*, 141124. [CrossRef]
57. Liu, B.; Li, M.; Wu, Y.; Ren, R.; Ge, Q.; Fan, L.; Qin, P.; Qian, J.; Ding, X. High-entropy BiSnSbCuAl nanoalloys conformed in carbon fibers as fast-charging and high-capacity anode material for sodium-ion batteries. *J. Power Sources* **2025**, *652*, 237600. [CrossRef]
58. Ujah, C.O.; Kallon, D.V.V.; Aigbodion, V.S. Study on properties of high entropy alloys reinforced with carbon nanotubes/graphene-A review. *J. Alloys Metall. Syst.* **2024**, *8*, 100117. [CrossRef]
59. Gao, X.; Chen, R.; Liu, T.; Fang, H.; Qin, G.; Su, Y.; Guo, J. High-entropy alloys: A review of mechanical properties and deformation mechanisms at cryogenic temperatures. *J. Mater. Sci.* **2022**, *57*, 6573–6606. [CrossRef]
60. Bolar, S.; Ito, Y.; Fujita, T. Future prospects of high-entropy alloys as next-generation industrial electrode materials. *Chem. Sci.* **2024**, *15*, 8664–8722. [CrossRef]
61. Wang, Y.; Zhang, Y.; Xing, P.; Li, X.; Du, Q.; Fan, X.; Cai, Z.; Yin, R.; Yao, Y.; Gan, W. Self-encapsulation of high-entropy alloy nanoparticles inside carbonized wood for highly durable electrocatalysis. *Adv. Mater.* **2024**, *36*, 2402391. [CrossRef]
62. Zhao, M.; Wu, T.; Liu, D.; Huang, Y.; Zhao, L.; Tang, Y.; Shen, M.; Hu, Y.; Zhang, J.; Li, J.; et al. Effect of carbon fiber on microstructure evolution and surface properties of FeCoCrNiCu high-entropy alloy coatings. *Mater. Corros.* **2019**, *71*, 430–439. [CrossRef]
63. Liu, Y.; Yuan, J.; Zhou, J.; Pan, K.; Zhang, R.; Zhao, R.; Li, L.; Huang, Y.; Liu, Z. Laser solid-phase synthesis of graphene shell-encapsulated high-entropy alloy nanoparticles. *Light Sci. App.* **2024**, *13*, 270. [CrossRef]
64. Wang, Z.; Ge, H.; Liu, S.; Li, G.; Gao, X. High-entropy alloys to activate the sulfur cathode for lithium-sulfur batteries. *Energy Environ. Mater.* **2022**, *6*, e12358. [CrossRef]
65. Pourghaz, A.; Rajabi, M.; Torabi, M.; Amirnejad, M. The impressive improvement of CoCrFeMnNi high entropy alloy mechanical properties and pitting corrosion resistance by incorporation of carbon nanotube reinforcement. *Intermetallics* **2023**, *63*, 108073. [CrossRef]

66. Khan, I.; Khan, S.; Wu, S.-Y.; Liu, L.; Alodhayb, A.N.; Mead, J.L.; Ali, S.; Hassan, S.U.; Chen, H.-T.; Ju, S.-P.; et al. Advanced fabrication of graphene-integrated high-entropy alloy@carbon nanocomposites as superior multifunctional electrocatalysts. *ACS Appl. Mater. Interfaces* **2025**, *17*, 21033–21052. [CrossRef] [PubMed]
67. Wang, J.; Yang, P.; Wang, Y.; Wang, S. High entropy alloy nanoparticles decorated carbon-based electrode as interfacial Li-ion localized accelerators for anode-free lithium metal batteries. *Carbon* **2024**, *228*, 119432. [CrossRef]
68. Xiao, W.; Xu, H.; Pei, Y.; Hu, L.; Yang, Z. High entropy alloy/graphite composite as a new strategy to enhance the performance of lithium-ion batteries. *Inorg. Chem. Commun.* **2025**, *178*, 114594. [CrossRef]
69. Vaidya, M.; Muralikrishna, G.M.; Murty, B.S. High-entropy alloys by mechanical alloying: A review. *J. Mater. Res.* **2019**, *34*, 664–686. [CrossRef]
70. Kumar, A.; Singh, A.; Suhane, A. Mechanically alloyed high entropy alloys: Existing challenges and opportunities. *J. Mater. Res. Technol.* **2022**, *17*, 2431–2456. [CrossRef]
71. Martin, P.; Aguilar, C.; Cabrera, J.M. A review on mechanical alloying and spark plasma sintering of refractory high-entropy alloys: Challenges, microstructures, and mechanical behavior. *J. Mater. Res. Technol.* **2024**, *30*, 1900–1928. [CrossRef]
72. Liu, C.; Jiang, X.; Sun, H.; Liu, T.; Wu, Z.; Yang, L. Steam oxidation properties of graphene reinforced bioinspired laminated CoCrFeNiMn high-entropy alloy matrix composites at 1000 °C. Mater. *Today Commun.* **2024**, *38*, 107962.
73. Singh, S.; Shaikh, S.M.; Kumar MK, P.K.; Murty, B.S.; Srivastava, C. Microstructural homogenization and substantial improvement in corrosion resistance of mechanically alloyed FeCoCrNiCu high entropy alloys by incorporation of carbon nanotubes. *Materialia* **2020**, *14*, 100917. [CrossRef]
74. Peng, Y.Z.; You, X.; Li, C.J.; Yang, C.M.Y.; Xu, Z.Y.; Lu, Q.; Xu, G.Y.; Wang, Y.R.; Feng, Z.X.; Bao, R.; et al. Strength-ductility synergy in a novel carbon nanotube-high entropy alloy co-reinforced aluminum matrix composite. *Compos. Part A Appl. Sci. Manuf.* **2024**, *181*, 108116. [CrossRef]
75. Han, C.X.; Zhi, J.Q.; Zeng, Z.; Wang, Y.S.; Zhou, B.; Gao, J.; Wu, Y.X.; He, Z.Y.; Wang, X.M.; Yu, S.W. Synthesis and characterization of nano-polycrystal diamonds on refractory high entropy alloys by chemical vapour deposition. *Appl. Surf. Sci.* **2023**, *623*, 157108. [CrossRef]
76. Lu, D.; Fu, X.; Guo, D.; Ma, W.; Sun, S.; Shao, G.; Zhou, Z. Challenges and opportunities in 2D high-entropy alloy electrocatalysts for sustainable energy conversion. *SusMat* **2023**, *3*, 730–748. [CrossRef]
77. Hassan, S.U.; Hou, L.; Yang, Y.; Qamar, T.H.; Wang, S. Optimizing the electromagnetic wave attenuation properties of carbon encapsulated $FeCoNiCuMn_x$ high entropy alloy nanoparticles. *Carbon* **2024**, *229*, 119502. [CrossRef]
78. Ciriminna, R.; Pagliaro, M. Open challenges in sol-gel science and technology. *J. Sol-Gel Sci. Technol.* **2022**, *101*, 29–36. [CrossRef]
79. Wang, Q.; Wang, J.; Jiang, S.; Li, P. Recent progress in sol-gel method for designing and preparing metallic and alloy nanocrystals. *Acta. Phys. Chim. Sin.* **2019**, *35*, 1186–1206. [CrossRef]
80. Niu, B.; Zhang, F.; Ping, H.; Li, N.; Zhou, J.; Lei, L.; Xie, J.; Zhang, J.; Wang, W.; Fu, Z. Sol-gel autocombustion synthesis of nanocrystalline high-entropy alloys. *Sci. Rep.* **2017**, *7*, 3421. [CrossRef]
81. Zhang, Q.; Ye, Y.; Sun, L.; Sun, P.; Wei, J.; Gan, Q. A heterojunction of high-entropy alloy and nitrogen-doped carbon nanospheres for efficient electromagnetic wave absorption. *J. Mater. Chem. C* **2025**, *13*, 7205–7218. [CrossRef]
82. Yu, Y.; Cui, W.; Xu, Z.; Wang, S.; Jiang, W.; Sun, R.; Qi, L.; Pan, K. High-entropy $Pt_{18}Ni_{26}Fe_{15}Co_{14}Cu_{27}$ nanocrystalline crystals in situ grown on reduced graphene oxide with excellent electromagnetic absorption properties. *J. Colloid Interf. Sci.* **2023**, *639*, 193–202. [CrossRef]
83. Ji, D.; Lin, Y.; Guo, X.; Ramasubramanian, B.; Wang, R.; Radacsi, N.; Jose, R.; Qin, X.; Ramakrishna, S. Electrospinning of nanofibres. *Nat. Rev. Methods Primers* **2024**, *4*, 1. [CrossRef]
84. Keirouz, A.; Wang, Z.; Reddy, V.S.; Nagy, Z.K.; Vass, P.; Buzgo, M.; Ramakrishna, S.; Radacsi, N. The history of electrospinning: Past, present, and future developments. *Adv. Mater. Technol.* **2023**, *8*, 2201723. [CrossRef]
85. Li, H.; Zhu, H.; Sun, S.; Hao, J.; Zhu, Z.; Xu, F.; Lu, S.; Duan, F.; Du, M. Thermodynamically driven metal diffusion strategy for controlled synthesis of high-entropy alloy electrocatalysts. *Chem. Commun.* **2021**, *57*, 10027–10030. [CrossRef] [PubMed]
86. Liu, W.; Zhang, B.; Wen, X.; Wang, L.; Jin, M.; Wu, H.; Qin, S.; Yang, X.; Liu, Y.; Yu, G. Bottom-up synthesis of high-entropy alloy/carbon nanofiber with magnetic-dielectric synergy for low-filler and broadband electromagnetic wave absorption. *ACS Appl. Nano Mater.* **2024**, *7*, 26706–26716. [CrossRef]
87. Wang, L.; Wang, C.; Mu, Y.; Fan, J.; Yang, X.; Yu, C.; Guo, B.; Zeng, G. Mesoporous high-entropy-alloy electrocatalysts via electrospinning for enhanced alkaline water electrolysis. *Fuel* **2025**, *391*, 134800. [CrossRef]
88. Wu, H.; Qi, Y.; Yin, R.; Long, Y.; Song, J.; Xie, P.; Zhong, J.; Wang, C.; Hou, Q.; Fan, R.; et al. Biocompatible radio frequency epsilon-near-zero materials for wearable electronics. *Adv. Compos. Hybrid Mater.* **2024**, *8*, 32. [CrossRef]
89. Haile, B.S.; Pal, V.; Pal, T.; Slathia, S.; Jigi, G.M.; Negedu, S.D.; Tiwari, N.; Singh, H.; Joseph, A.; Olu, F.E.; et al. Direct ink writing (3D printing) of robust, highly efficient, double-half-heusler thermoelectric high-entropy alloy. *Adv. Eng. Mater.* **2025**, *27*, 2402283. [CrossRef]

90. Ai, J.; Liu, S.; Zhang, Y.; Han, Y.; Liu, B.; Yin, Y.; Ma, H.; Feng, J. 3D-printed high-entropy alloy nanoarchitectures. *Small* **2025**, *21*, 2409900. [CrossRef]
91. Bajaj, D.; Feng, A.; Qu, S.; Chen, Z.; Li, D.; Chen, D.L. Deformation behavior of 3D-printed high-entropy alloys: A critical review. *Adv. Eng. Mater.* **2023**, *26*, 2300615. [CrossRef]
92. Yan, T.; Ye, X.; He, E.; Gao, Q.; Wang, Y.; Ye, Y.; Wu, H. 3D-Printed $(CoCrFeMnNi)_3O_4$@C-GR dual core-shell composites: Multilevel control and mechanisms of microwave absorption performance. *Chem. Eng. J.* **2024**, *497*, 154777. [CrossRef]
93. Yi, J.; Deng, Q.; Cheng, H.; Zhu, D.; Zhang, K.; Yang, Y. Unique hierarchically structured high-entropy alloys with multiple adsorption sites for rechargeable Li-CO_2 batteries with high capacity. *Small* **2024**, *20*, 2401146. [CrossRef]
94. Wen, Z.; Mu, X.; Sun, X.; Xu, Z.; Zheng, M.; Zhou, H.; He, P. Boosting Li-CO_2 battery performance via high-entropy alloy catalysts: Insights into configurational entropy effect. *Angew. Chem. Int. Edit.* **2025**, *64*, e202424121. [CrossRef]
95. Wang, J.; Wang, Y.; Lu, X.; Qian, J.; Yang, C.; Manke, I.; Song, H.; Liao, J.; Wang, S.; Chen, R. Ultra-sleek high entropy alloy tights: Realizing superior cyclability for anode-free battery. *Adv. Mater.* **2023**, *36*, 2308257. [CrossRef]
96. Wang, P.; Guo, S.; Xu, Y.; Yuan, X.; Tian, Y.; Xu, B.; Zhao, Z.; Wang, Y.; Li, J.; Wang, X.; et al. Upcycling spent cathodes from Li-ion batteries into a high-entropy alloy catalyst with reverse electron transfer for Li-O_2 batteries. *ACS Nano* **2025**, *19*, 17589–17605. [CrossRef] [PubMed]
97. Wu, R.; Zhang, Q.; Yang, Q.; Hu, Z.; Zhao, Y. High-entropy alloy nanoparticles functionalized with reduced graphene oxide as a high-performance cathode for lithium–oxygen batteries. *Nanoscale* **2025**, *17*, 10581–10588. [CrossRef] [PubMed]
98. Tian, J.; Rao, Y.; Shi, W.; Yang, J.; Ning, W.; Li, H.; Yao, Y.; Zhou, H.; Guo, S. Sabatier relations in electrocatalysts based on high-entropy alloys with wide-distributed d-band centers for Li-O_2 batteries. *Angew. Chem. Int. Ed.* **2023**, *62*, e202310894. [CrossRef] [PubMed]
99. Bai, M.; Tang, X.; Liu, S.; Wang, H.; Liu, Y.; Shao, A.; Zhang, M.; Wang, Z.; Ma, Y. An anodeless, mechanically flexible and energy/power dense sodium battery prototype. *Energy Environ. Sci.* **2022**, *15*, 4686–4699. [CrossRef]
100. Zhang, R.; Xue, H.; Du, D.; Shen, Y.; Zheng, J.; Li, C.; Feng, Z.; Sun, T. Regulation of configurational entropy to realize long cycle lifespan of high entropy alloy anodes for potassium batteries. *Adv. Funct. Mater.* **2025**, *35*, 2422218. [CrossRef]
101. Chang, C.B.; Tseng, Y.Y.; Lu, Y.R.; Yang, Y.C.; Tuan, H.Y. High entropy induced local charge enhancement promotes Frank-Van der merwe growth for dendrite-free potassium metal batteries. *Adv. Funct. Mater.* **2024**, *34*, 2411193. [CrossRef]
102. Zheng, D.; Zha, J.; Wang, Y.; Wei, Z.; Qi, J.; Wei, F.; Meng, Q.; Xue, X.; Zhao, D.; Li, Y.; et al. $Co_{0.2}Sb_{0.2}Fe_{0.2}Mn_{0.2}Ni_{0.2}$ high-entropy alloy carbon nanofiber as anode for lithium/potassium ion batteries. *J. Mater. Sci.* **2025**, *60*, 3926–3939. [CrossRef]
103. Han, F.; Zhang, L.; Jin, Q.; Ma, X.; Zhang, Z.; Sun, Z.; Zhang, X.; Wu, L. High rate and long-cycle life of lithium-sulfur battery enabled by high d-band center of high-entropy alloys. *ACS Nano* **2025**, *19*, 9182–9195. [CrossRef] [PubMed]
104. Ma, Y.; Ren, Y.; Sun, D.; Wang, B.; Wu, H.; Bian, H.; Cao, J.; Cao, X.; Ding, F.; Lu, J.; et al. High entropy alloy nanoparticles dual-decorated with nitrogen-doped carbon and carbon nanotubes as promising electrocatalysts for lithium-sulfur batteries. *J. Mater. Sci. Technol.* **2024**, *188*, 98–104. [CrossRef]
105. Cao, X.; Gao, Y.; Wang, Z.; Zeng, H.; Song, Y.; Tang, S.; Luo, L.; Gong, S. FeNiCrCoMn high-entropy alloy nanoparticles loaded on carbon nanotubes as bifunctional oxygen catalysts for rechargeable zinc-air batteries. *ACS Appl. Mater. Interfaces* **2023**, *15*, 32365–32375. [CrossRef]
106. Han, J.; Zhang, W.; Liu, K.; Zheng, H.; Li, Y.; Luo, L.; Gong, S.; Jia, Y.; Liang, X. PtFeCoNiMoY high-entropy alloy nanoparticles as bifunctional oxygen catalysts for zinc air batteries. *Appl. Surf. Sci.* **2025**, *687*, 162238. [CrossRef]
107. Yao, Y.; Li, Z.; Dou, Y.; Jiang, T.; Zou, J.; Lim, S.Y.; Norby, P.; Stamate, E.; Jensen, J.O.; Zhang, W. High entropy alloy nanoparticles encapsulated in graphitised hollow carbon tubes for oxygen reduction electrocatalysis. *Dalton Trans.* **2023**, *52*, 4142–4151. [CrossRef]
108. Shen, E.; Song, X.; Chen, Q.; Zheng, M.; Bian, J.; Liu, H. Spontaneously forming oxide layer of high entropy alloy nanoparticles deposited on porous carbons for supercapacitors. *ChemElectroChem* **2021**, *8*, 260–269. [CrossRef]
109. Mohanty, G.C.; Verma, A. Supercapacitive study of a high entropy alloy and green carbon composite electrode with a wide potential window (1.3 V) and its application towards an aqueous symmetric device. *New J. Chem.* **2025**, *49*, 521–529. [CrossRef]

batteries

MDPI

Article

Electrochemical Performances of Li-Ion Batteries Based on $LiFePO_4$ Cathodes Supported by Bio-Sourced Activated Carbon from Millet Cob (MC) and Water Hyacinth (WH)

Wend-Waoga Anthelme Zemane and Oumarou Savadogo *

Laboratory of New Materials for Energy and Electrochemistry, Polytechnique Montreal, Montreal, QC H3T 1J4,Canada; wend-waoga-anthelme.zemane@polyml.ca

* Correspondence: osavadogo@polymtl.ca; Tel.: +1-514-340-4725

Abstract

The electrochemical performance of Li-ion batteries employing $LiFePO_4$ (LFP) cathodes supported by bio-sourced activated carbon derived from millet cob (MC) and water hyacinth (WH) were systematically investigated. Carbon activation was carried out using potassium hydroxide (KOH) at varying mass ratios of KOH to precursor material: 1:1, 2:1, and 5:1 for both WH and MC-derived carbon. The physical properties (X-ray diffraction patterns, BET surface area, micropore and mesopore volume, conductivity, etc.) and electrochemical performance (specific capacity, discharge at various current rates, electrochemical impedance measurement, etc.) were determined. Material characterization revealed that the activated carbon derived from MC exhibits an amorphous structure, whereas that obtained from WH is predominantly crystalline. High specific surface areas were achieved with activated carbons synthesized using a low KOH-to-carbon mass ratio (1:1), reaching 413.03 $m^2 \cdot g^{-1}$ for WH and 216.34 $m^2 \cdot g^{-1}$ for MC. However, larger average pore diameters were observed at higher activation ratios (5:1), measuring 8.38 nm for KOH/WH and 5.28 nm for KOH/MC. For both biomass-derived carbons, optimal electrical conductivity was obtained at a 2:1 activation ratio, with values of 14.7×10^{-3} $S \cdot cm^{-1}$ for KOH/WH and 8.42×10^{-3} $S \cdot cm^{-1}$ for KOH/MC. The electrochemical performance of coin cells based on cathodes composed of 85% $LiFePO_4$, 8% of these activated carbons, and 7% polyvinylidene fluoride (PVDF) as a binder, with lithium metal as the anode were studied. The $LiFePO_4$/C (LFP/C) cathodes exhibited specific capacities of up to 160 $mAh \cdot g^{-1}$ at a current rate of C/12 and 110 $mAh \cdot g^{-1}$ at 5C. Both LFP/MC and LFP/WH cathodes exhibit optimal energy density at specific values of pore size, pore volume, charge transfer resistance (R_{ct}), and diffusion coefficient (D_{Li}), reflecting a favorable balance between ionic transport, accessible surface area, and charge conduction. Maximum energy densities relative to active mass were recorded at 544 $mWh \cdot g^{-1}$ for LFP/MC 2:1, 554 $mWh \cdot g^{-1}$ for LFP/WH 2:1, and 568 $mWh \cdot g^{-1}$ for the reference LFP/graphite system. These performance results demonstrate that the development of high-performing bio-sourced activated carbon depends on the optimization of various parameters, including chemical composition, specific surface area, pore volume and size distribution, as well as electrical conductivity.

Keywords: $LiFePO_4$; lithium; activated carbon; water hyacinth; millet cob; porosity; coin cells; cycling performance; specific capacity; electrochemical impedance spectroscopy

1. Introduction

The Lithium iron phosphate has emerged as a cost-effective and promising positive electrode for lithium-ion batteries, particularly suited for applications in electronics, telecommunications, transportation, and beyond. Since its introduction as a cathode [1], $LiFePO_4$ has demonstrated numerous advantages, including a highly theoretical capacity (170 mAh/g), exceptional longevity, an excellent stable discharge profile, the ability to handle high current rates, moderate or low cost, natural presence of its constituent elements on earth, excellent thermal and chemical stability and more environmentally friendly utilization than the other cathode-based materials [2–4]. Its thermal stability, attributed to its phospho-olivine structure, is rooted in a unique crystalline configuration: oxygen atoms form a hexagonal close-packed lattice, iron ions create zigzag chains of octahedra connected by phosphate (PO_4) groups, and Li^+ ions occupy octahedral sites arranged in one-dimensional tunnels along the [010] axis [1–3,5]. The P^{5+} ions create strong covalent bonds with oxygen and confer high structural stability, enhancing thermal resistance and ensuring reliable performance at elevated temperatures, a critical advantage for demanding applications. Additionally, it has been shown that [6] the combination of phosphorus and iron optimizes redox energy, generating a voltage of 3.4 V relative to Li/Li^+, a key factor for stable electrochemical performance. The excellent reversibility of $LiFePO_4$ is linked to the structural similarity between $LiFePO_4$ and $FePO_4$ [6]. During the phase transition between these two states during de-lithiation, changes in crystal parameters remain minimal, with a volume reduction of only 6.8% and a density increase of 2.6%, reducing structural stress during charge-discharge cycles [3]. Composed of naturally abundant elements such as iron, lithium, and phosphorus, $LiFePO_4$ could reduce cathode costs by 10–50% of the total lithium-ion battery cost, enhancing its competitiveness for large-scale commercial applications [3,7].

Despite its many advantages, certain limitations hinder the widespread adoption of $LiFePO_4$. Its low ionic diffusivity ($\sim10^{-14}$ cm^2/s) and poor electronic conductivity ($\sim10^{-9}$ S/cm) limit its performance during rapid charge and discharge cycles. This low electronic conductivity was attributed to the bonding structure of iron atoms: FeO_6 octahedra share vertices rather than edges or faces, increasing the distance between iron atoms and complicating electron transport [2,8]. Another limitation of this cathode material is its low operating voltage which is not higher than 3.4 V vs. Li/Li^+ whereas the other commercial cathodes exhibit potentials closer to 4 V vs. Li/Li^+. To address these drawbacks, various approaches have been developed, including particle size reduction [9–11], doping with metals [9,12,13], and coating $LiFePO_4$ particles with carbon [9,14–17]. These methods improve electronic conductivity and overall performance, though they may incur additional costs. Issues are also related to the long-term stability of the dopant. Ongoing research aims to identify innovative and cost-effective solutions to optimize $LiFePO_4$ performance and stability while controlling expenses [18].

On the particle size reduction, it has been noted [19] that nanomaterials exhibit excellent mechanical stress resistance, facilitating adaptation to volume changes during charge-discharge cycles while improving lithium-ion diffusion and electron transfer due to reduced diffusion distances. However, their highly specific surface area makes them susceptible to undesirable side reactions, such as electrolyte degradation, which can affect chemical stability. Moreover, excessively small particle sizes may reduce electronic conductivity and lead to agglomeration, complicating processing and incorporation into composite materials [20–22]. Metal doping has proven effective in enhancing Li^+ ion diffusion and cathode conductivity [22–25]. It has been shown that integrating super-valent ions [22], such as Mg^{2+}, Al^{3+}, or Ti^{4+}, can increase electronic conductivity by orders of magnitude by creating p- and n-type semiconducting regions. It has been shown that partial replacement of Fe^{2+} with Mn^{2+} improves conductivity, achieving a specific capacity exceeding 140 mAh/g with minimal performance loss over cycles [3,8]. Additionally, it has been confirmed that doping with precious metals like platinum and palladium significantly enhances electrochemical properties [13,23–25]. On the contrary, it has been noted that aliovalent doping is poorly tolerated by the olivine structure of $LiFePO_4$, as dopant ions are not always well integrated into the crystal lattice, limiting their effectiveness in improving electronic conductivity [26,27]. To overcome the conductivity limitations of $LiFePO_4$ cathodes, it has been suggested [28] to use conductive carbon-based materials as supports, promoting optimal lithium-ion flow and enhancing electrochemical performance. Several studies [19,29–34] have explored the application of carbon coatings at varying percentages to strengthen conductivity and optimize overall cathode performance.

More recently, several studies have focused on the improvement of the performances of $LiFePO_4$ cathodes. A study explored the effect of holes created by a picosecond pulsed laser on the high-rate performance of a hybrid $LiFePO_4$/activated carbon electrode [35]. Another research effort involved wrapping $LiFePO_4$ particles with a combination of activated carbon and reduced graphene oxide, yielding cathodes with remarkably high capacities and specific energy densities [36]. A separate investigation prepared LFP/C composites with varying carbon contents via the carbothermal reduction method. Accordingly, it was found that the amount of carbon significantly influences the material properties: increasing the carbon content led to smaller and more uniformly distributed particles. Among the tested samples, the LFP/C-15 composite exhibited the best electrochemical performance, achieving a discharge capacity of 160.7 mAh g^{-1} after 100 cycles at 0.1 C, and maintaining a capacity retention of 82.1% [37].

This innovative study explores the use of additives bio-sourced carbon derived from agricultural waste, such as millet cob (MC) and water hyacinth (WH), to develop $LiFePO_4$ (LFP) cathodes for lithium batteries. These novel materials, applied in this context for the first time, are expected to provide new scientific insights on the correlation which might exist between the physicochemical properties and the improvements in the electrochemical parameters of LFP cathodes enhanced with additives bio-derived carbon from MC and WH. This research will help to advance in the understanding of bio-sourced carbon additives influence the performance of LFP-based cathodes in lithium-ion batteries.

2. Materials and Methods

2.1. Materials, Electrodes, Li-Ion Battery Cell Elaboration and Fabrications

The millet cob and water hyacinth samples were weighed using an analytical balance and placed in refractory crucibles. A furnace operating under an argon atmosphere was used both for drying and for determining the volatile matter and ash content. The obtained ash of millet cob and water hyacinth samples were chemically activated using potassium hydroxide (KOH) as the activating agent. Three different concentrations were investigated: 10% KOH (KOH/CM 1:1), 20% KOH (KOH/CM 2:1), and 50% KOH (KOH/CM 5:1). Accordingly, the activated carbon samples derived from millet cob were labeled KOH/MC 1:1, KOH/MC 2:1, and KOH/MC 5:1, while those from water hyacinth were referred to as KOH/WH 1:1, KOH/WH 2:1, and KOH/WH 5:1. The synthetic graphite powder (99.99% purity), used as a reference material, was supplied by Sigma Aldrich, Oakville, ON, Canada.

The working anode electrodes were prepared by mixing 85 wt% synthesized activated carbon, 8 wt% carbon black, and 7 wt% polyvinylidene fluoride (PVDF) binder dissolved in N-methyl-2-pyrrolidone (NMP) to form a homogeneous slurry. The slurry was uniformly coated onto copper foil current collectors and dried under vacuum at 120 °C for 12 h to remove residual solvent. Circular electrodes (12 mm diameter) were punched from the coated foil prior to assembly. CR2032-type coin half-cells were assembled in an argon-filled glovebox using lithium metal foil as the counter/reference electrode.

Galvanostatic charge–discharge tests of the anodes were conducted at a current rate of C/24 (based on theoretical graphite capacity) within a potential window of 0.01–3.0 V vs. Li^+/Li, using a potentiostat/galvanostat. This low rate minimizes parasitic side reactions and enables accurate probing of the intrinsic Li-storage behavior of the carbons.

Cathode Electrodes were fabricated using LFP powder, with a purity of more than 99% and particle size less than 5 microns (Sigma Aldrich) and lithium metal discs (MSE Supplies). A polymeric binder, polyvinylidene fluoride (PVDF), dissolved in dimethylacetamide (DMAC), was used to homogenize and coat the active $LiFePO_4$/carbon (LFP/C) mixture onto an aluminum current collector. The electrode composition consisted of 85% LFP, 8% activated carbon or graphite, and 7% PVDF dissolved in DMAC at 60 °C. The cathode electrode assembly was also carried out in CR2032 coin cell using standard components, including casings, springs, spacers, and gaskets.

All assembly steps were performed in an argon-filled glove box to prevent exposure to moisture and oxygen.

For the separate characterization of both anode or cathode electrodes, the electrolyte, supplied by Sigma Aldrich, consisted of a 1 M solution of $LiPF_6$ dissolved in an equimolar mixture of ethylene carbonate (EC), dimethyl carbonate (DMC), and ethyl methyl carbonate (EMC), ensuring high ionic conductivity and thermal stability for lithium-ion battery operation. A 25 μm-thick microporous polypropylene membrane (Celgard® 2400) was used as the separator, cut into 1.90 cm discs to fit the dimensions of the coin cells.

2.2. Methods

2.2.1. X-Ray Diffraction (XRD)

The crystalline phase, structural characteristics, and purity of the synthesized powder were analyzed using a Bruker D8 X-ray diffractometer equipped with a CuKα radiation source (λ = 0.15406 nm, E = 8.04867 keV). The measurements were conducted in θ–2 θ geometry, with a scanning range from 5° to 80° (2 θ), a step size of 0.02°, and an angular sweep rate of 0.02°/s.

2.2.2. Specific Surface Area Determination Using BET Technic

The specific surface area (S_{BET}), pore volume, and pore size distribution (PSD) of the various samples were characterized using the Brunauer–Emmett–Teller (BET) method. These measurements were conducted with a Quantachrome AS1Win™ automated gas sorption analyzer, which utilizes acquisition and reduction software for data collection and analysis.

Prior to analysis, the LFP powder was degassed under dynamic vacuum ($\leq 10^{-3}$ mbar) at 300 °C for 3 H to eliminate adsorbed impurities. A mass of 791.50 mg was then loaded into the BSD-PM sample tube. The sample was cooled to 77.3 K (liquid nitrogen temperature), and nitrogen adsorption isotherms were measured as a function of relative pressure (P/P_0), with P_0 representing the initial pressure (1.0258 bars). BET surface area analysis was performed in the range of P/P_0 = 0.05–0.30. Subsequently, the pressure was increased to near saturation ($P/P_0 \approx 0.99$) and then gradually decreased to obtain the desorption branch for Barrett–Joyner–Halenda (BJH) mesopore analysis (pore sizes 2–50 nm). Finally, low-pressure adsorption data ($P/P_0 < 0.01$) were collected for Horváth–Kawazoe (HK) micropore characterization (<2 nm).

2.2.3. Pore Size Distribution Techniques

Several techniques are available in the literature for characterizing carbonaceous pore systems from BET measurement, including the BJH method [38], HK method [38], SF

(Saito-Foley) method [39], NLDFT (non-local density functional theory) method [40] and MP (Micropore) method [41]. These analytical models provide meaningful insights into the pore architecture of synthesized activated carbons [42]. The HK and SF methods are widely employed for quantifying micropore size distributions, whereas the BJH method is commonly used to assess meso-porosity. NLDFT offers a more comprehensive analysis, capable of simultaneously resolving micropores, mesopores, and macropores [40]. In the present study, micropores were characterized using the HK method, while mesopores and macropores were analyzed via NLDFT, a robust, multi-parametric technique for detailed pore size and pore distribution evaluation.

2.2.4. Surface Morphology Characterisation and Qualitative Chemical Composition Analysis

Scanning electron microscopy (SEM) was used to investigate the surface morphology and microstructure of activated carbon (WH and MC), including particle size and surface characteristics. Imaging was performed using a JEOL JSM-7600F thermal field-emission scanning electron microscope (FE-SEM) from JEOL North America, Peabody, MA, USA, operated at an acceleration voltage of 15–30 kV and a probe current ranging from 0.1 pA to 200 nA. The equipment includes an energy-dispersive X-ray spectroscopy (EDS) detector for elemental mapping and analyzing each component. EDS, a qualitative technique, was employed to identify and characterize the elementary constituents of the synthesized carbon samples [43–45].

2.2.5. Conductivity Measurement of the Activated Carbons

Electrical conductivity was measured using the four-point probe method using a S-302 Four Point Resistivity Probing Equipment from Lucas Labs, Gilroy, CA, USA. This test assesses the conductivity of the activated carbon, a critical property of this electrode material for battery applications.

2.2.6. Electrochemical Studies

The electrochemical performance evaluation was based on the utilization of the following various technics: galvanostatic charge–discharge tests, cycling stability measurements, and electrochemical impedance spectroscopy (EIS). These techniques were used to determine key performance metrics, including specific capacity, Coulombic efficiency, Peukert coefficient, cycling durability of the cathodes, and lithium-ion diffusion coefficients. The results were analyzed to assess the influence of activated carbon properties on cathode behavior and compared to those obtained using graphite, to evaluate the advantages of bio-based carbon materials.

Electrochemical testing of Li-ion coin cell with LFP/C was performed using a Princeton Applied Research 273A potentiostat, a Solartron CellTest 1470 multichannel potentiostat (Scribner, Southern Pines, NC, USA), and a Solartron 1255B frequency response analyzer (Scribner). The 273A potentiostat was employed for cyclic voltammetry analysis, while the Solartron 1470 multichannel potentiostat was used for galvanostatic charge–discharge cycling. The Solartron 1255B analyzer, interfaced with the 273A system, enabled the acquisition of EIS data.

3. Results and Discussion

3.1. Morphological and Structural Characterization of Pure $LiFePO_4$

Figure 1a displays the XRD pattern of the synthesized $LiFePO_4$ powder. For comparison, the reference diffraction pattern of pure $LiFePO_4$, based on standard card 01-090-1862, is shown in Figure 1b [46]. All diffraction peak positions in Figure 2 closely match those of the $LiFePO_4$ reference) [46], confirming the successful formation of the $LiFePO_4$ phase.

Moreover, the sharp and well-defined peaks indicate a high degree of crystallinity and support the presence of the stable triphylite phase. Well-ordered crystalline structures helps for optimizing ionic conductivity and facilitate lithium-ion transport by providing well-defined diffusion pathways, which is indispensable for efficient charge/discharge cycling.

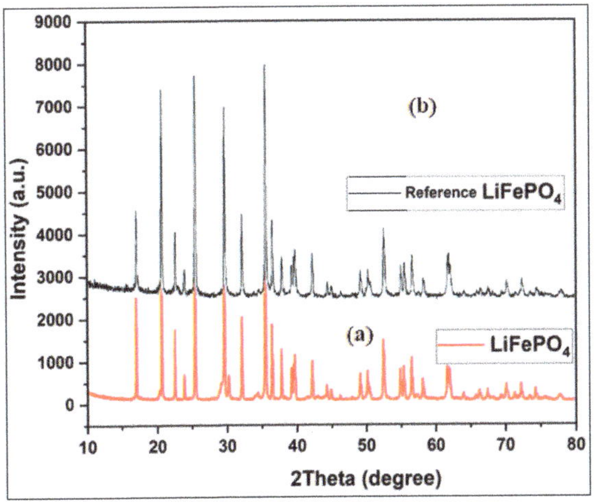

Figure 1. X-ray Diffractogram of synthesized $LiFePO_4$ (**a**) and the standard $LFPO_4$ (**b**) [46].

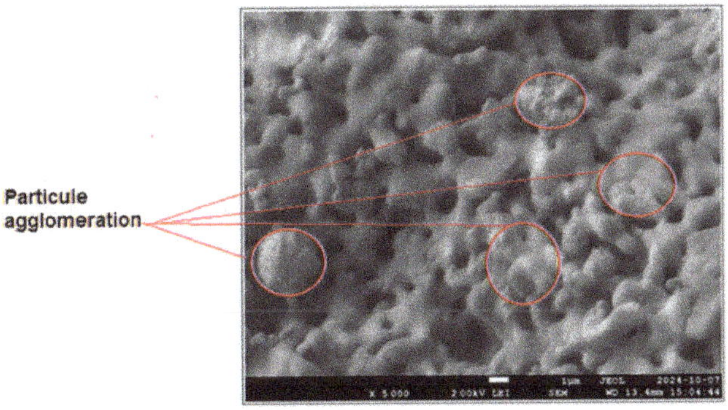

Figure 2. SEM Image of the synthesized $LiFePO_4$ Powder: ×5000.

SEM analysis (Figure 2) reveals that the $LiFePO_4$ particles exhibit relatively uniform morphology, ranging from spherical to ellipsoidal shapes, with particle sizes between 100 nm and 500 nm. However, SEM micrographs also show evidence of particle agglomeration (highlighted by red circles), which could negatively impact cathode performance by limiting electrical conductivity, increasing ion diffusion path lengths, and ultimately reducing overall electrochemical efficiency.

Figure 3 illustrates the nitrogen adsorption–desorption isotherms of $LiFePO_4$, showing the variation of the adsorbed volume (cm^3/g) as a function of relative pressure (P/P_0). The hysteresis loop corresponds to type H4 according to IUPAC classification, indicating the coexistence of micropores and mesopores in the material. The initial rise in nitrogen adsorption at low relative pressures suggests the presence of micropores (<2 nm), as these are typically filled first. This is followed by a more pronounced increase in adsorption at higher relative pressures, characteristic of mesoporous structures (2–50 nm). Thus, the isotherm supports the presence of both pore types, with mesopores likely predominating, as inferred from the overall shape of the curve. This porosity suggests a hierarchical pore structure that may enhance electrolyte accessibility and facilitate ion transport within the electrode material [47].

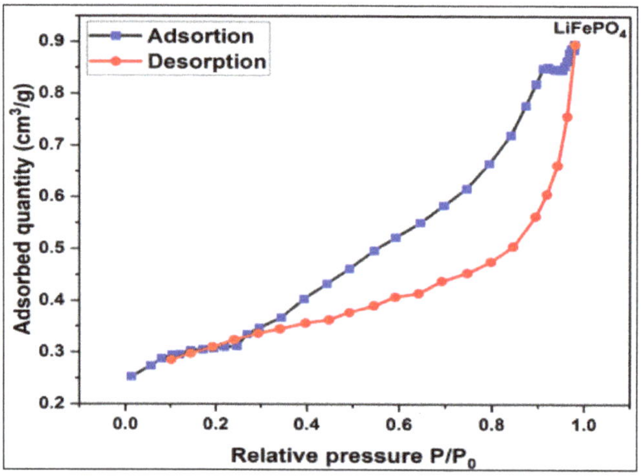

Figure 3. Variation of the nitrogen adsorption–desorption volume of $LiFePO_4$ (cm^3/g) as a function of the relative pressure P/Po.

The PSD, specifically for mesoporous structures, was evaluated using the Barrett–Joyner–Halenda (BJH) method, which is based on the Kelvin equation (Equation (1)) [48]. This model provides a quantitative framework for converting nitrogen adsorption data into PSD profiles. According to the BJH approach, the radius of the condensate with a hemispherical meniscus (r_i), or the curvature radius of the adsorbent surface (r_i), is expressed in nanometers and defined as follows:

$$r_i = -\frac{2\sigma V \cos\theta}{RT \ln\left(\frac{pi}{P_0}\right)} \quad (1)$$

where σ is the liquid surface tension of the adsorbate in N/cm, V is the molar volume of the liquid phase in ml/mol; θ is the contact angle between the meniscus and the solid wall in radians; R is the ideal gas constant (8.31 J/(mol·K); T is the temperature in Kelvin (K); P_i is the applied pressure in atm; P_0 is the reference pressure in atm; P_i/P_0 is the partial pressure when pressure P_i is applied to the sample.

From this relation, a set of algorithms is employed to estimate pore structure characteristics at each relative pressure P_i/P_0, based on nitrogen adsorption–desorption isotherms. These include the pore anisotropy factor b_i, defined as the ratio of pore length (Li) to pore diameter (D_i), that is, $b_i = L_i/D_i$. The method also enables the determination of pore surface area (Sp_i), pore volume (Vp_i), and pore radius (r_i) = Di/2 at each relative pressure. By plotting log (Sp_i3/Vpi_2) vs. log r_i, a linear relationship is obtained with slope $s_i = \alpha_i - 1$ where α_i is defined as the pore length parameter. This slope is further related to the pore length (l_i) via the power law relation: $l_i = kr_i\alpha_i - 1$, where k is a proportionality constant (in grams). Thus, α serves as a parameter for assessing pore elongation.

For pores narrower than 2 nm, the fluid–solid interaction becomes dominant, necessitating alternative models. In such cases, the pores are classified as micropores, and their filling occurs in the pre-capillary condensation region of the physisorption isotherm. The Horváth–Kawazoe (HK) method offers a specialized approach to determining micropore size distributions. Initially developed for slit-shaped carbonaceous pores, it has since been adapted to account for cylindrical geometries such as those found in silica-based materials. The HK model is founded on the principle that the relative pressure required to fill micropores of a specific shape and size is directly correlated with the adsorbate–adsorbent interaction energy [49].

Based on these energies of interaction, the following equation (Equation (2)) was developed to relate variations in partial pressure to pore dimensions and relevant physical parameters [49].

$$RT\ln\left(\frac{P}{P_0}\right) = K\frac{N_a A_a + N_A + A_A}{\sigma^4(l-d)} \times \left[\frac{\sigma^4}{3(l-d/2)^3} - \frac{\sigma^{10}}{9(l-d/2)^9} - \frac{\sigma^4}{3(d/2)^3} + \frac{\sigma^{10}}{9(d/2)^9}\right] \quad (2)$$

where: K is Avogadro's constant ($K = 6.02 \times 10^{23}$ atoms); σ is the distance between a gas atom and the surface at zero energy of interaction, l is the distance from the surface; N_a is the number of atoms per unit area of the surface, A_a is a constant; $d = d_a + d_A$, where d_a is the diameter of an adsorbent atom, and d_A is the diameter of the adsorbate molecule. The condition $l > d$ must be satisfied for the interaction model to remain valid.

A comprehensive analysis of $LiFePO_4$'s textural properties as a cathode material was conducted using BET surface area measurements in conjunction with BJH and HK pore structure analyses (Figures 4 and 5, Table 1). The resulting data detailing specific surface area, PDS, and pore diameter profiles offer key insights into the structure property relationships governing the material's suitability for lithium-ion battery applications.

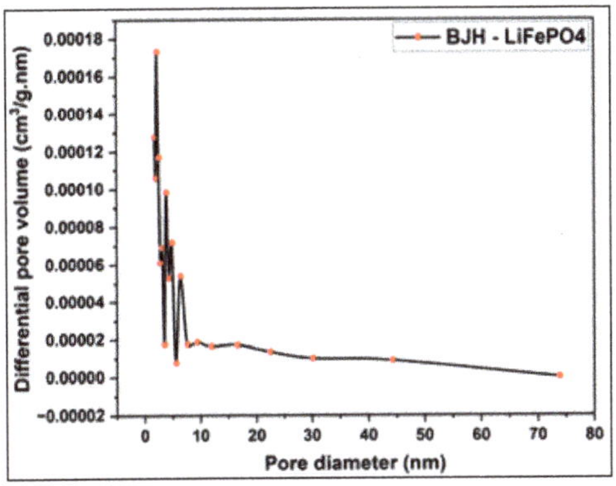

Figure 4. Variation of the differential pore volume ($cm^3/g{\cdot}nm$) vs. pore diameter of the pore size distribution of $LiFePO_4$ using BJH method.

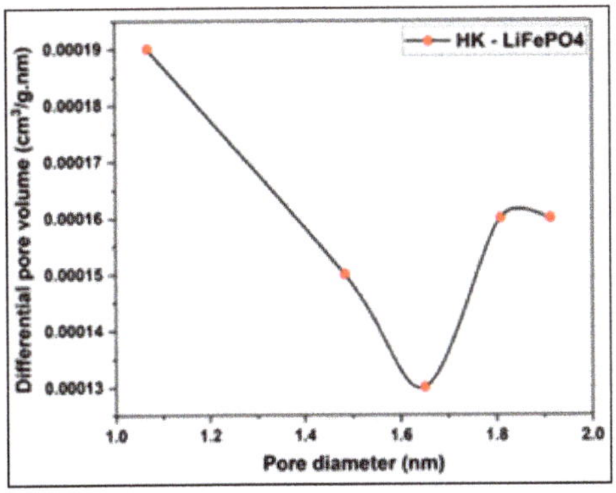

Figure 5. Variation of the differential pore volume ($cm^3/g{\cdot}nm$) vs. pore diameter of the pore size distribution of synthesized $LiFePO_4$ using HK method.

Table 1. Specific surface area and pore characteristics of $LiFePO_4$ powder.

Sample	BET Surface Area (m^2/g) (±2–5%)	Total Pore Volume (cm^3/g) (±2–5%)	Vmicropore (cm^3/g) (±2–5%)	Vmesopore (cm^3/g) (±2–5%)	Average Pore Diameter (nm) (±0.5 nm)
$LiFePO_4$	1.2	1.4×10^{-3}	5×10^{-4}	9×10^{-4}	4.7

The results indicate that the $LiFePO_4$ powder possesses a low BET surface area of 1.2 m^2/g and a limited total pore volume of 1.4×10^{-3} cm^3/g, with mesopores accounting for the majority (9×10^{-4} cm^3/g) and an average pore diameter of 4.7 nm above the 2 nm threshold that defines microporosity. This low pore volume suggests limited capacity for electrochemical reactions and restricted transport of lithium ions within the pristine $LiFePO_4$ structure. To mitigate these limitations, the incorporation of activated carbon as a conductive support is proposed. Activated carbon can significantly increase the surface area and porosity of the composite, thereby improving electrolyte–electrode contact and facilitating more efficient lithium-ion diffusion.

The pore size distribution derived using the BJH method, exhibits prominent peaks at approximately 2 nm, 4 nm, 6 nm, and 42 nm. This confirms that mesopores dominate the pore structure—an attribute favorable for enhancing lithium-ion diffusion and improving electrochemical performance.

3.2. Chemical and Physical Characterization of Activated Carbons at 700 °C from Millet Cob (MC) and Water Hyacinth (WH)

3.2.1. BET Surface, Pore Size and Pore Distribution Determination of Activated Carbons at 700 °C from MC and WH

Adsorption and Desorption Isotherms of Activated Carbons from MC and WH

Figure 6a,b present, respectively, the nitrogen adsorption/desorption isotherms for activated carbons derived from MC and WH. These curves distinctly reveal the characteristic pore structures of each material, reflecting differences in surface area, pore volume, and distribution resulting from their respective activation and precursor properties.

These isotherm diagrams vary depending on the KOH-to-biomass mass ratios used during activation (KOH/MC 1:1, 2:1, and 5:1). For both MC and WH–derived activated carbons, a decreasing trend in maximum nitrogen uptake is observed with increasing KOH content. This behavior may reflect changes in pore structure and surface chemistry induced by the higher activating agent concentrations.

The nitrogen adsorption/desorption isotherms of functionalized carbon samples from MC (KOH/MC 1:1, 2:1, and 5:1) and WH (KOH/WH 1:1, 2:1, and 5:1) exhibit type IV profiles with H4-type hysteresis loops, indicative of a dominant mesoporous structure [50]. The isotherm curves can be interpreted in three distinct regions:

Region I ($P/P_0 < 0.2$): The overlap of adsorption and desorption branches corresponds to monolayer adsorption, characteristic of micropore filling [51].

Region II ($0.2 < P/P_0 < 0.5$): The initial divergence between the branches suggests the presence of partially developed pores and progressive multilayer formation within the micropore–mesopore transition region [38,52].

Region III ($P/P_0 > 0.5$): A pronounced hysteresis loop appears, attributed to capillary condensation in open mesopores and larger macropores [38].

The presence of all three regions confirms a hierarchical pore system composed of micropores, mesopores, and macropores. This structural complexity is further supported by the pore size distribution data, which aligns with the trends observed in the isotherm profiles.

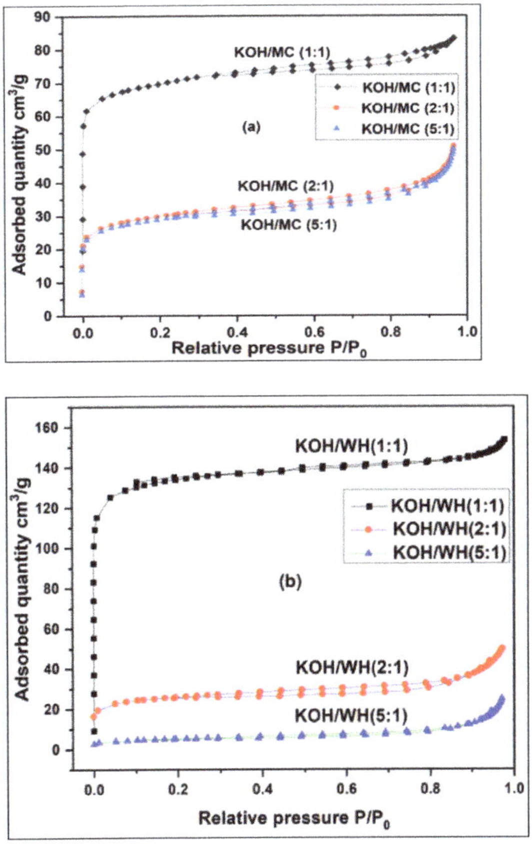

Figure 6. (**a**) Nitrogen adsorption /desorption isotherms of millet cob (MC) at the ratios KOH/MC of: KOH/MC= 1:1; KOH/MC =2:1; KOH/MC = 5:1; (**b**) Nitrogen adsorption/desorption isotherms of water hyacinth (WH) at the ratios KOH/WH of: KOH/WH= 1:1; KOH/WH =2:1; KOH/WH = 5:1.

Distribution of the Pore Dimensions of the Activated Carbons

Pore size distribution encompasses both micropores and mesopores, each contributing distinct functional advantages. Micropores offer a high specific surface area, thereby enhancing lithium-ion adsorption capacity, while mesopores facilitate more efficient lithium-ion diffusion within the cathode structure.

Case of the Water Hyacinth (WH) Sample

The micropore size distributions of WH–derived activated carbon samples, synthesized at different KOH/WH mass ratios, were estimated using the HK method and are shown in Figure 7a–c. Complementary analysis of mesopores and macropores was conducted using the NLDFT model, as presented in Figure 7d–f.

In Figure 7a, the HK method reveals a dominant micropore (D < 2 nm) peak centered at 0.58 nm. A comparable result is obtained in Figure 7d using the NLDFT method, which shows a peak at approximately 0.50 nm, confirming consistency between the two models. For the samples corresponding to Figure 7b,c (KOH/WH 2:1 and 5:1), broader and weaker peaks are observed near 1.8 nm, indicating a shift in pore size distribution with increased activation ratio. These findings align with the NLDFT profiles in Figure 7e,f, where peaks appear at 1.65 nm, suggesting the presence of wider micropores or narrow mesopores.

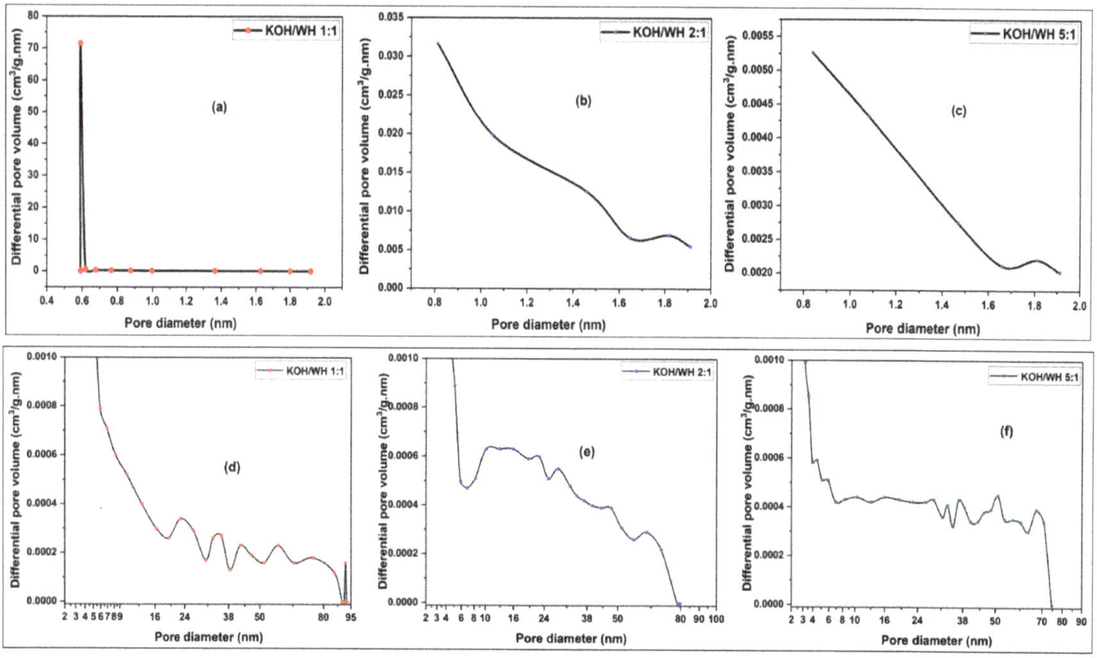

Figure 7. KOH/WH pore size distribution using HK method: (**a**) KOH/WH 1:1 (**b**) KOH/WH 2:1 and (**c**) KOH/WH 5:1 at 700 °C; KOH/WH pore size distribution using NLDFT method: (**d**) KOH/WH 1:1 (**e**) KOH/WH 2:1 and (**f**) KOH/WH 5:1 at 700 °C.

Notably, the micropore diameter for the KOH/WH 1:1 sample (0.58 nm) is at least three times smaller than those of the KOH/WH 2:1 and 5:1 sample (1.8 nm), highlighting the influence of activation intensity on pore development. This progressive shift in pore size distribution may directly impact on the electrochemical accessibility and ion transport behavior of the resulting carbon materials.

Figure 7d shows that the KOH/WH 1:1 sample exhibits mesopore peaks at 2.15, 5.1, 23, 35, and 42 nm, with additional macropore features at 58, 75, and 92 nm. In contrast, the KOH/WH 2:1 sample (Figure 7e) displays a broader mesopore distribution, with peaks at 2.15, 2.40, 2.83, 3.96, 10, 22, 28, and 47 nm, along with a single macropore peak at 63 nm. Similarly, the KOH/WH 5:1 sample (Figure 7f) reveals mesopores at 2.15, 2.55, 4.45, 5.82, 10.3, 16, 29, and 37 nm, accompanied by weak macropore ($D > 50$ nm) signals at 51 and 67 nm. Across all samples, mesopores are more prevalent than macropores. Notably, the 2:1 and 5:1 sample exhibit a greater number and broader distribution of mesopores compared to the 1:1 sample, suggesting the development of a more complex and well-interconnected mesoporous network.

This richer mesopore architecture is expected to enhance electrolyte accessibility, as a broader range of mesopore sizes, within an optimal window facilitates more efficient lithium-ion diffusion in Li-ion battery systems. Furthermore, the improved mesopore distribution may promote better surface utilization, thereby increasing charge storage capacity. The presence of interconnected pores of varying sizes can also reduce ionic transport resistance by optimizing electrolyte penetration and ion flow pathways. Conversely, the KOH/WH 1:1 sample, characterized by fewer but larger mesopores, may favor bulk ion transport yet offer lower active surface area. The weak macropore signatures observed in all cases suggest that macropores contribute minimally to the electrochemical behavior of these materials.

Case of the Millet cob (MC) sample

Figure 8a displays a prominent micropore peak centered at approximately 0.64 nm. A comparable result is observed in the NLDFT model (Figure 8d), where the micropore peak is located around 0.51 nm. Similarly, the HK model results shown in Figure 8b,c also exhibit dominant peaks near 0.64 nm, while the corresponding NLDFT profiles in Figure 8e,f confirm micropore sizes of approximately 0.5 nm. Unlike the WH–based samples, the micropore size of the MC–derived activated carbon remains largely unaffected by variations in the KOH/MC activation ratio. This indicates that, for this biomass precursor, the micropore dimensions are more dependent on the intrinsic structural characteristics of the raw material than on the activation intensity.

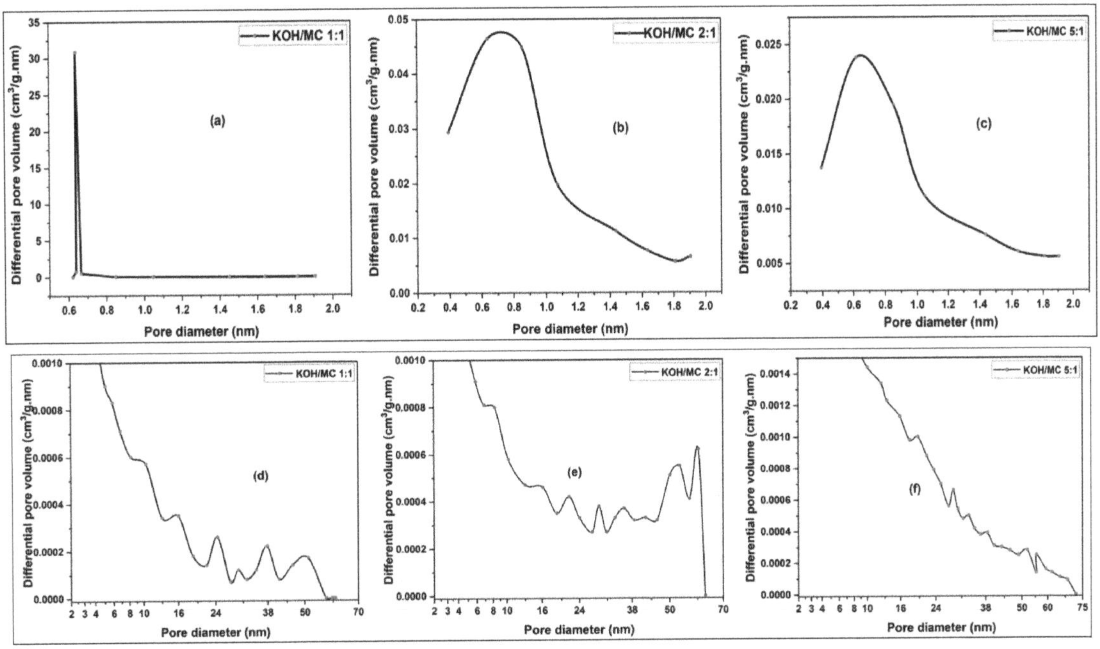

Figure 8. KOH/MC pore size distribution using HK method: (**a**) KOH/MC 1:1 (**b**) KOH/MC 2:1 and (**c**) KOH/MC 5:1 at 700 °C; KOH/MC pore size distribution using NLDFT method: (**d**) KOH/MC 1:1 (**e**) KOH/MC 2:1 and (**f**) KOH/MC 5.1 at 700 °C.

The pore structure of activated carbon derived from MC varies with the KOH/MC mass ratio. For the KOH/MC 1:1 sample (Figure 8d), mesopore peaks are observed at 2.28, 3.96, 16, 24.5, and 38 nm, along with a weak macropore feature at 52 nm. The KOH/MC 2:1 sample (Figure 8e) exhibits mesopores at 2.3, 3.94, 16.2, 21.7, 28.88, 35.72, and 42 nm, accompanied by minor macropore signals at 53.6 and 60.4 nm. In the KOH/MC 5:1 sample (Figure 8f, NLDFT model), mesopore (2 nm < D < 50 nm) peaks occur at 2.40, 4.45, 19.75, 28.67, and 38.52 nm, with additional macropores detected at 52 and 55 nm.

The micropore size distribution remains largely consistent across all KOH/MC ratios, suggesting that micropore formation is relatively insensitive to activation intensity in MC-derived carbons. In contrast, mesopore characteristics are clearly influenced by the KOH ratio, with higher ratios generating broader and more complex mesopore distributions. Macropore size, however, remains relatively unchanged. This behavior contrasts with that of WH–derived samples, where micropore, mesopore, and macropore features all vary significantly with activation ratio.

BET Surface Area of Functionalized Carbons from MC and at WH

Table 2 summarizes the key textural properties of all activated carbon samples, including BET surface area, total pore volume, micropore volume, mesopore volume, and average pore diameter. For both WH and MC–derived carbons, an increase in the KOH activation ratio leads to two consistent trends: (1) a reduction in BET surface area and total pore volume, and (2) an increase in average pore diameter. This shift toward larger pore sizes is primarily attributed to a decrease in micropore volume, indicating that more intensive chemical activation promotes pore expansion at the expense of microporosity.

Table 2. Surface parameters of activated carbons from MC and WH at 700 °C.

Sample	BET Surface Area (m^2/g) (±2–5%)	Total Pore Volume (10^{-3} cm^3/g) (±2–5%)	Vmicropore (10^{-3} cm^3/g) (±2–5%)	V mesopore (10^{-3} cm^3/g) (±2–5%)	Average Pore Diameter (nm) (±0.5 nm)
KOH/WH 1:1	413.0	237.6	203.1	34.3	2.3
KOH/WH 2:1	83.2	77.4	39.1	38.2	3.7
KOH/WH 5:1	18.53	38.9	7.5	12.9	8.4
KOH/MC 1:1	216.3	128.6	104.3	24.3	2.4
KOH/MC 2:1	95.8	78.7	46.9	31.1	3.3
KOH/MC 5:1	57.8	76.3	28.0	48.0	5.3
Graphite	45.1	75	21	53	6.6

The observed reduction in BET surface area for both MC and WH–derived activated carbon samples at higher KOH activation ratios can be attributed to two primary mechanisms. First, excessive KOH concentrations may promote excessive pore widening during chemical activation, resulting in a decline in micropore density and, consequently, a lower overall surface area. Second, the formation and retention of residual KOH or potassium-containing byproducts may lead to partial pore blockage, further limiting surface accessibility. These structural alterations, particularly the loss of microporosity and potential occlusion of diffusion pathway, may adversely affect the electrochemical performance of Li-ion batteries by reducing lithium-ion accessibility.

3.2.2. SEM Characterization of Activated Carbons at 460 °C and 700 °C from Millet Cob and Water Hyacinth

SEM analysis of samples carbonized at 460 °C (KOH/MC and KOH/WH 2:1) (Figure 9) revealed irregular, blocky morphologies with large pore sizes of approximately 2600 nm and 1600 nm, and no evidence of mesopores or micropores. These structures, indicative of incomplete carbonization, are further corroborated by the low carbon content (~30 wt%) measured in these samples (see Section 3.2.3). Due to their poorly developed porosity and limited carbon content, these materials are not suitable for use as electrode components in Li-ion battery systems.

In contrast, SEM micrographs of activated carbons synthesized at 700 °C (Figure 10) exhibit well-developed, homogeneous porous morphologies across all KOH/MC and KOH/WH mass ratios. The elevated carbonization temperature promotes complete thermal decomposition and efficient chemical activation by KOH, leading to structurally robust carbon matrices with high carbon content (≥85 wt%; see Section 3.2.3). At this temperature, the evolution of volatile species and enhanced etching reactions facilitate the generation of interconnected pore networks unique to each activation condition.

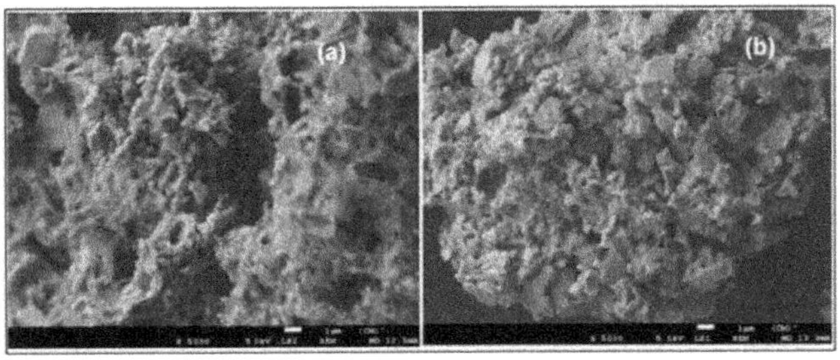

Figure 9. SEM images of millet cob activated carbon at 460 °C: (**a**) KOH/MC 2:1; (**b**) KOH/WH 2:1.

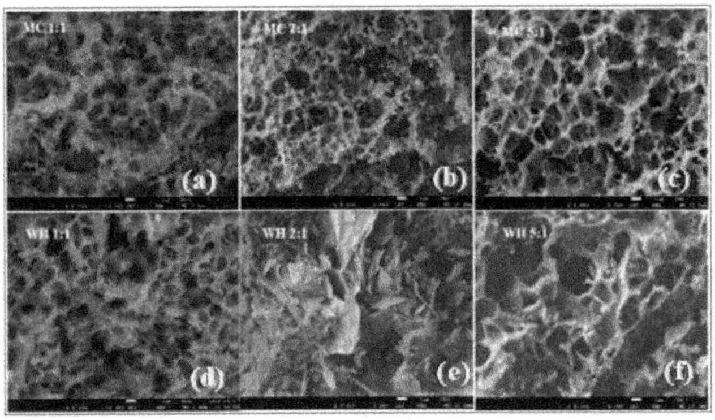

Figure 10. SEM images of various activated carbon at 700 °C: (**a**) KOH/MC 1:1; (**b**) KOH/MC 2:1; (**c**) KOH/MC 5:1; (**d**) KOH/WH 1:1; (**e**) KOH/WH 2:1; (**f**) KOH/WH 5:1.

Samples prepared with lower KOH concentrations (KOH/MC and KOH/WH 1:1) display a high density of small, uniform pores. This is attributed to the moderate reactivity of KOH at this ratio, which limits excessive carbon wall degradation while favoring the development of fine porosity. As illustrated in Figure 10a,d, these pore structures are advantageous for applications requiring high surface area, particularly in energy storage systems where micropore presence plays an important role in enhancing adsorption capacity and lithium-ion storage performance.

Samples activated with higher KOH ratios (KOH/MC and KOH/WH 2:1 and 5:1) exhibit increased pore sizes compared to those prepared at a KOH 1:1 ratio. Notably, the KOH/WH 2:1 sample reveals a graphite-like layered morphology, in which the interlayer spacing forms a porous network that may facilitate lithium-ion diffusion. For both biomass at the KOH ratio 5:1, the pore structures (Figure 10c,f) appear more open and less compact, suggesting that excessive KOH concentrations can lead to partial degradation of the carbon framework—enlarging pore size while diminishing pore density and potentially reducing the specific surface area.

While higher KOH content enhances chemical activation by promoting the removal of volatile matter and increasing porosity, the 2:1 ratio appears to strike a balance between pore development and structural integrity—making it the most favorable condition for Li-ion battery applications among the studied ratios. In contrast, further increasing the KOH ratio to 5:1 induces structural collapse, resulting in fewer, oversized pores that may hinder electrochemical performance.

These observations underscore the necessity of optimizing both the KOH/carbon ratio and carbonization temperature (with 700 °C proving effective) to tailor the porosity of biomass-derived carbons for energy storage. The KOH/WH 2:1 sample carbonized at

700 °C demonstrates a particularly promising pore architecture, aligning well with the structural and functional demands of lithium-ion battery electrodes.

3.2.3. Elemental Composition of Porous Carbons from MC and WH Using EDS

Table 3 presents the EDS results for samples activated at 460 °C with a KOH/carbon 2:1 ratio. The carbon content in both samples does not exceed 30 wt%, indicating incomplete carbonization of the organic precursors. Additionally, the EDS profiles reveal higher relative concentrations of oxygen and potassium compared to carbon, further supporting the presence of residual activating agents and unconverted biomass. These findings confirm insufficient thermal decomposition at this activation temperature.

Table 3. Chemical composition (%mass with an error range of 0.1–0.5%) using EDS of activated carbon from MC and WH at 460 °C.

Samples	KOH/CM	C (%)	O (%)	K (%)	Na (%)	Cl (%)	Al (%)	Si (%)	Mg (%)	P (%)
MC 460 °C	2:1	30	33	35	-	-	0.2	1.7	0.1	-
WH 460 °C	2:1	26	41	30	0.3	1.4	0.1	0.4	0.3	0.1

Table 4 summarizes the EDS results of activated carbon samples derived from various KOH/carbon mass ratios, all carbonized at 700 °C. Compared to those carbonized at 460 °C, these samples consistently show carbon contents exceeding 70 wt%, indicating successful carbonization at elevated temperatures. The data also reveal a systematic decrease in carbon content with increasing KOH ratios. For MC-derived samples, the carbon content decreases from 85.5% at KOH/MC 1:1 to 71% both KOH/MC 2:1 and 5:1. Similarly, for water hyacinth-based samples, carbon content declines from 80% (KOH/WH 1:1) to 70% (KOH/WH 2:1 or 5:1).

Table 4. Chemical Composition using EDS Activated Carbon from MC and WH at 700 °C.

Samples	KOH/C	Chemical Elements (% Mass with an Error Range of 0.1–0.5%)											
		C	O	K	Na	Cl	Al	Si	Fe	Ca	Mg	S	P
Millet cob	1:1	85.5	12	2	-	0.3	-	0.2	-	-			
	2:1	71.5	19	6	0.1	0.8	0.8	1.1	0.1	0.1	0.1	0.4	-
	5:1	71.0	20	8.3	-	0.1	0.2	0.3	-	-	-	0.1	-
Water hyacinth	1:1	80.0	13	2	-	2	0.2	0.5	0.4	0.8	0.1	0.9	0.1
	2:1	70.0	18	8	0.1	1.8	0.2	0.7	0.1	0.6	0.2	0.2	0.1
	5:1	70.0	18	8	0.1	2.4	0.1	0.4	0.2	0.2	0.1	0.5	-

This reduction in carbon content is attributed to intensified chemical activation at higher KOH concentrations, which promotes gasification reactions that convert carbon into CO_2 or potassium-based compounds such as K_2CO_3. The resulting loss of carbon during activation leads to lower carbon yields in the final product. Additionally, the increasing levels of oxygen and potassium detected in samples with higher KOH/carbon ratios reflect the retention of potassium species—either as residual unreacted KOH or as activation byproducts such as potassium carbonates.

The observed increase in oxygen content from 12% to 20% for MC and from 13% to 18% for WH indicates an accumulation of oxygenated functional groups and oxygen-containing inorganic residues (e.g., carbonates). Similarly, the potassium content rises significantly, from 2% to 8.3% for MC and from 2% to 8% for WH, likely due to the formation of potassium-based compounds such as K_2CO_3 and K_2O. These species may become embedded within the porous carbon matrix, particularly at elevated KOH concentrations.

Increasing the KOH/MC or KOH/WH mass ratio intensifies the activation process, promoting the development of larger pores and a more open structure. However, excessive KOH can also trigger gasification reactions that convert a portion of the carbon into gaseous by-products, thereby reducing carbon yield and introducing additional potassium residues. The concurrent rise in oxygen content suggests the persistence of residual oxygen-containing functional groups, further reflecting incomplete carbonization or the retention of oxidized species.

3.2.4. X-Ray Diffraction Studies of Porous Carbons from Millet Cob and Water Hyacinth

Figure 11 presents the XRD patterns of activated carbon samples prepared from millet cob (KOH/MC 2:1) and water hyacinth (KOH/WH 2:1) at an activation temperature of 460 °C. Both patterns exhibit distinct crystalline reflections, suggesting the presence of ordered structural components. However, the observed peaks do not correspond to that characteristic of typical carbonaceous phases such as graphite or crystallin/amorphous carbon. Instead, the diffraction features are likely associated with residual inorganic species originating from the activating agent (KOH) or undecomposed mineral constituents intrinsic to the biomass precursors. These results indicate that at 460 °C, the activation temperature is insufficient to achieve complete carbonization and removal of non-carbonaceous phases.

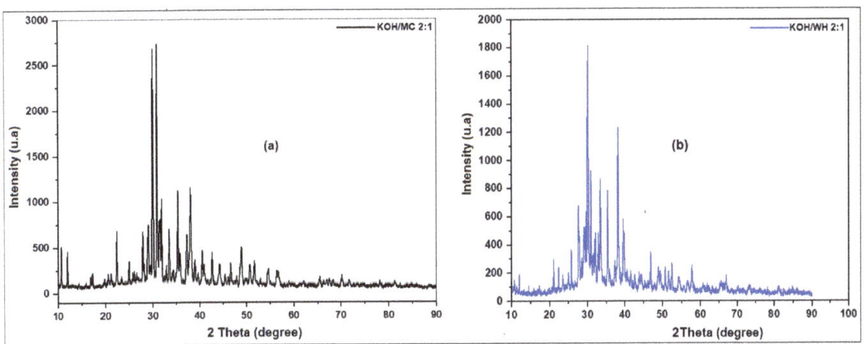

Figure 11. XRD patterns of activated carbon samples prepared from millet cob (KOH/MC 2:1) (**a**) and water hyacinth (KOH/WH 2:1) (**b**) at an activation temperature of 460 °C. Both patterns exhibit distinct crystalline reflections, suggesting the presence of ordered structural components.

The very low intensity of diffraction features attributable to carbonaceous phases suggest that the more prominent sharp peaks in the XRD patterns may arise from crystalline inorganic contaminants. This interpretation is supported by the EDS data presented in Table 3, which indicate the presence of silicon in the activated carbon—1.7 wt% for MC–derived samples and 0.4 wt% for WH–derived samples. The absence of well-defined carbon peaks further implies that carbonization and chemical activation were incomplete at 460 °C. This insufficient thermal treatment likely resulted in the retention of residual inorganic species originating from the precursor biomass and the activating agent (KOH), including metal oxides, potassium salts, or other processing-related impurities [53]. The low carbon content detected via EDS in these samples further corroborates the predominance of non-carbonaceous components.

Figure 12 presents the XRD patterns of activated carbon samples synthesized at 700 °C for KOH/carbon mass ratios of 1:1, 2:1, and 5:1, alongside the reference pattern for pure graphite (PDF 00-056-0159). Activation at this elevated temperature enhances the decomposition of organic constituents and facilitates the development of a more ordered and graphitized carbon matrix, as evidenced by the emergence of broader yet distinguishable carbon diffraction peaks.

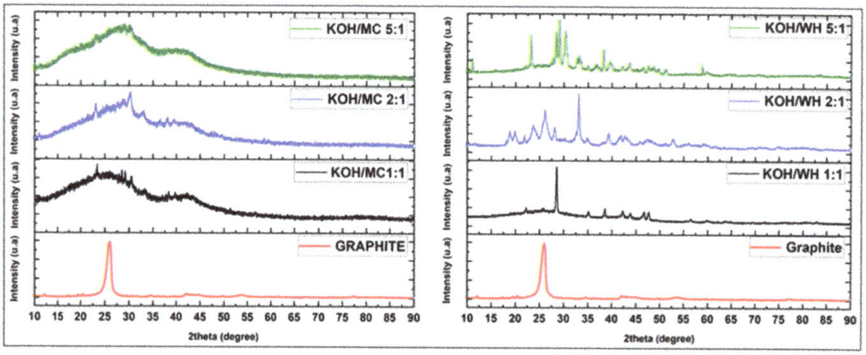

Figure 12. XRD patterns of activated carbon at 700 °C from Millet Cob (MC): KOH/MC 1:1, KOH/MC 2:1, KOH/MC 5:1, pure graphite and from Water Hyacinth: KOH/WH 1:1, KOH/WH 2:1, KOH/WH 5:1, pure graphite (this work).

The XRD patterns of KOH-activated carbon samples derived from MC (KOH/MC 1:1, 2:1, and 5:1) exhibit two broad diffraction features: a more intense peak centered around $2\theta = 26°$ and a weaker one at $2\theta = 43°$. The broadness of these peaks and the absence of sharp reflections confirm the predominantly amorphous nature of the carbonaceous framework, which is characteristic of chemically activated carbon materials [54–57]. The peak at $2\theta = 26°$ is attributed to the (002) diffraction plane, while the signal at $2\theta = 43°$ corresponds to the (100) plane, both typically associated with graphitic carbon structures [54,56,58–60]. The XRD patters of pure graphite in this work (Figure 12) are in agreement of the results obtained elsewhere on this material [61]. EDS analysis supports this structural characterization, revealing decreasing carbon contents of 85.5 wt% (1:1), 71.5 wt% (2:1), and 71 wt% (5:1). However, for the 1:1 and 2:1 sample, the presence of weak sharp peaks superimposed on the broad humps suggests a minor crystalline phase embedded within the largely amorphous matrix, potentially arising from residual inorganic species or structural impurities.

A comparative analysis of the three MC-derived samples shows a gradual decrease in crystallinity with an increasing KOH ratio, indicating a transition toward a more disordered structure. Quantitative crystallinity indices reveal a decline from 12.7% (KOH/MC 1:1) to 11.9% (KOH/MC 2:1), and down to 8.6% (KOH/MC 5:1), underscoring the impact of elevated KOH concentrations disrupting graphitic domains. This degradation is attributed to intensified chemical etching by KOH, which generates porosity at the expense of crystalline order by disintegrating graphitic planes.

In contrast, XRD patterns of WH–derived activated carbon (Figure 12) exhibit features of more crystalline carbon across all mass ratios (KOH/WH 1:1, 2:1, and 5:1). Each sample displays a sharp peak at $2\theta = 26°$, corresponding to the (002) plane of graphitic carbon as indexed in PDF reference 00-056-0159, alongside a second sharp peak at $2\theta = 43°$, attributed to the (111) plane of carbon (PDF 00-060-0053).

These patterns reveal a strong dependency of crystallinity on the KOH/WH ratio. The intensity of both graphite-associated peaks increases from KOH/WH 1:1 to 2:1, with a slight decline observed at 5:1. Quantitative analysis of crystallinity confirms this trend, with values rising from 40.5% (1:1) to 51.8% (2:1), before decreasing slightly to 50.6% (5:1). These results suggest that while moderate KOH concentrations enhance graphitic domain development, excessive activation may disrupt structural order.

Additional sharp peaks observed in the XRD patterns for all KOH/WH samples are attributed to residual inorganic elements—such as potassium (K), sodium (Na), iron (Fe), and chlorine (Cl)—which are known as byproducts of incomplete rinsing or activation residue [53]. EDS analysis corroborates their presence, indicating that these extraneous species influence diffraction behavior and reflect the imperfect removal of chemical byproducts during the post-activation washing process [53].

3.2.5. Electric Conductivity of Activated Carbons from MC and WH

The fabrication of carbon platelets from powdered activated carbon using a hydraulic press reveals that electrical conductivity is sensitive to the applied pressure during compression [62]. Additionally, morphological characteristics and pore distribution are known to significantly influence electrical conductivity in porous carbons [63,64]. To ensure consistency and eliminate pressure-induced variability, all platelets were produced under identical conditions: a uniaxial pressure of 600 kPa and constant sample mass for each type of activated carbon.

Electrical conductivity measurements indicate that samples derived from MC and WH, activated at 460 °C, exhibit negligible conductivity. This result is consistent with insufficient carbonization at this lower activation temperature. In contrast, samples activated at 700 °C demonstrate markedly improved conductivity, as presented in Table 5.

Table 5. Evolution of electrical conductivity as a function of the mixing ratio of potassium hydroxide and activated carbon materials KOH/CM: CM= Carbon Materials.

Sample	Mixing Ration (KOH/CM)	Electrical Conductivity (10^{-3} S.cm^{-1}) (±0.5–1.5%)	Total Pore Volume (10^{-3} cm^3/g) (±2–5%)	Average Pore Diameter (nm) (±0.5)
Graphite	Reference	672	75	6.6
Millet cob	1:1	6.74	237.60	2.30
	2:1	8.42	77.40	3.72
	5:1	4.99	38.90	8.38
Water hyacinth	1:1	8.92	128.60	2.38
	2:1	14.70	78.70	3.29
	5:1	6.40	76.30	5.28

The electrical conductivity of both MC and WH–derived activated carbon samples vary as a function of the KOH/carbon mass ratio. The highest conductivity for each biomass type is observed at a KOH/CM 2:1 ratio, with the KOH/WH 2:1 sample exhibiting a significantly higher conductivity value (1.47×10^{-2} S·cm^{-1}) compared to KOH/MC 2:1 (8.42×10^{-3} S·cm^{-1}). Additionally, the samples prepared with a 1:1 mass ratio (KOH/MC 1:1 and KOH/WH 1:1) show higher conductivity than their 5:1 counterpart. These results suggest that electrical conductivity is maximized in samples with moderate total pore volumes, and equilibrium in micro/mesopore as reported in Table 5 for the KOH/WH 2:1 sample. Conversely, the formation of overly large pores in the KOH/WH 5:1 sample appears to impede conductivity, likely due to a reduction in the density of effective conductive pathways within the carbon matrix. This is attributed to the need of electrons to travel through carbon materials with large pores on longer distances which increases the electric resistance. Distances. In comparison, moderate total pore volume creates small pathways which allow fast electrons transfer, leading to higher conductivity.

The commercial graphite exhibited more high conductivity than the activated bioresources carbons. This might be due to differences in crystal structures (graphite is more crystal-like structure than activated bio-resources carbons and/or the difference in surface treatment. The commercial graphite surface was already treated by oxidation and by fluorination to improve its electrochemical performance [65]. These findings indicate that specific surface area alone may not be the most reliable probe of the variation in the electrical conductivity. As reported in [66], electrical resistivity may increases with the surface area. But excessive pore enlargement—arising from the transformation of micropores into meso- and macropores—can diminish conductivity by disrupting the continuity of the conductive network [66,67]. The optimised pore size between 3 and 4 nm may help in this improvement of the conductivity Accordingly, the reduced conductivity observed in the highly activated KOH/MC 5:1 and KOH/WH 5:1 sample could be attributed to the decreases of carbon content, nonconductive activation by-products and /or the increased degree of activation.

Overall, WH-derived samples consistently outperform MC-derived ones in electrical conductivity. As illustrated in Figure 12, this behavior is likely linked to the higher crystallinity of WH-based carbons, as evidenced by their more pronounced graphitic peaks in the XRD patterns. This observation is consistent with the findings of [68], which report enhanced electrical performance in carbonaceous materials possessing well-ordered graphite-like structures. Finally, KOH/CM 2:1 appears optimal for both precursors, striking a balance between pore development and retention of conductive domains, and thereby enabling superior electrical conductivity.

3.3. Electrochemical Characterization

3.3.1. Specific Capacity of Anodes Based on Synthesized Activated Carbon

Figure 13 presents the charge/discharge profiles of the anodes based on MC and WH activated and Figure 14 displays the corresponding electrochemical curves of a commercial graphite (reference) electrode under identical conditions. The intersection points between the charge and discharge curves were used to determine the specific capacity (mAh/g) of each sample, as summarized in Table 6.

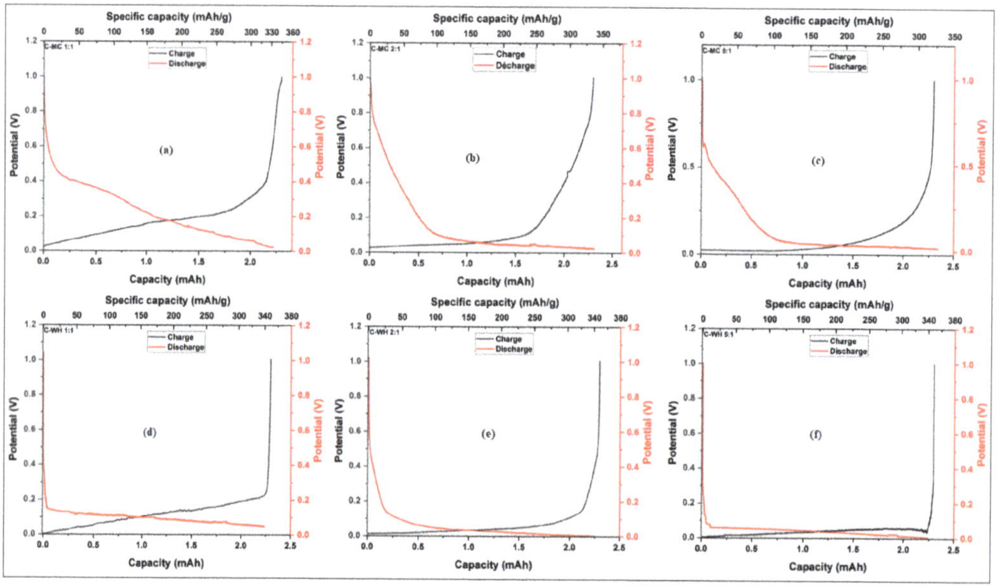

Figure 13. Charge/discharge at C/24, of Li-ion anodes based on KOH/MC 1:1 (**a**), 2:1 (**b**). 5:1 (**c**) and KOH/WH 1:1 (**d**), 2:1 (**e**), 5:1 (**f**).

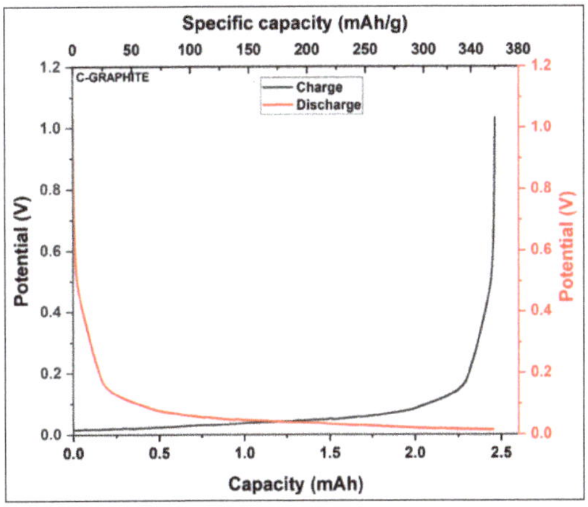

Figure 14. Charge/discharge (Li-ion anode based on commercial Graphite) at C/24.

Table 6. Respective specific capacity (mAh/g), anodes based on activated carbons based on Millet Cob (MC) and Water Hyacinth (WH) obtained from different ratios of KOH/CM and KOH/WH for a current rate of C/24.

	Mixing Ratio (KOH/MCor KOH/WH) (±2 mAh/g)	Mixing Ratio (KOH/MC or KOH/WH) (±2 mAh/g)	Mixing Ratio (KOH/MC or KOH/WH) (±2 mAh/g)
Ok. Fine. No bolt also	1.1	2:1	5:1
Carbon of Millet cob	333	335	330
Water hyacinth	336	339	332

For comparison, the specific capacity obtained for commercial graphite based anode under C/24 cycling conditions, as shown in Figure 14, is 362 mAh·g^{-1}, which closely aligns with the widely reported theoretical literature value of 372 mAh/g corresponding to the LiC_6 anode [69]. This high value of the specific capacity of graphite is agreement with his high conductivity, optimized total pore volume and the diameter size. The specific capacities of the various activated carbon samples vary only slightly, ranging from a minimum of 330 mAh·g^{-1} for the KOH/MC 5:1 sample (Figure 15c) to a maximum of 339 mAh·g^{-1} for the KOH/WH 2:1 sample (Figure 13e). No substantial difference in specific capacity is observed between samples derived from millet cob and water hyacinth, or across different activation ratios.

According to the data presented in Table 6, samples prepared with KOH/carbon mass ratios of 1:1 and 5:1 exhibit marginally lower capacities, which is likely attributable to differences in pore architecture. Although micropores contribute significantly to the overall specific surface area, an excess of ultra-micropores (<2 nm) can hinder lithium-ion penetration, imposing diffusion limitations. In contrast, a reasonable higher proportion of mesopores facilitates improved ion accessibility, enabling more efficient lithium-ion transport and deeper intercalation into the carbon structure.

The determination of the specific capacity of anodes based on new active carbons is important for carbon-based Li-ions batteries development because it is the limiting factor of the anode specific capacity.

The development of additive carbons with optimized parameters may play an important role in obtaining the best electrochemical performance of electrodes based on $LiFePO_4$/biomass-derived activated carbons.

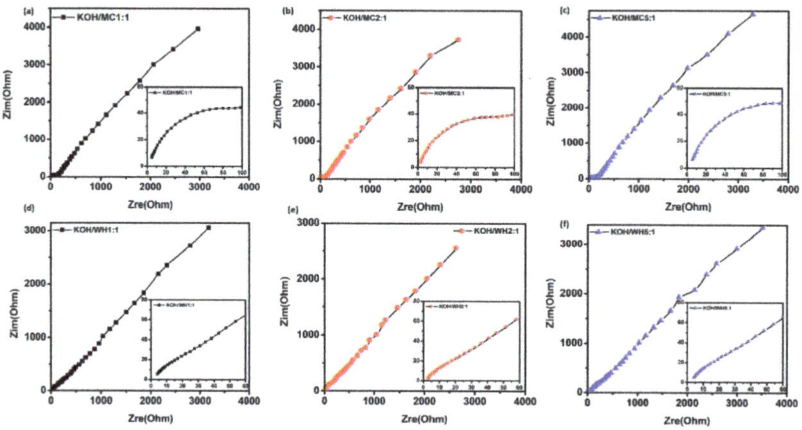

Figure 15. Nyquist diagram of EIS of anodes based on MC (**a–c**) and WH (**d–f**) carbon.

3.3.2. EIS Study of Synthesized Activated Carbon

To assess the battery characteristics based on the synthesized activated carbons, EIS measurements were conducted using Li/C half-cells. In this configuration, the cathode consisted of activated carbon coated onto an aluminum current collector, the anode comprised lithium metal, and the electrolyte was a solution of lithium hexafluorophosphate ($LiPF_6$). Impedance spectra were recorded over a frequency range from 100 kHz to 0.1 Hz with a root mean square (RMS) amplitude of 5 mV. Measurements were benchmarked against those obtained from a standard Li/graphite half-cell. A Solartron 1255B frequency response analyzer coupled with a 273A potentiostat was employed for EIS acquisition.

The resulting Nyquist plots, along with their equivalent circuit models generated using ZSimpWin software, version 3.6, are presented in Figure 15 for activated carbon samples derived from MC (a, b, c) and WH (d, e, f). Activated carbon samples of KOH/CM 2:1 ratio, derived from both MC and WH, exhibit the most favorable electrochemical parameters, characterized by significantly lower R_{ct} compared to other samples. Samples synthesized at a KOH/CM 1:1 ratio rank next in performance, while those prepared at a KOH/CM 5:1 ratio show the highest R_{ct} values, indicating less favorable electrochemical behavior.

These trends are consistent with prior findings, where a combination of moderately enhanced crystallinity, optimized micro- and mesoporous structure, and elevated electrical conductivity synergistically contribute to improved impedance characteristics. The KOH/CM 2:1 sample, in particular, benefits from a balanced pore architecture and preserved graphitic domains, which collectively facilitate efficient charge transport and reduced interfacial resistance.

3.3.3. Analysis of the Electrochemical Characteristics of $LiFePO_4$/C Electrode

Study of Coulombic Efficiency

Coulombic efficiency, defined as the ratio of discharge capacity to charge capacity, is a metric for assessing electrochemical reversibility. Table 7 summarizes the Coulombic efficiency values for $LiFePO_4$/C half-cells. The LFP/MC 1:1, 2:1, and 5:1 configuration exhibit first-cycle. Similarly, the LFP/WH 1:1, 2:1, and 5:1 cell achieves efficiencies near 96%, with the LFP/WH 2:1 sample presenting slightly higher values compared to the others in the MC series.

Table 7. Coulombic efficiency in % of the first three cycles of the $LiFePO_4/C$ based cell.

Samples	Coulombic Efficiency (%) ± 1%			
	Cycle 1	Cycle 2	Cycle 3	Cycle 4
LFP/MC 1:1	95.0	100	99.8	99.6
LFP/MC 2:1	95.9	100	99.8	99.6
LFP/MC 5:1	95.5	100	99.7	99.5
LFP/WH 1:1	96.0	100	99.9	99.7
LFP/WH 2:1	96.4	100	100	99.9
LFP/WH 5:1	95.9	100	99.8	99.6

The improvement in Coulombic efficiency observed during the second cycle is indicative of the formation of a solid electrolyte interphase (SEI) at the electrode/electrolyte interface. This nanometric layer forms on the surface of the anode during the initial charge/discharge cycles of Li-ion batteries and results from the reaction between lithium ions and electrolyte components [70]. During the first charge cycle, a portion of lithium ions is irreversibly consumed to form this passivating layer, which explains why the initial capacity is not representative of the battery's true reversible capacity in subsequent cycles [71]. To investigate SEI formation and evaluate Coulombic efficiency, the cells were cycled at a C/12 rate for three cycles. A high Coulombic efficiency reflects minimal parasitic side reactions and efficient lithium-ion cycling [72]. Once the SEI stabilizes, subsequent cycles demonstrate improved charge retention, highlighting enhanced electrode–electrolyte compatibility and reduced side reactions [72,73]. While the first-cycle SEI formation dominates, long-term growth contributes to additional lithium and electrolyte consumption, ultimately degrading cell conductivity, capacity, and Coulombic efficiency over time [73,74]. This might be the reason why after cycle 2, the current efficiency is going slightly down as shown for cycle 3 and cycle 4 (Table 7).

Study of the Discharge of $LiFePO_4$ (LFP)/C at Different Current Rates

Using Lithium metal as the anode, the charge/discharge performance of LFP/C (MC, WH) cathodes was conducted at various current rates C/12, C/6, 1C, and 5C to assess the rate capability and overall electrochemical behavior of the composite electrodes under different operating conditions. The constant cycling current at each applied rate is calculated based on the active mass of the cathode, the C-rate current value, and the theoretical specific capacity of LFP (170 mAh/g).

The corresponding discharge profiles, presented in the respective figures, illustrate the reversibility and rate-dependent behavior of the electrodes. Figures 16 and 17 depict the discharge curves for LFP/C cathodes incorporating, respectively, an additive of 8% mass of MC-derived activated carbon at KOH/MC or KOH/WH, mass ratios of 1:1, 2:1, and 5:1.

Figure 16 shows discharge curves at different current rates for LFP/C samples from millet cob. All samples display a well-defined voltage plateau near 3.4 V, characteristic of the Fe^{3+}/Fe^{2+} redox transition and associated with the biphasic reaction between $FePO_4$ and $LiFePO_4$.

Figure 17 presents the discharge profiles of $LiFePO_4/C$ (WH) cathodes with samples KOH/WH1:1, 2:1, and 5:1. All samples also display a well-defined voltage plateau near 3.4 V, characteristic of the Fe^{3+}/Fe^{2+} redox transition and associated with the biphasic reaction between $FePO_4$ and $LiFePO_4$. This behavior closely agrees with those observed for MC-derived carbon cathode.

Figure 18 shows the discharge curves for LFP with 8%mass of graphite. These curves exhibit a well-defined voltage plateau at 3.4 V, corresponding to the Fe^{3+}/Fe^{2+} redox couple and characteristic of the biphasic transition between $FePO_4$ and LFP [75].

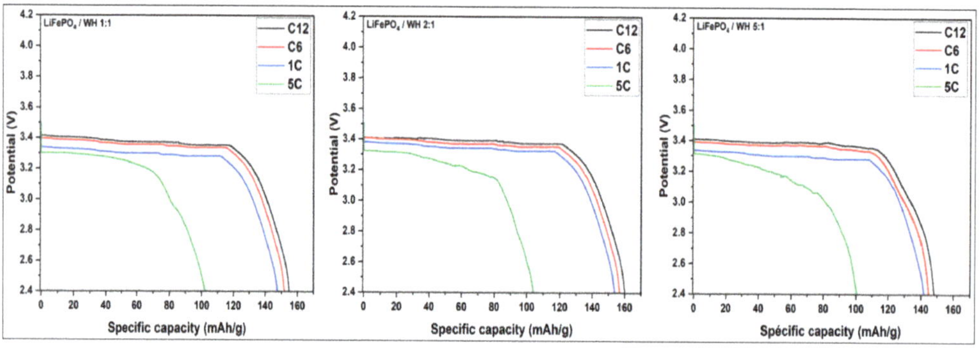

Figure 16. Discharge curves at different current rates for LFP/C samples from millet cob.

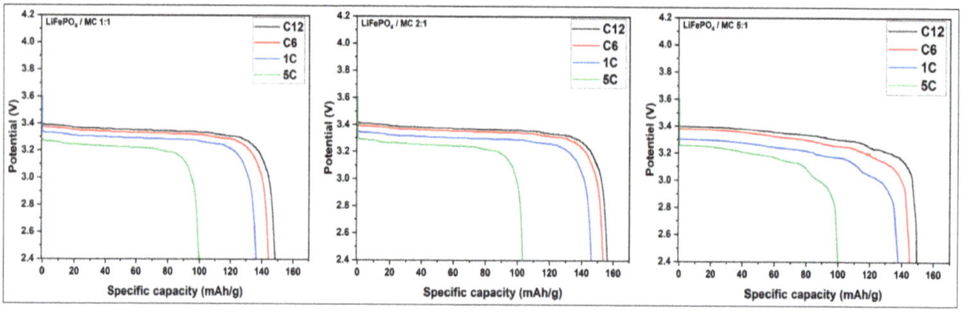

Figure 17. Discharge curves at different current rates for LFP/C samples from water hyacinth.

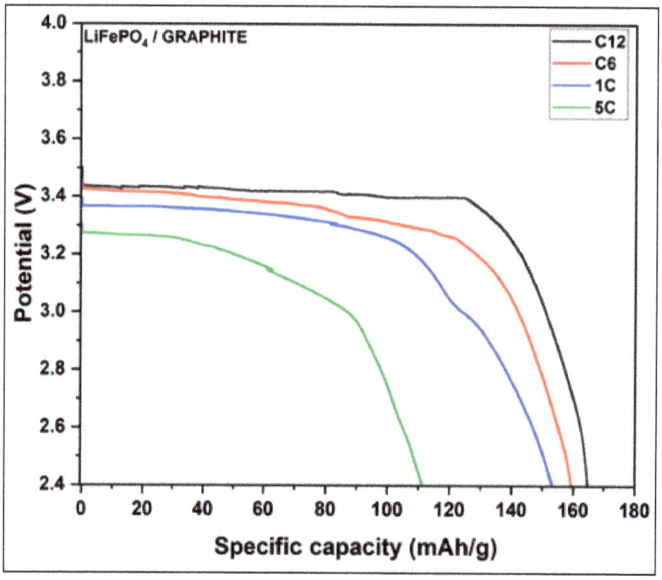

Figure 18. Discharge curves at different current rates of LFP/Graphite.

Using Figures 16–18, the variation of the specific capacity with the current rate of Li-ion batteries based on LFP/C with the additive of 8% mass of respective activated carbon based on MC (LFP/MC), WH (LFP/WH) and graphite (LFP/graphite) are shown in Table 8. It should be noted that the theoretical specific capacity of LFP is 170 mAh·g^{-1}.

Table 8. Variation of the specific capacity with the current rate of Li-ion batteries based on LFP/C with an additive of 8% mass of different activated carbon based on MC (LFP/MC) and WH (LFP/WH). Those of LFP/graphite is also shown.

Sample	Specific Capacity (mAh/g) (± 2 mAh/g)			
	Current Rate (C/12)	Current Rate (C/6)	Current Rate (1C)	Curent Rate (5C)
LFP/Graphite	167	163	161	120
LFP/MC 1:1	152	148	140	102
LFP/MC 2:1	160	160	157	106
LFP/MC 5:1	149	145	138	100
LFP/WH 1:1	158	155	151	108
LFP/WH 2:1	163	160	157	110
LFP/WH 5:1	153	149	145	106

The results demonstrate, respectively, that activated carbon derived from MC or WH at a KOH/MC 2:1 or KOH/WH 2:1 ratio, exhibits the highest specific capacity and electrical conductivity, surpassing those of the KOH/MC 1:1 and 5:1 or KOH/WH 1:1 and 5:1 samples. This trend is supported by the influence of the KOH activation ratio on the resulting electrochemical and electrical properties of the carbon material.

Among the WH-based cathodes, the LFP/WH 2:1 sample delivers the best specific capacity across multiple current rates. This enhanced performance is attributed to its superior electrical conductivity and well-optimized porous architecture, which jointly facilitates efficient lithium-ion diffusion and improved charge transport kinetics.

These results corroborate the electrical conductivity measurements (Table 5), which had already identified the KOH/WH 2:1 sample as the most effective ratio. Owing to its high electrical conductivity and optimized pore architecture, this material exhibits exceptional electrochemical performance, particularly under high current densities where an increase of efficient ion transport becomes necessary.

Cycling Performance of $LiFePO_4$ (LFP)/C Sample

The electrochemical performance of LFP/C cathodes was assessed through galvanostatic charge/discharge cycling at multiple current rates. The corresponding discharge curves, shown in Figure 19, enable a comparative evaluation of the rate-dependent behavior of each carbon source, highlighting differences in lithium-ion transport efficiency and capacity retention.

$LiFePO_4$/C cathodes incorporating activated carbon derived from MC and water WH exhibit a typical decline in specific capacity with increasing current rates of well-documented behavior in lithium-ion battery systems. For MC-based electrodes, this capacity fade becomes more pronounced at elevated rates (1C and 5C), particularly for the LFP/MC 1:1 and LFP/MC 5:1 sample, likely due to kinetic limitations that hinder effective ion and electron transport. In contrast, the LFP/MC 2:1 sample demonstrates superior performance across all tested current rates, benefiting from higher electrical conductivity and improved accessibility to electrochemically active sites.

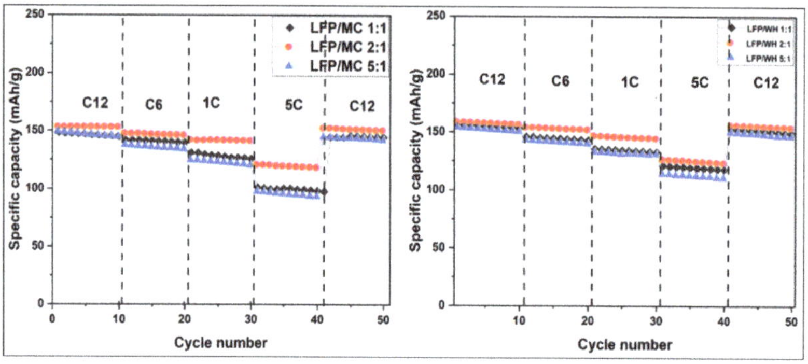

Figure 19. Cycling charge/discharge curves at different rates for $LiFePO_4/C$ samples.

A similar trend is observed among WH-based samples, where the LFP/WH 2:1 formulation outperforms both LFP/WH 1:1 and LFP/WH 5:1 counterpart, maintaining higher specific capacities even under high-rate conditions. This enhanced rate capability is attributed to a well-optimized pore structure and elevated conductivity, which together promotes efficient lithium-ion diffusion and facilitates rapid charge transfer processes. This is supported by the results on the study on the interaction of carbon with silicon as anodes for $LiFePO_4$ Li-ion battery applications [76]. It was shown that the appropriate design the edge-surface-inter nanocarbon on silicon ca be used to form the vertical conductive channels for rapid Li^+ transport [76].

EIS Analysis of $LiFePO_4/C$ Cathodes

EIS analysis of $LiFePO_4/C$ (MC, WH) is illustrated in the Nyquist plots (Figures 20 and 21), which represent the relationship between the real (Z_{re}) and imaginary (Z_{im}) components of impedance. These spectra offer valuable insights into the R_{ct} and lithium-ion diffusion kinetics. All three spectra corresponding to the LFP/MC and LFP/WH samples exhibit a similar profile: an absent or poorly defined semicircle in the high-frequency region followed by an inclined linear segment at low frequencies (Warburg region), indicative of lithium-ion diffusion within the electrode material, a typical behavior for Li-ion battery systems [77].

Among the MC-based samples, the LFP/MC 2:1 electrode demonstrates the lowest values of both R_e and R_{ct}, suggesting enhanced electronic conductivity and more efficient lithium-ion transport. In contrast, the LFP/MC 5:1 sample exhibits the highest R_{ct}, indicating less favorable interfacial kinetics, while the LFP/MC 1:1 electrode displays intermediate behavior. Figure 20 presents the Nyquist plots and the corresponding equivalent circuit models for the LFP/MC cathodes at different KOH/MC mass ratios.

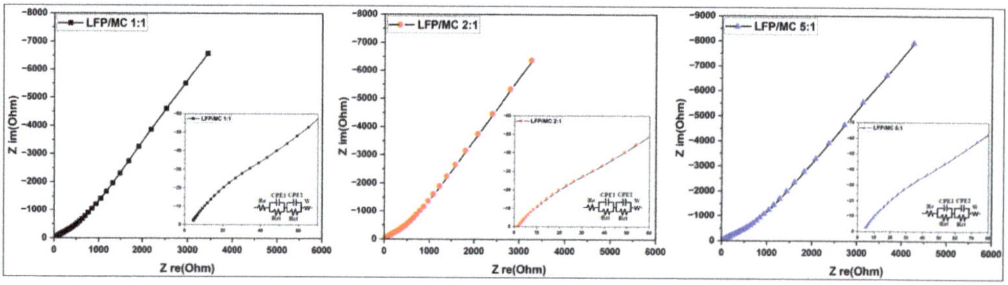

Figure 20. Nyquist plot of EIS for $LiFePO_4/C$ (millet cob carbon).

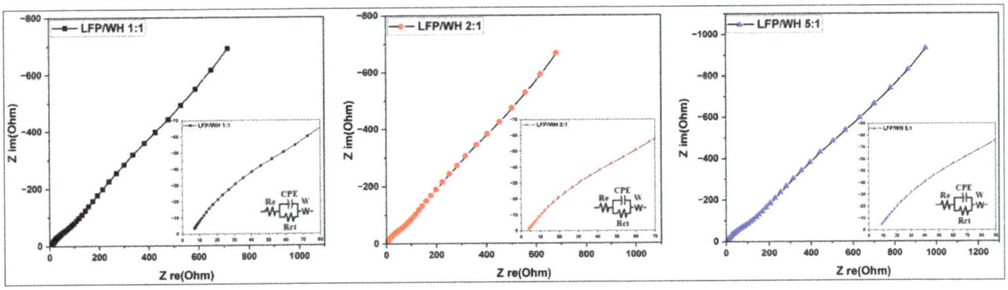

Figure 21. Nyquist plot of EIS for LiFePO4/C (water hyacinth carbon).

Among the $LiFePO_4$/C cathodes (LFP/WH), EIS reveals that the sample prepared with a KOH/WH 2:1 mass ratio exhibits the lowest resistance values. This reflects enhanced electronic conductivity and more efficient lithium-ion transport at the electrode–electrolyte interface. Like the trend observed in MC–derived samples, the LFP/WH 5:1 configuration demonstrates elevated charge transfer resistance, while the LFP/WH 1:1 sample exhibits intermediate electrochemical behavior.

Figure 21 also presents the Nyquist plot for the LFP samples using water hyacinth-derived activated carbon as a support.

The analysis of the influence of KOH/CM mass ratio reveals that a 2:1 ratio offers the optimal balance between electronic conductivity and lithium-ion transport. In contrast, excessive KOH (5:1 ratio) leads to significant degradation of electrochemical performance, likely resulting from structural overactivation and morphological disruption of the carbon framework. Such structural alteration reduces interparticle contact and increases R_{ct}. Conversely, a lower ratio (KOH/CM 1:1) promotes excessive microporosity, which impedes charge carrier mobility by restricting ion diffusion during cycling.

Comparative characterization of LFP/C cathodes from LFP/WH and LFP/MC reveals that the WH-based composites exhibit intrinsically lower impedance. This enhancement in electrochemical conductivity may be attributed to differences in the crystalline organization of the carbon matrix. The WH–derived carbon appears to promote more uniform dispersion of LFP particles and facilitate improved electronic percolation pathways.

EIS, analyzed through equivalent circuit fitting, enables extraction of the Warburg coefficient using Equation (3) [77]. This parameter provides quantitative insight into lithium-ion diffusion kinetics. The application of this method highlights the distinct ionic transport behaviors of MC- and WH-based composites, with WH samples exhibiting superior diffusion characteristics. Equation (3) is defined as follows [77]:

$$Z_{re} = R_e + R_{ct} + \sigma_\omega \omega^{-1/2} \quad (3)$$

where: Z_{re} is the real impedance, R_e is the electrolyte resistance, R_{ct} is the charge transfer resistance, ω is the angular frequency in the low-frequency range, and σ_ω represents the slope of Z_{re} curve as a function of $\omega^{-1/2}$.

To determine the slope of the Z_{re} ($\omega^{-1/2}$) σ_ω in Equation (1), we employed a systematic approach by plotting the linear relationship between the real impedance component (Z_{re}) and the inverse square root of low-frequency angular frequencies (ω) for the LFP/MC and LFP/WH composites. This linear regression analysis provides a quantitative assessment of the materials' electrochemical behavior, particularly their ionic diffusion dynamics. The resulting fitting lines, displayed in Figure 22, reveal distinct trends for each cathode system, highlighting the superior Warburg-type diffusion characteristics of the WH-derived carbon (LFP/WH) compared to the MC-based counterpart (LFP/MC). Building on the linear correlations demonstrated in Figure 22, the lithium-ion diffusion coefficients (D_{Li}) for the LFP/MC and LFP/WH composites were calculated using Equation (4) given in [78,79]:

$$D_{Li} = 0.5 \left(\frac{R * T}{A * C * \sigma_{\omega} * F^2} \right)^2 \tag{4}$$

where: R = universal gas constant (8.314 J·mol^{-1}·K^{-1}), T = absolute temperature (298.5 K), A = electrode surface area (experimental parameter, m^2), C is the molar concentration of lithium ions (mol·m^{-3}), F = Faraday constant (96,500 C·mol^{-1}).

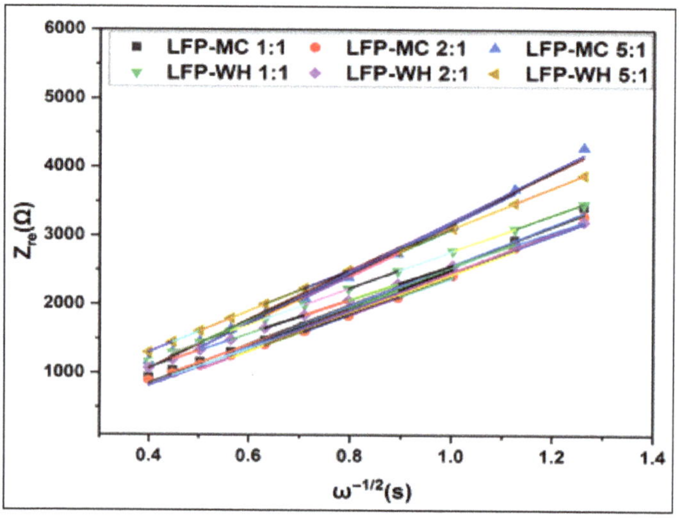

Figure 22. Relationship between Zre and ω-1/2 at low frequencies of $LiFePO_4$/C.

This methodology enables an evaluation of ionic mobility within the cathode materials, directly linking structural properties to electrochemical kinetic performance. The derived D_{Li} values underscore the superior ion transport efficiency of the WH-based composite, consistent with its enhanced charge transfer dynamics observed in earlier analyses (Figure 22).

The key electrochemical parameters derived from EIS analysis, including R_e, R_{ct}, and D_{Li}, are shown in Table 9 for the various LFP/MC and LFP/WH cathodes.

Table 9. EIS-derived electrochemical parameters for LFP/MC and LFP/WH cathodes.

Samples	Re (Ω) (±0.5)	Rct (Ω) (±0.5)	D_{Li}(10^{-13} cm^2/s)(± 5%)
LFP/MC 1:1	4.4	99.9	1.7
LFP/MC 2:1	2.7	95.9	1.8
LFP/MC 5:1	5.7	124.9	1.0
LFP/WH 1:1	2.7	98.6	1.9
LFP/WH 2:1	2.5	91.1	2.3
LFP/WH 5:1	3.0	110.4	1.5
LFP/graphite	2.3	73.8	3.5

Comparative analysis reveals that the LFP/WH 2:1 composite exhibits the lowest ohmic resistance (R_e = 2.49 Ω) and charge transfer resistance (R_{ct} = 91.12 Ω) among all tested samples. This synergistic improvement corresponds to a notably higher lithium-ion diffusion coefficient (D_{Li} = 2.28 × 10^{-13} $cm^2 \cdot s^{-1}$), exceeding values obtained for other configurations. The enhanced ionic transport is attributed to the optimized carbon architecture of the LFP/WH 2:1 electrode, which reduces diffusion path length through a hierarchically porous network, thereby promoting lithium-ion mobility. In contrast, the LFP/MC 5:1 composite, whether derived from millet cob or water hyacinth, exhibits the lowest D_{Li} values. This limitation is likely due to deficient microporosity and the dominance of poorly connected macropores, which hinder efficient ion percolation. The best values of these respective parameters are obtained with LFP/Graphite.

Graphite, used as a reference material and tested under identical conditions, displays superior electrochemical properties compared to synthesize activated carbon sample.

Figure 23 displays, Nyqiest and Z_{rc} curves of LFP/Graphite. It shows the analysis of the discharge profiles reveals that the LFP/Graphite cathode consistently delivers slightly higher specific capacities across all tested current rates compared to the LFP/WH 2:1 sample. The EIS parameters for this cathode, summarized in Table 9 indicate notably lower values of both R_e and R_{ct} relative to the biomass-derived activated carbon electrodes. As a result, the LFP/Graphite cell exhibits the highest D_{Li}, with the LFP/WH 2:1 sample ranking second. These findings underscore the excellent electrochemical kinetics of graphite while highlighting the promising performance of WH–derived carbon as a viable alternative electrode material.

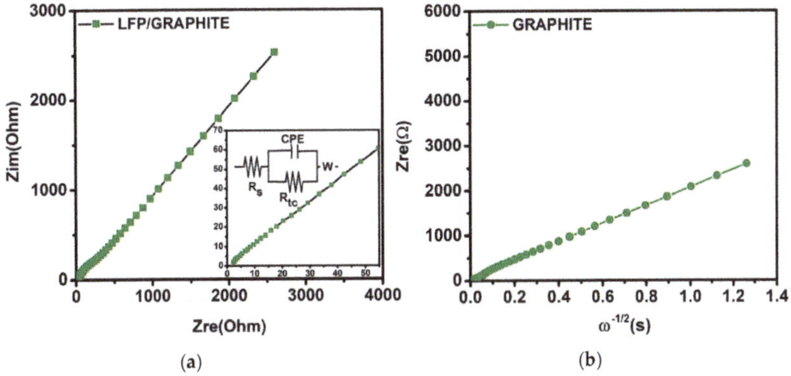

Figure 23. Nyquist plot (**a**), Z_{re} ($\omega^{-1/2}$) (**b**) curves of LFP/Graphite.

3.4. Correlation Between the Various Results

Galvanostatic charge/discharge tests demonstrate that, using Lithium anode, the optimum capacities are obtained for the carbons activated at the KOH/CM 2:1, corresponding to the optimum pore structure and pore diameters (Table 6). On the other hand, $LiFePO_4$ cathodes supported by bio-derived activated carbon achieve specific capacities approaching the theoretical limit of $LiFePO_4$ (170 $mAh \cdot g^{-1}$), with optimal performance observed in samples activated at a mass ratio KOH/CM 2:1 (Table 8). This ratio offers a favorable balance of pore structure, pore diameter and conductivity, providing pathways for rapid lithium-ion diffusion and efficient electron transport. Notably, carbon derived from WH with its relatively higher crystallinity contributes to superior conductive network formation, enabling enhanced capacity retention across a range of current rates.

Furthermore, Coulombic efficiency stabilizes after the initial cycles (Table 7), indicating successful formation of a stable SEI, which is essential for suppressing parasitic reactions and preserving electrode integrity. In particular, the KOH/WH 2:1 sample display rapid SEI stabilization and sustained high Coulombic efficiency from the second cycle onward, highlighting their potential as effective conductive additives in $LiFePO_4$/C systems. On the other hand, the chemical composition the chemical composition the sample evaluated by EDS (Table 4) and the results on their electrochemical parameters (Tables 6–8, Figures 18 and 23–25) indicate that the best electrochemical performances are obtained with activated biomass containing around 70% of carbon, high content of potassium and high content of silicon. But no evident correlation can be deduced from these results. Further studies are underway to make the possible correlation between the chemical composition and the electrochemical performances.

Cycling stability evaluations of activated KOH/CM carbons reveal excellent capacity retention exceeding 96% of the initial value over multiple cycles (Table 7, Figure 17) for the, KOH/MC or KOH/WH 2:1 sample derived from both MC and WH. These findings underscore the viability of bio-derived activated carbons as cost-effective and environmentally friendly alternatives in lithium-ion battery cathodes. The improved stability is attributed to the synergistic effect of enhanced electrical conductivity and expanded electrochemically accessible surface areas through the crystallinity, pore distribution and size imparted by the biomass carbon.

EIS further supports these observations, indicating relatively low R_{ct} carbon. Figure 24 presents the correlation between specific capacity and R_{ct} as a function value for $LiFePO_4/C$ cathodes incorporating KOH/MC 2:1 and KOH/WH 2:1 activated of the KOH activation ratio.

Among all samples, the KOH/WH 2:1 exhibits the lowest overall resistance, attributed to its well-ordered carbon structure and hierarchical porosity. While a high specific surface area dominated by micropores may limit lithium-ion accessibility, the presence of mesopores is essential for facilitating efficient transport. For example, the KOH/MC 1:1 and KOH/WH 1:1 sample, despite their elevated surface areas, but with small pores sizes, display a higher charge transfer resistance compared to the KOH/CM 2:1 counterpart. These results emphasize the importance of pore size distribution, not just surface area, in determining electrochemical performance. Beyond the optimal KOH/CM 2:1 activation ratio, excessive KOH levels lead to structural degradation manifested as pore over-widening which compromises both capacity and conductivity.

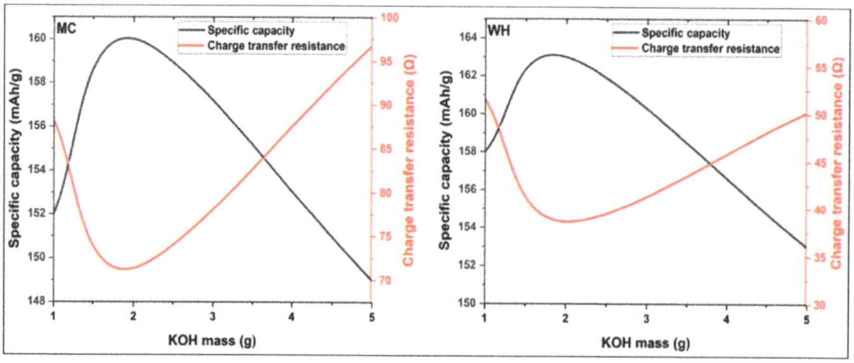

Figure 24. Evolution of capacity and charge transfer resistance as a function of KOH Mass.

EIS enabled precise determination of the lithium-ion diffusion coefficients D_{Li} for $LiFePO_4/C$ cathodes. Among the samples, the LFP/WH 2:1 composite exhibited the highest diffusion coefficient ($D_{Li} \approx 2.28 \times 10^{-13}$ $cm^2 \cdot s^{-1}$), indicating superior lithium-ion transport kinetics. Interestingly, the KOH/WH 2:1 sample achieved this performance despite having a lower specific surface area than the KOH/WH 1:1 counterpart, emphasizing that D_{Li} is not solely governed by surface area. Instead, it reflects a confluence of factors, including the electrical conductivity of the carbon matrix, elemental composition affecting charge transfer dynamics, and a well-balanced micro/mesoporous structure that promotes ion mobility.

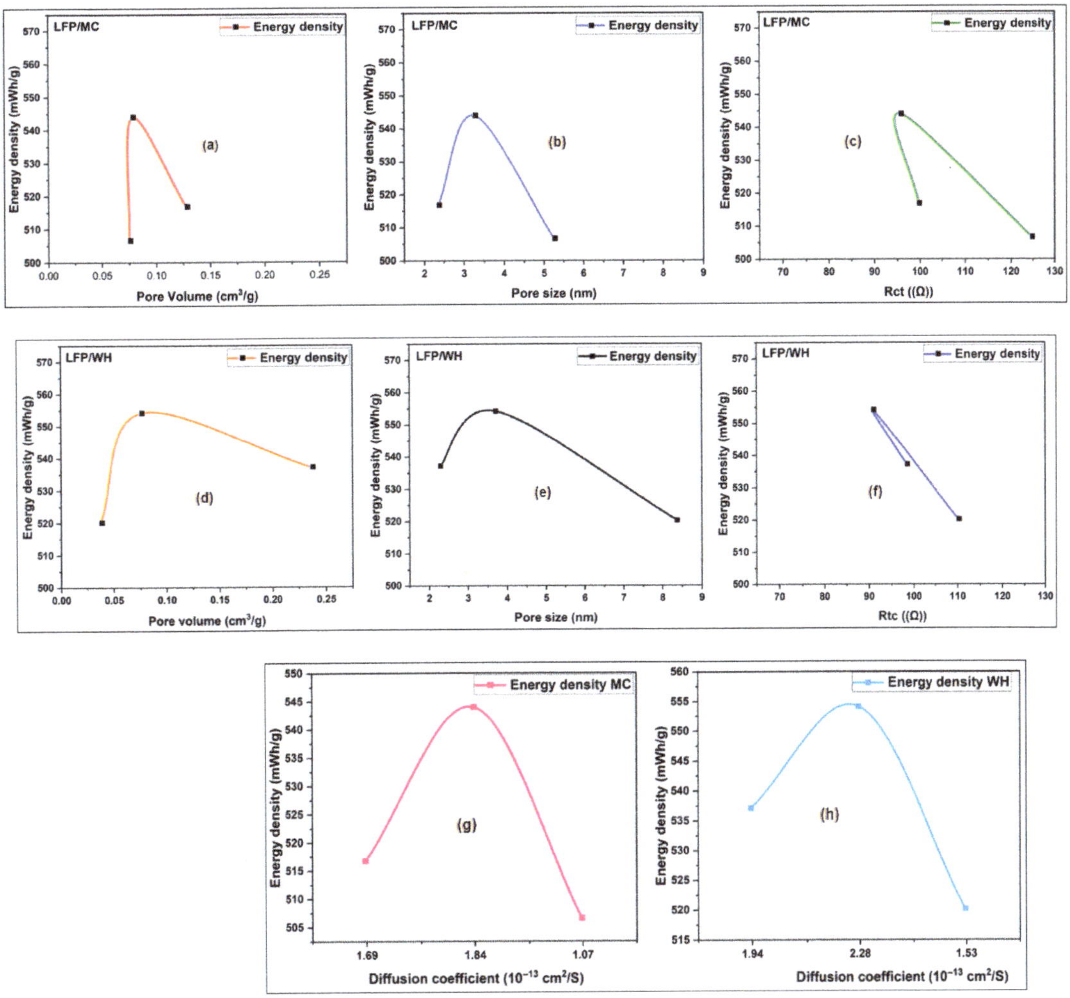

Figure 25. Energy density curves as a function: pore volume (**a**), pore size (**b**), charge transfer resistance (**c**) for millet cob, pore volume (**d**), pore size (**e**), charge transfer resistance (**f**) for water hyacinth, diffusion coefficient for millet cob (**g**) and water hyacinth (**h**).

In cases where microporosity dominates (>80% of total pore volume), diffusion pathways become constricted, impeding charge carrier mobility and lowering D_{Li}. Conversely, a hierarchical pore distribution with an approximately balance micro/mesopore, as observed in the LFP/WH 2:1 electrode, facilitates continuous and accessible ion channels. This structural optimization enhances lithium-ion flux within the composite and reinforces the conclusion that balanced porosity—not maximal surface area—is essential for achieving high-performance lithium-ion battery cathodes.

Furthermore, the energy density curves support and enrich the previous findings. Figure 25 presents, respectively, the energy density curves as a function of pore size, pore volume, charge transfer resistance, and diffusion coefficient for KOH/MC For both studied materials (LFP/MC and LFP/WH), a local maximum in energy density is observed as a function of each of the four key electrochemical parameters: pore size, pore volume, charge transfer resistance (R_{ct}), and diffusion coefficient. These curves highlight the presence of optimal conditions where the balance between ionic accessibility, active surface area, and charge transport is most favorable. The energy density curves show a maximum energy density relative to the active mass of 544 mWh/g for the LFP/MC 2:1 sample, 554 mWh/g for the LFP/WH 2:1 sample, and 568 mWh/g for the reference LFP/graphite sample. These energy densities were calculated using the capacities obtained from Tables 8 and 9 by the operating potential of 3.4 Volts of Li-ion battery based respectively on KOH/MC, KOH/WH and graphite.

Considering these results, it becomes evident that, to obtain high-performance bio-based activated carbon as a $LiFePO_4$ cathode additives, several criteria must be met: a high carbon content, both for amorphous or highly crystalline structure. The balanced distribution of micropores and mesopores, and an average pore size greater than 2 nm may lead to good carbon additives for $LiFePO_4$ cathodes.

4. Conclusions

This study presents a comprehensive evaluation of $LiFePO_4$ cathodes supported by activated carbon derived from renewable biomass sources, millet cob and water hyacinth. Electrochemical characterizations, including galvanostatic charge/discharge testing, cycling stability assessments, and EIS, collectively revealed outstanding performance, particularly for samples activated at a mass ratio of KOH/carbon 2:1. The LFP/WH 2:1 cathode achieved a specific capacity of 163 $mAh \cdot g^{-1}$ at a C/12 rate and retained 110 $mAh \cdot g^{-1}$ at 5 C, while the LFP/MC 2:1 sample demonstrated comparable behavior with 160 $mAh \cdot g^{-1}$ at C/12 and 106 $mAh \cdot g^{-1}$ at 5 C. Across all current densities, these samples exhibited excellent cycling stability, maintaining high capacity retention at elevated rates (1 C and 5 C), further reinforcing their electrochemical robustness.

EIS analysis showed that WH-derived cathodes exhibited lower R_{ct} and R_e compared to MC-based counterparts, confirming the superior electronic conductivity of crystalline carbon obtained from water hyacinth relative to the predominantly amorphous carbon structure in millet cob. This enhanced conductivity, coupled with a more favorable porous architecture, contributes to the superior performance of WH-based cathodes.

Furthermore, activated carbons with moderate specific surface areas and a balance between micropore/mesopore demonstrated optimal performance as conductive support. The presence of interconnected mesopores facilitates efficient lithium-ion diffusion and minimizes charge transport resistance, thereby enhancing overall electrode kinetics. These findings underscore the promise of biomass-derived activated carbon based on millet cob or water hyacinth as sustainable and alternatives to conventional carbon materials, offering a compelling pathway toward high-performance, environmentally friendly lithium-ion batteries.

Author Contributions: W.-W.A.Z.: Data curation, all experimental data collections, Writing—original draft. O.S.: Conceptualization, Formal analysis, Supervision, Validation, Writing—review & editing. All authors have read and agreed to the published version of the manuscript.

Funding: The research was funded by the National Science and Engineering Research Council of Canada (NSERC) (RGP20002705), UNESCO Chair (FRQ-CUN-338927).

Data Availability Statement: Data sets generated during the current study are available from the corresponding author on a reasonable request.

Acknowledgments: We thank the National Science and Engineering Research Council of Canada (NSERC) (RGP20002705) for the financial and the Fonds de recherche du Québec, UNESCO Chair (FRQ-CUN-338927).

Conflicts of Interest: The authors declare that they have no known competing financial interests or personal relationships that could have appeared to influence the work reported in this paper. The authors declare the following financial interests/personal relationships which may be considered as potential competing interests: Oumarou Savadogo reports administrative support was provided by Polytechnique Montreal. Oumarou Savadogo reports a relationship with Polytechnique Montreal that includes employment. If there are other authors, they declare that they have no known competing financial interests or personal relationships that could have appeared to influence the work reported in this paper.

Abbreviations

BET	Brunauer, Emmett, and Teller
BJH	Barrett, Joyner, and Halenda
C	Carbon, Current Rate
CS	Specific Capacitances
D	Diameter
DMAC	Dimethylacetamide
XRD	X-ray Diffraction
EIS	Electrochemical Impedance Spectroscopy
MC	Millet Cob
EDS	Energy Dispersive Spectroscopy
HCl	Hydrogen Chloride
HK	Horváth-Kawazoe
H_3PO_4	Phosphoric Acid
WH	Water Hyacinth
KOH	Potassium Hydroxide
$LiFePO_4$ or LFP	Lithium Iron Phosphate
Li+	Lithium Ion
$LiPF_6$	Lithium Hexafluorophosphate
CM	Carbonaceous Materials
SEM	Scanning Electron Microscopy
N_2	Nitrogen
NLDFT	Non-Local Density Functional Theory
PVDF	Polyvinylidene Fluoride
R_{ct}	Charge Transfer Resistance
R_e	Electrolyte Resistance

References

1. Padhi, A.K.; Nanjundaswamy, K.S.; Goodenough, J.B. Phospho-olivines as positive-electrode materials for rechargeable lithium batteries. *J. Electrochem. Soc.* **1997**, *144*, 1188. [CrossRef]
2. Lee, S.; Jang, I.; Lim, H.a.; Aravindan, V.; Kim, H.; Lee, Y. Preparation and electrochemical characterization of $LiFePO_4$ nanoparticles with high rate capability by a sol–gel method. *J. Alloys Compd.* **2010**, *491*, 668–672. [CrossRef]
3. Li, Z.; Zhang, D.; Yang, F. Developments of lithium-ion batteries and challenges of $LiFePO_4$ as one promising cathode material. *J. Mater. Sci.* **2009**, *44*, 2435–2443. [CrossRef]
4. Balakrishnan, N.T.; Paul, A.; Krishnan, M.; Das, A.; Raphaez, L.R.; Ahn, J.-H.; Jabeen Fatima, M.; Prasanth, R. Lithium Iron Phosphate (LiFePO4) as High-Performance Cathode Material for Lithium Ion Batteries. In *Metal, Metal-Oxides and Metal Sulfides for Batteries, Fuel Cells, Solar Cells, Photocatalysis and Health Sensors*; Springer: Berlin/Heidelberg, Germany, 2021; pp. 35–73.
5. Kim, D.-H.; Kim, J. Synthesis of LiFePO4 nanoparticles in polyol medium and their electrochemical properties. *Electrochem. Solid-State Lett.* **2006**, *9*, A439. [CrossRef]
6. Yun, N.J.; Ha, H.-W.; Jeong, K.H.; Park, H.-Y.; Kim, K. Synthesis and electrochemical properties of olivine-type $LiFePO_4/C$ composite cathode material prepared from a poly (vinyl alcohol)-containing precursor. *J. Power Sources* **2006**, *160*, 1361–1368.
7. Ritchie, A.; Howard, W. Recent developments and likely advances in lithium-ion batteries. *J. Power Sources* **2006**, *162*, 809–812. [CrossRef]

8. Nakamura, T.; Miwa, Y.; Tabuchi, M.; Yamada, Y. Structural and surface modifications of $LiFePO_4$ olivine particles and their electrochemical properties. *J. Electrochem. Soc.* **2006**, *153*, A1108. [CrossRef]
9. Guan, Y.; Shen, J.; Wei, X.; Zhu, Q.; Zheng, X.; Zhou, S.; Xu, B. $LiFePO_4$/activated carbon/graphene composite with capacitive-battery characteristics for superior high-rate lithium-ion storage. *Electrochim. Acta* **2019**, *294*, 148–155. [CrossRef]
10. Zhang, H.; Liu, D.; Qian, X.; Zhao, C.; Xu, Y. A novel nano structured LiFePO4/C composite as cathode for Li-ion batteries. *J. Power Sources* **2014**, *249*, 431–434. [CrossRef]
11. Wang, H.; Wang, R.; Liu, L.; Jiang, S.; Ni, L.; Bie, X.; Yang, X.; Hu, J.; Wang, Z.; Chen, H. In-situ self-polymerization restriction to form core-shell LiFePO4/C nanocomposite with ultrafast rate capability for high-power Li-ion batteries. *Nano Energy* **2017**, *39*, 346–354. [CrossRef]
12. Tu, J.; Wu, K.; Tang, H.; Zhou, H.; Jiao, S. Mg–Ti co-doping behavior of porous $LiFePO_4$ microspheres for high-rate lithium-ion batteries. *J. Mater. Chem. A* **2017**, *5*, 17021–17028. [CrossRef]
13. Talebi-Esfandarani, M. Synthesis, Characterization and Modification of $LiFePO_4$ by Doping with Platinum and Palladium for Lithium-Ion Batteries. Ph.D. Thesis, École Polytechnique de Montréal, Montreal, QC, Canada, 2013.
14. Wang, B.; Al Abdulla, W.; Wang, D.; Zhao, X. A three-dimensional porous $LiFePO_4$ cathode material modified with a nitrogen-doped graphene aerogel for high-power lithium ion batteries. *Energy Environ. Sci.* **2015**, *8*, 869–875. [CrossRef]
15. Jinli, Z.; Jiao, W.; Yuanyuan, L.; Ning, N.; Junjie, G.; Feng, Y.; Wei, L. High-performance lithium iron phosphate with phosphorus-doped carbon layers for lithium ion batteries. *J. Mater. Chem. A* **2015**, *3*, 2043–2049. [CrossRef]
16. Zhou, Y.; Lu, J.; Deng, C.; Zhu, H.; Chen, G.Z.; Zhang, S.; Tian, X. Nitrogen-doped graphene guided formation of monodisperse microspheres of $LiFePO_4$ nanoplates as the positive electrode material of lithium-ion batteries. *J. Mater. Chem. A* **2016**, *4*, 12065–12072. [CrossRef]
17. He, Z.; Zhang, C.; Zhu, Z.; Yu, Y.; Zheng, C.; Wei, F. Advances in carbon nanotubes and carbon coatings as conductive networks in silicon-based anodes. *Adv. Funct. Mater.* **2024**, *34*, 2408285. [CrossRef]
18. Ali, A.; Cai, R.; Wang, T.; Zhang, C.; Liu, C.; Ma, D.; Hou, H.; Lei, Q.; Xia, Y.; Wang, S. Emerging flexible battery technology: Innovative structural design, material integration and wearable applications and beyond. *Chem. Eng. J.* **2025**, *519*, 165358. [CrossRef]
19. Wang, Y.; Wang, Y.; Hosono, E.; Wang, K.; Zhou, H. The design of a $LiFePO_4$/carbon nanocomposite with a core–shell structure and its synthesis by an in situ polymerization restriction method. *Angew. Chem. Int. Ed.* **2008**, *47*, 7461–7465. [CrossRef]
20. Huang, H.; Yin, S.-C.; Nazar, L.s. Approaching theoretical capacity of $LiFePO_4$ at room temperature at high rates. *Electrochem. Solid-State Lett.* **2001**, *4*, A170. [CrossRef]
21. Shahid, R.; Murugavel, S. Synthesis and characterization of olivine phosphate cathode material with different particle sizes for rechargeable lithium-ion batteries. *Mater. Chem. Phys.* **2013**, *140*, 659–664. [CrossRef]
22. Chung, S.-Y.; Bloking, J.T.; Chiang, Y.-M. Electronically conductive phospho-olivines as lithium storage electrodes. *Nat. Mater.* **2002**, *1*, 123–128. [CrossRef]
23. Talebi-Esfandarani, M.; Savadogo, O. Effects of palladium doping on the structure and electrochemical properties of $LiFePO_4$/C prepared using the sol-gel method. *J. New Mater. Electrochem. Syst.* **2014**, *17*, 91–97. [CrossRef]
24. Talebi-Esfandarani, M.; Savadogo, O. Enhancement of electrochemical properties of platinum doped $LiFePO_4$/C cathode material synthesized using hydrothermal method. *Solid State Ion.* **2014**, *261*, 81–86. [CrossRef]
25. Talebi-Esfandarani, M.; Savadogo, O. Synthesis and characterization of Pt-doped $LiFePO_4$/C composites using the sol–gel method as the cathode material in lithium-ion batteries. *J. Appl. Electrochem.* **2014**, *44*, 555–562. [CrossRef]
26. Ravet, N.; Abouimrane, A.; Armand, M. On the electronic conductivity of phospho-olivines as lithium storage electrodes. *Nat. Mater.* **2003**, *2*, 702. [CrossRef]
27. Shi, M.; Li, R.; Liu, Y. In situ preparation of $LiFePO_4$/C with unique copolymer carbon resource for superior performance lithium-ion batteries. *J. Alloys Compd.* **2021**, *854*, 157162. [CrossRef]
28. Ravet, N.; Besner, S.; Simoneau, M.; Vallée, A.; Armand, M.; Magnan, J.-F. Electrode Materials with High Surface Conductivity. US Patents US6855273B2, 15 February 2005.
29. Chen, Z.; Dahn, J. Reducing carbon in $LiFePO_4$/C composite electrodes to maximize specific energy, volumetric energy, and tap density. *J. Electrochem. Soc.* **2002**, *149*, A1184. [CrossRef]
30. Dominko, R.; Bele, M.; Gaberscek, M.; Remskar, M.; Hanzel, D.; Pejovnik, S.; Jamnik, J. Impact of the carbon coating thickness on the electrochemical performance of $LiFePO_4$/C composites. *J. Electrochem. Soc.* **2005**, *152*, A607. [CrossRef]
31. Doeff, M.M.; Wilcox, J.D.; Kostecki, R.; Lau, G. Optimization of carbon coatings on $LiFePO_4$. *J. Power Sources* **2006**, *163*, 180–184. [CrossRef]
32. Doeff, M.M.; Wilcox, J.D.; Yu, R.; Aumentado, A.; Marcinek, M.; Kostecki, R. Impact of carbon structure and morphology on the electrochemical performance of $LiFePO_4$/C composites. *J. Solid State Electrochem.* **2008**, *12*, 995–1001. [CrossRef]
33. Talebi-Esfandarani, M.; Savadogo, O. Improvement of electrochemical and electrical properties of $LiFePO_4$ coated with citric acid. *Rare Met.* **2016**, *35*, 303–308. [CrossRef]

34. Zhi, X.; Liang, G.; Wang, L.; Ou, X.; Gao, L.; Jie, X. Optimization of carbon coatings on $LiFePO_4$: Carbonization temperature and carbon content. *J. Alloys Compd.* **2010**, *503*, 370–374. [CrossRef]
35. Tsuda, T.; Ando, N.; Utaka, T.; Kojima, K.; Nakamura, S.; Hayashi, N.; Soma, N.; Gunji, T.; Tanabe, T.; Ohsaka, T. Improvement of high-rate performance of $LiFePO_4$ cathode with through-holed $LiFePO_4$/Activated carbon hybrid electrode structure fabricated with a pico-second pulsed laser. *Electrochim. Acta* **2019**, *298*, 827–834. [CrossRef]
36. Mwizerwa, J.P.; Liu, C.; Xu, K.; Zhao, N.; Li, Y.; Ndagijimana, P.; Chen, Z.; Shen, J. Activated carbon/reduced graphene oxide wrapped $LiFePO_4$ cathode for Li-ion batteries with ultrahigh capacities and high specific energy density. *FlatChem* **2022**, *34*, 100393. [CrossRef]
37. Liu, X.; Zhao, R.; Xia, Y.; Li, Q. Improved electrochemical performance of $LiFePO_4$/carbon cathode for lithium-ion batteries. *Ionics* **2022**, *28*, 4579–4585. [CrossRef]
38. Han, W.; Zhou, G.; Gao, D.; Zhang, Z.; Wei, Z.; Wang, H.; Yang, H. Experimental analysis of the pore structure and fractal characteristics of different metamorphic coal based on mercury intrusion-nitrogen adsorption porosimetry. *Powder Technol.* **2020**, *362*, 386–398. [CrossRef]
39. Saito, A.; Foley, H. Curvature and parametric sensitivity in models for adsorption in micropores. *AIChE J.* **1991**, *37*, 429–436. [CrossRef]
40. Ravikovitch, P.I.; Haller, G.L.; Neimark, A.V. Density functional theory model for calculating pore size distributions: Pore structure of nanoporous catalysts. *Adv. Colloid Interface Sci.* **1998**, *76*, 203–226. [CrossRef]
41. Mikhail, R.S.; Brunauer, S.; Bodor, E. Investigations of a complete pore structure analysis: I. Analysis of micropores. *J. Colloid Interface Sci.* **1968**, *26*, 45–53. [CrossRef]
42. Rouquerol, F.; Rouquerol, J.; Sing, K. Assessment of mesoporosity. In *Adsorption by Powders and Porous Solids*; Academic Press: Oxford, UK, 1999; pp. 191–217.
43. Sohouli, E.; Adib, K.; Maddah, B.; Najafi, M. Manganese dioxide/cobalt tungstate/ nitrogen-doped carbon nano-onions nanocomposite as new supercapacitor electrode. *Ceram. Int.* **2022**, *48*, 295–303. [CrossRef]
44. Sohouli, E.; Adib, K.; Maddah, B.; Najafi, M. Preparation of a supercapacitor electrode based on carbon nano-onions/manganese dioxide/iron oxide nanocomposite. *J. Energy Storage* **2022**, *52*, 104987. [CrossRef]
45. Albo Hay Allah, M.A.; Alshamsi, H.A. Green synthesis of ZnO NPs using Pontederia crassipes leaf extract: Characterization, their adsorption behavior and anti-cancer property. *Biomass Convers. Biorefinery* **2024**, *14*, 10487–10500. [CrossRef]
46. Wilson, G.; Zilinskaite, S.; Unka, S.; Boston, R.; Reeves-McLaren, N. Establishing operando diffraction capability through the study of Li-ion (de) intercalation in $LiFePO_4$. *Energy Rep.* **2020**, *6*, 174–179. [CrossRef]
47. Rouquerol, J.; Rouquerol, F.; Llewellyn, P.; Maurin, G.; Sing, K. *Adsorption by Powders and Porous Solids: Principles, Methodology and Applications*; Academic Press: Cambridge, MA, USA, 2013.
48. Barrett, E.P.; Joyner, L.G.; Halenda, P.P. The determination of pore volume and area distributions in porous substances. I. Computations from nitrogen isotherms. *J. Am. Chem. Soc.* **1951**, *73*, 373–380. [CrossRef]
49. Horváth, G.; Kawazoe, K. Method for the calculation of effective pore size distribution in molecular sieve carbon. *J. Chem. Eng. Jpn.* **1983**, *16*, 470–475. [CrossRef]
50. Sing, K.S. Reporting physisorption data for gas/solid systems with special reference to the determination of surface area and porosity (Recommendations 1984). *Pure Appl. Chem.* **1985**, *57*, 603–619. [CrossRef]
51. Yu, S.; Bo, J.; Fengli, L.; Jiegang, L. Structure and fractal characteristic of micro-and meso-pores in low, middle-rank tectonic deformed coals by CO2 and N2 adsorption. *Microporous Mesoporous Mater.* **2017**, *253*, 191–202. [CrossRef]
52. Tang, J.; Feng, L.; Li, Y.; Liu, J.; Liu, X. Fractal and pore structure analysis of Shengli lignite during drying process. *Powder Technol.* **2016**, *303*, 251–259. [CrossRef]
53. Rahma, N.A.; Kurniasari, A.; Pambudi, Y.D.S.; Bintang, H.M.; Zulfia, A.; Hudaya, C. Characteristics of Corncob-Originated Activated Carbon Using Two Different Chemical Agent. *IOP Conf. Ser. Mater. Sci. Eng.* **2019**, *622*, 012030. [CrossRef]
54. Wang, S.; Lu, G. A comprehensive study on carbon dioxide reforming of methane over Ni/γ-Al_2O_3 catalysts. *Ind. Eng. Chem. Res.* **1999**, *38*, 2615–2625. [CrossRef]
55. Kamath, S.R.; Proctor, A. Silica gel from rice hull ash: Preparation and characterization. *Cereal Chem.* **1998**, *75*, 484–487. [CrossRef]
56. Omri, A.; Benzina, M. Characterization of activated carbon prepared from a new raw lignocellulosic material: *Ziziphus spina-christi* seeds. *J. Société Chim. Tunis.* **2012**, *14*, 175–183.
57. Ahmed, S.; Parvaz, M.; Johari, R.; Rafat, M. Studies on activated carbon derived from neem (*Azadirachta indica*) bio-waste, and its application as supercapacitor electrode. *Mater. Res. Express* **2018**, *5*, 045601. [CrossRef]
58. Chen, Y.; Zhu, Y.; Wang, Z.; Li, Y.; Wang, L.; Ding, L.; Gao, X.; Ma, Y.; Guo, Y. Application studies of activated carbon derived from rice husks produced by chemical-thermal process—A review. *Adv. Colloid Interface Sci.* **2011**, *163*, 39–52. [CrossRef] [PubMed]
59. De, B.; Karak, N. A green and facile approach for the synthesis of water soluble fluorescent carbon dots from banana juice. *RSC Adv.* **2013**, *3*, 8286–8290. [CrossRef]
60. Qiu, T.; Yang, J.-G.; Bai, X.-J.; Wang, Y.-L. The preparation of synthetic graphite materials with hierarchical pores from lignite by one-step impregnation and their characterization as dye absorbents. *RSC Adv.* **2019**, *9*, 12737–12746. [CrossRef]

61. Phillips, R.; Jolley, K.; Zhou, Y.; Smith, R. Influence of temperature and point defects on the X-ray diffraction pattern of graphite. *Carbon Trends* **2021**, *5*, 100124. [CrossRef]
62. Barroso-Bogeat, A.; Alexandre-Franco, M.; Fernández-González, C.; Sánchez-González, J.; Gómez-Serrano, V. Electrical conductivity of metal (hydr) oxide–activated carbon composites under compression. A comparison study. *Mater. Chem. Phys.* **2015**, *152*, 113–122. [CrossRef]
63. Celzard, A.; Marêché, J.; Payot, F.; Furdin, G. Electrical conductivity of carbonaceous powders. *Carbon* **2002**, *40*, 2801–2815. [CrossRef]
64. Adinaveen, T.; Vijaya, J.J.; Kennedy, L.J. Comparative study of electrical conductivity on activated carbons prepared from various cellulose materials. *Arab. J. Sci. Eng.* **2016**, *41*, 55–65. [CrossRef]
65. Zhao, H.; Zuo, H.; Wang, J.; Jiao, S. Practical application of graphite in lithium-ion batteries: Modification, composite, and sustainable recycling. *J. Energy Storage* **2024**, *98*, 113125. [CrossRef]
66. Ryu, S.; Kim, S.; Gallego, N.; Edie, D. Physical properties of silver-containing pitch-based activated carbon fibers. *Carbon* **1999**, *37*, 1619–1625. [CrossRef]
67. Hashisho, Z.; Rood, M.J.; Barot, S.; Bernhard, J. Role of functional groups on the microwave attenuation and electric resistivity of activated carbon fiber cloth. *Carbon* **2009**, *47*, 1814–1823. [CrossRef]
68. Huang, Y. Electrical and thermal properties of activated carbon fibers. In *Activated Carbon Fiber and Textiles*; Elsevier: Amsterdam, The Netherlands, 2017; pp. 181–192.
69. Dahn, J.R.; Zheng, T.; Liu, Y.; Xue, J. Mechanisms for lithium insertion in carbonaceous materials. *Science* **1995**, *270*, 590–593. [CrossRef]
70. Ciosek Högström, K.; Malmgren, S.; Hahlin, M.; Rensmo, H.; Thébault, F.; Johansson, P.; Edstrom, K. The influence of PMS-additive on the electrode/electrolyte interfaces in $LiFePO_4$/graphite Li-ion batteries. *J. Phys. Chem. C* **2013**, *117*, 23476–23486. [CrossRef]
71. Ekström, H.; Lindbergh, G. A model for predicting capacity fade due to SEI formation in a commercial graphite/$LiFePO_4$ cell. *J. Electrochem. Soc.* **2015**, *162*, A1003. [CrossRef]
72. Yang, F.; Song, X.; Dong, G.; Tsui, K.-L. A coulombic efficiency-based model for prognostics and health estimation of lithium-ion batteries. *Energy* **2019**, *171*, 1173–1182. [CrossRef]
73. Goodenough, J.B.; Park, K.-S. The Li-ion rechargeable battery: A perspective. *J. Am. Chem. Soc.* **2013**, *135*, 1167–1176. [CrossRef] [PubMed]
74. Goodenough, J.B.; Kim, Y. Challenges for rechargeable Li batteries. *Chem. Mater.* **2010**, *22*, 587–603. [CrossRef]
75. Prabaharan, S.S.; Michael, M.S. *Nanotechnology in Advanced Electrochemical Power Sources*; CRC Press: Boca Raton, FL, USA, 2014.
76. Yang, Y.; Wang, J.; Sun, D.; Li, Y.; Xiao, T.; Zhang, C.; Lu, C.; Gao, J.; Xu, C.; Li, Y.; et al. Edge-Surface-Inter Carbon Nanoarchitecture on Silicon. *ACS Nano* **2025**, *19*, 16597–16610. [CrossRef]
77. Deleebeeck, L.; Veltzé, S. Electrochemical impedance spectroscopy study of commercial Li-ion phosphate batteries: A metrology perspective. *Int. J. Energy Res.* **2020**, *44*, 7158–7182.
78. Cui, Y.; Zhao, X.; Guo, R. Improved electrochemical performance of $La_{0.7}Sr_{0.3}MnO_3$ and carbon co-coated $LiFePO_4$ synthesized by freeze-drying process. *Electrochim. Acta* **2010**, *55*, 922–926. [CrossRef]
79. Nikgoftar, K.; Madikere Raghunatha Reddy, A.K.; Reddy, M.V.; Zaghib, K. Carbonaceous Materials as Anodes for Lithium-Ion and Sodium-Ion Batteries. *Batteries* **2025**, *11*, 123. [CrossRef]

batteries

MDPI

Correction

Correction: Zeng et al. Sustainable Synthesis of a Carbon-Supported Magnetite Nanocomposite Anode Material for Lithium-Ion Batteries. *Batteries* 2024, *10*, 357

Hui Zeng, Jiahui Li, Haoyu Yin, Ruixin Jia, Longbiao Yu, Hongliang Li and Binghui Xu *

Institute of Materials for Energy and Environment, College of Materials Science and Engineering, Qingdao University, Qingdao 266071, China; zenghui@qdu.edu.cn (H.Z.); lijiahui1@qdu.edu.cn (J.L.); yinhaoyu@qdu.edu.cn (H.Y.); jiaruixin01@163.com (R.J.); 13406366724@163.com (L.Y.); lhl@qdu.edu.cn (H.L.)
* Correspondence: xubinghuiqdu@qdu.edu.cn or xubinghuicsu@163.com

The authors would like to make the following corrections to their published paper [1].

The changes are as follows: References [35,54,56] were removed due to insufficient alignment with the discussed context. Reference [39] was deleted as it duplicates content with [25]. References [5–7,17,19,24,25,27,30] have been updated to the following:

5. Chen, Y.X.; He, L.H.; Shang, P.J.; Tang, Q.L.; Liu, Z.Q.; Liu, H.B.; Zhou, L.P. Micro-sized and nano-sized Fe_3O_4 particles as anode materials for lithium-ion batteries. *J. Mater. Sci. Technol.* **2011**, *27*, 41–45. https://doi.org/10.1016/S1005-0302(11)60023-6.
6. Zhao, P.X.; Jiang, L.; Li, P.S.; Xiong, B.; Zhou, N.; Liu, C.Y.; Jia, J.B.; Ma, G.Q.; Zhang, M.C. Tailored engineering of Fe_3O_4 and reduced graphene oxide coupled architecture to realize the full potential as electrode materials for lithium-ion batteries. *J. Colloid Interface Sci.* **2022**, *634*, 737–746. https://doi.org/10.1016/j.ensm.2019.03.030.
7. Tu, J.W.; Tong, H.G.; Wang, P.C.; Wang, D.D.; Yang, Y.; Meng, X.F.; Hu, L.; Wang, H.; Chen, Q.W. Octahedral/tetrahedral vacancies in Fe_3O_4 as k-storage sites: A case of anti-spinel structure material serving as high-performance anodes for PIBs. *Small* **2023**, *19*, 2301606. https://doi.org/10.1002/smll.202301606.
17. Liu, Z.H.; Li, B.Z.; Yuan, J.S.; Yuan, Y.J. Creative biological lignin conversion routes toward lignin valorization. *Trends Biotechnol.* **2022**, *40*, 1550–1566. https://doi.org/10.1016/j.tibtech.2022.09.014.
19. Tu, M.Y.; Wang, K.K.; Bao, S.C.; Zhang, R.; Tan, Q.K.; Kong, X.L.; Yu, L.B.; Wu, G.L.; Xu, B.H. Sodium carboxymethylcellulose induced engineering a porous carbon and graphene immobilized magnetite composite for lithium-ion storage. *J. Colloid Interface Sci.* **2022**, *608*, 1707–1717. https://doi.org/10.1016/j.jcis.2021.10.068.
24. Gao, X.J.; Xiao, Z.P.; Jiang, L.L.; Wang, C.; Lin, X.R.; Sheng, L.Z. Yolk-shell porous Fe_3O_4@C anchored on graphene as anode for Li-ion half/full batteries with high rate capability and long cycle life. *J. Colloid Interface Sci.* **2023**, *641*, 820–830. https://doi.org/10.1016/j.jcis.2023.03.121.
25. Li, M.; Ma, W.S.; Tan, F.Q.; Yu, B.; Cheng, G.H.; Gao, H.; Zhang, Z.H. Fe_3O_4@C-500 anode derived by commercial ammonium ferric citrate for advanced lithium ion batteries. *J. Power Sources* **2023**, *574*, 233146. https://doi.org/10.1016/j.jpowsour.2023.233146.
27. Salunkhe, T.T.; Kadam, A.N.; Hur, N.; Kim, I.T. Green and sustainably designed intercalation-type anodes for emerging lithium dual-ion batteries with high energy density. *J. Energy Chem.* **2023**, 80, 466–478. https://doi.org/10.1016/j.jechem.2023.01.051.

30. Yan, Z.Q.; Sun, Z.H.; Liu, H.S.; Guo, Z.H.; Wang, P.; Zhao, L.L.; Qian, L.; Wu, X.L. Heterogeneous interface in hollow ferroferric oxide/iron phosphide@carbon spheres towards enhanced Li storage. *J. Colloid Interface Sci.* **2022**, *617*, 442–453. https://doi.org/10.1016/j.jcis.2022.03.030.

With these corrections, the order of some references has been adjusted accordingly. The authors state that the scientific conclusions are unaffected. This correction was approved by the Academic Editor. The original publication has also been updated.

Reference

1. Zeng, H.; Li, J.; Yin, H.; Jia, R.; Yu, L.; Li, H.; Xu, B. Sustainable Synthesis of a Carbon-Supported Magnetite Nanocomposite Anode Material for Lithium-Ion Batteries. *Batteries* **2024**, *10*, 357. [CrossRef]

https://doi.org/10.3390/batteries12020038